t for difference between two independent means $= \dfrac{(\overline{Y}_1 - \overline{Y}_2) - (\mu_1 - \mu_2)}{\hat{s}_{\overline{Y}_1 - \overline{Y}_2}}$

Sum of squares between $= SS_{\text{BETWEEN}} = \Sigma\, n_j\, T_j^2$

Sum of squares total $= SS_{\text{TOTAL}} = \Sigma(Y - G)^2 = \Sigma Y^2 - (\Sigma Y)^2 / N$

Sum of squares within $= SS_{\text{WITHIN}} = SS_{\text{TOTAL}} - SS_{\text{BETWEEN}}$

Variance ratio $= F = \dfrac{MS_{\text{BETWEEN}}}{MS_{\text{WITHIN}}}$

Correlation coefficient $= r = \dfrac{CP}{\sqrt{(SS_X)(SS_Y)}} = \dfrac{\Sigma XY - \dfrac{(\Sigma X)(\Sigma Y)}{N}}{\sqrt{(SS_X)(SS_Y)}}$

Slope $= b = \dfrac{CP}{SS_X} = \dfrac{N(\Sigma XY) - (\Sigma X)(\Sigma Y)}{N\Sigma X^2 - (\Sigma X)^2}$

Intercept $= a = \overline{Y} - b\overline{X}$

t for correlation coefficient $= \dfrac{r - \rho}{\sqrt{\dfrac{1 - r^2}{N - 2}}}$

Chi square $= \chi^2 = \displaystyle\sum_{i=1}^{k} \left(\dfrac{(O_i - E_i)^2}{E_i} \right)$

Cramer's statistic $= V = \sqrt{\dfrac{\chi^2}{N(L - 1)}}$

Eta squared $= 1 - \dfrac{SS_{\text{ERROR}}}{SS_{\text{TOTAL}}}$

$\phantom{\text{Eta squared}} = \dfrac{t^2}{t^2 + df}$

$\phantom{\text{Eta squared}} = r^2$

$\phantom{\text{Eta squared}} = \dfrac{(df_{\text{BETWEEN}})F}{(df_{\text{BETWEEN}})F + df_{\text{WITHIN}}}$

Statistics for the Behavioral Sciences

James Jaccard
State University of New York at Albany

Wadsworth Publishing Company
Belmont, California
A Division of Wadsworth, Inc.

This book is dedicated to the city of Victor, Colorado.

Credits

American Psychological Association: Table 9.3, page 171, from C. Boneau, "The Effects of Violations of Assumptions Underlying the T Test," *Psychological Bulletin*, 1960. Copyright 1960 by the American Psychological Association. Reprinted by permission of the publisher and author.

Harcourt Brace Jovanovich: Figure 7.1, page 130, from R. B. McCall, *Fundamental Statistics for Psychology*, Third Edition, Harcourt Brace Jovanovich, 1980. Used with permission.

Harper & Row: Box 8.1, page 153, from Schuyler W. Huck and Howard M. Sandler, *Rival Hypotheses: Alternative Interpretations of Data Based Conclusions*. Copyright © 1979 by Schuyler W. Huck and Howard M. Sandler. By permission of Harper & Row, Publishers, Inc.

Prentice-Hall, Inc.: Figure 6.1, page 116, and Figure 17.1, page 348, from M. K. Johnson and R. M. Liebert, *Statistics: Tool of the Behavioral Sciences*, © 1977, reprinted by permission of Prentice-Hall, Inc., Englewood Cliffs, New Jersey. Table 9.3, page 171, from Glass and Stanley, *Statistical Methods in Education and Psychology*, © 1970, reprinted by permission of Prentice-Hall, Inc., Englewood Cliffs, New Jersey.

John Wiley & Sons, Inc.: Table 6.4, page 115, and Figure 13.7, page 271, from E. W. Minium, *Statistical Reasoning in Psychology and Education*, John Wiley & Sons, 1978. Used with permission.

Psychology Editor: *Ken King*
Production Editor: *Gary Mcdonald*
Designer: *Patricia Girvin Dunbar*
Copy Editor: *Joanne Cuthbertson*
Technical Illustrator: *Evanell Towne*

Printed in the United States of America

2 3 4 5 6 7 8 9 10—87 86 85 84 83

ISBN 0-534-01247-7

Library of Congress Cataloging in Publication Data

Jaccard, James.
  Statistics for the behavioral sciences.

  Bibliography: p.
  Includes index.
  1. Psychometrics. I. Title.
BF39.J28 1983   519.5'0243   82-13544
ISBN 0-534-01247-7

Contents

1
Statistical
Preliminaries
1

2
The Analysis of Bivariate Relationships 144

Chapter 14: Chi Square Test of Independence   288

Chapter 15: Nonparametric Statistical Tests   308

Chapter 16: Selecting the Appropriate Statistical Technique to Analyze Bivariate Relationships   330

**3
Additional
Topics
341**

Preface

To the Instructor

While developing the outline for a new statistics text, I was haunted by my own reaction to new statistics books: "Oh no, not *another* introductory statistics text." There are hundreds of introductory statistics books available, many of which are excellent. Several have been on the market for a decade and have had the benefit of going through several editions. Basic introductory statistics, unlike content areas in the social sciences, does not become quickly dated. Many of the concepts taught ten years ago are still, for most purposes, taught today. So why one more text?

Despite the existence of some excellent texts, I have been unable to locate any that accomplish my own personal goals in teaching introductory statistics at the undergraduate level. What follows is, in part, an elaboration of some of these goals and of how the present text differs from those currently available.

Applications and Integration. In my opinion, most introductory statistics texts fail to integrate sufficiently the subject matter of statistics with what students will experience in other social science courses and in social science journals. A statistics course should not only teach a student basic skills for analyzing data, but also help make him or her an intelligent consumer of scientific information. Students need to know how to make sense of research reports and journal articles, and a firm grasp of statistics is a major step in this direction.

In my view, there is a large gap between what a student learns in statistics classes and the way statistics are presented in research reports. Accordingly, one goal of the present book is to teach the reader how to present the results of statistical analyses when writing research reports. By definition, this also conveys to the reader the form in which he or she will encounter statistical information when *reading* research reports. Most statistics texts emphasize the importance of stating null and alternative hypotheses, critical values, and formal decision rules for drawing inferences with respect to the null and alternative hypotheses. Yet these are rarely explicitly stated in research reports and students often find this confusing. In fact, phrases like "the difference was statistically significant ($t = 2.89$, df $= 18$, $p < .05$)" have usually never been encountered by students who have had introductory statistics. The present book attempts to confront this discrepancy. A special section entitled Method of Presentation will be found in most chapters. This section discusses examples of how statistical analyses are typically presented in research reports and provides the rationale underlying the form of such presentation.

Learning When to Use a Statistical Test. Because of the way chapters and exercises are organized in most texts, students are essentially told which statistical procedure to use on a given set of data. This state of affairs is simply unrealistic. It is easy to teach a student *how to compute* a correlation coefficient, just as it is easy to teach a student *how to interpret* a correlation coefficient. It is much more difficult, but just as important, to teach the student *when* to use a correlation coefficient and *why* it should be used. Given a set of data, many students can't determine where to begin in answering their research questions. I have attempted to address this problem. To be sure, it is impossible to state any hard and fast rules for data analysis. It is always possible to find an exception or suggest a procedure that might give better insights into the data. But some rough guidelines can be given and the issues involved in selecting a form of data analysis can be explicated. In the present text, each statistical technique is introduced by stating instances where the test is most typically applied and an interesting research example is then given. Chapter 16 develops in detail issues to consider when selecting a statistical test to analyze one's data.

Relevance. A common complaint among students who take statistics courses is that statistics are irrelevant and boring. This view is fostered, in part, by our tendency to use examples and exercises that *are* irrelevant and boring. Sometimes it is necessary to use mundane examples to best illustrate concepts. I found many instances where this was necessary. But at the same time, it is possible to provide students with interesting applications of statistics. I have chosen two ways in which to accomplish this. First, in most chapters I have included boxed material that presents interesting research applications of the statistical method being developed. Second, in later chapters (9–17), I have presented example studies in the exercise section where students are to take representative data from research that has actually been conducted in the social science literature and analyze the data using what they have learned. Supplementing these are exercises that should help students to learn and understand the material covered in the chapter.

Unifying Themes. During the years I have taught statistics, I have found that students do not easily understand the common focus of the different statistical tests. Most students simply cannot appreciate the conceptual relationship between the *t* test, analysis of variance, correlation, and chi square tests. As a result, no unifying focal point is developed and the student loses sight of the purpose of the various tests. In the present book, a unifying structure is provided. Each of the major statistical techniques concerns the relationship between variables. The *t* test and analysis of variance are usually (but not always) applied when analyzing the relationship between a qualitative independent variable and a quantitative dependent variable, correlation is applied when analyzing the relationship between two quantitative variables, and the chi square test is applied when analyzing the relationship between two qualitative variables. Three questions serve as the organizing framework for each technique: (1) Given sample data, can we infer that a relationship exists between the variables?, (2) What is the strength of the relationship?, and (3) What is the nature of the relationship? As an example, in analysis of variance, the first question is addressed by the test of the null hypothesis, the second by an index such as eta squared, and the third by Tukey's HSD test. By continually relating each technique to these three questions, a unified framework emerges.

Variance Extraction Approach. Although I have used traditional presentations of most statistical methods, I have employed a few nontraditional modes of presentation as well. This has involved the development of variance extraction techniques in selected chapters. The chapters where this is most conspicuous are those on the independent groups t test, the correlated groups t test, and repeated measures analysis of variance. In the chapter on the independent groups t test, variance extraction techniques are used to give an *intuitive* understanding of statistics such as eta squared. Such techniques are also used to make explicit the logic of the correlated groups t test. One particular advantage of this approach, as developed in the context of the correlated groups t test, is that generalizations to repeated measures analysis of variance are direct. Whereas most introductory courses do not even treat repeated measures analysis of variance, I have found that I can cover the basics of the topic in one lecture, given mastery of the variance extraction approach. Recognizing that not everyone will care to use this approach, I have included the more traditional presentations as well. Advanced students can benefit by comparing the two approaches. Beginning students need only consider one of the two methods of development, whichever the instructor cares to emphasize and directs them to read.

Conceptual versus Computational Emphasis. The emphasis of the present book is on a conceptual understanding of statistics. With rapid progress in the computer field and with the widespread use of hand calculators geared toward statistics, it seems unwise to spend considerable time on computational formulas and methods of calculation. Very few students who take introductory statistics ever end up calculating statistics. Rather, they read about them in a research report or learn to program a computer to do the calculations. Because of this, most of the computations and formulas used in this book are conceptually rather than computationally based. Although they are not computationally efficient, my teaching experience has demonstrated that they are well worth the extra effort in terms of fostering an understanding of statistics.

Although computational formulas are downplayed, they are not ignored. Chapters 1–7 provide the student with the relevant background for reading statistical notation they will encounter. Computational formulas are provided, largely in appendixes, and serve as a reference that can be used in case the student ever needs to compute a statistic as quickly and efficiently as possible. However, I personally encourage my students to avoid these procedures in favor of the conceptual formulas when trying to *learn about* statistics (as opposed to *computing* statistics).

Research Design. Another unique characteristic of the present text is a chapter on research methods. I have always believed that statistics and experimental design are highly intertwined and that statistics should be placed in the context of research design. For example, how can students really grasp the meaning of error variance without some elementary understanding of disturbance variables? Chapter 8 is intended to provide an appropriate research context. In addition, each research example used to develop a statistical technique is discussed in the context of its methodological constraints. I hope this will encourage the student to consider the results of statistical analysis in a broader sense than most statistic books convey.

Advanced Students. I have included a special feature for advanced students who are especially interested and inquisitive with regard to statistics. This is the presence of appendixes to several chapters that explain in more detail certain advanced concepts. These appendixes are generally written at a higher level and should be skipped by introductory students.

Material Covered. At first glance, the table of contents suggests that this book is more advanced than the typical introductory statistics book. This is not the case. I recognize that different instructors like to emphasize different material. It is not expected that an introductory class could even begin to cover all seventeen chapters of the present book. Some instructors will wish to emphasize some topics at the expense of others. The chapters included are intended to provide the instructor with a useful set of topics from which to choose. The order of chapters is flexible, except for natural progressions (for example, no one would cover t tests before they covered means and standard deviations). The material not covered in class will be available as reference for students who pursue graduate work or advanced undergraduate research projects. Also, when students in my classes ask questions about topics not typically covered in an introductory text, I have found it useful to be able to give them appropriate reading material. The comprehensive coverage of the present text helps in this regard.

In talking with different statistics instructors, I have found one of the main differences in teaching statistics is how the topic of probability is treated. Some instructors prefer to cover it in some detail (as has been done in Chapter 5 of this book), while others prefer to give it less emphasis. *All* instructors recognize that probability is a key concept in most statistical procedures. However, some feel that topics such as conditional probabilities, joint probabilities, and sampling with and without replacement have little practical relevance in applied research settings (for example, for computing t tests, analysis of variance, and so on). Because of this, I have written Chapter 5 so that it can be omitted without disrupting succeeding chapters. The concept of probability is discussed in Chapters 1–4 in sufficient detail to give students the necessary appreciation for later statistical tests.

In my own one-semester courses (which are *very* introductory) I cover Chapters 1–16, omitting Chapters 5, 15, and 17. I try to give my students a brief, one to two lecture overview of what is in the remaining chapters and encourage them to read what I could not cover. Also, within certain chapters, I choose to skip selected sections (for example, percentiles) so I can emphasize later material that I think is more appropriate *for my particular students*. I have tried to structure sections within chapters so that instructors who want to skip a topic can easily do so.

I have also tried to focus discussion on those techniques that are most common in the social science literature. This is not to say that the omitted concepts are not important. The decision to exclude these reflects space demands and a cost-benefit analysis of what the student needs from the course more than anything else.

Chapter Structure. Chapters 9–17 develop the major statistical tests typically introduced in beginning statistics courses. I have imposed a common structure on each of these chapters to underscore the common focus of the tests.

Each chapter begins with a discussion of the conditions under which the test is typically applied. Attention then turns to inferring whether a relationship exists between the independent variable and the dependent variable, the strength of the relationship, and the nature of the relationship. These issues are developed in the context of a research example and critical computational stages are highlighted with study exercises that occur within the chapter. The statistics are then placed in context via a section on methodological considerations. This section underscores the importance of interpreting statistics relative to research design considerations. The Method of Presentation section discusses how the statistical technique will typically be reported in journal articles. This is followed by a numerical example that takes the student through an application of the statistical test from start to finish. Finally, a discussion of planning an investigation using the test is presented, with explicit consideration of power and sample size selection. The exercises for each chapter are of two types: (1) exercises designed to review and reinforce concepts the student has learned in the chapter, and (2) exercises designed so that the student can apply these concepts to a real research situation.

Acknowledgments

Many individuals have helped in the development of this book. First, I would like to thank the following publishers for permission to quote or adapt material from previously published works: Addison Wesley; the Trustees of *Biometrika;* Brooks/Cole; Harcourt Brace Jovanovich; Harper & Row; Holt, Rinehart & Winston; John Wiley & Sons; the Literary Executor of the late Sir Ronald Fisher; McGraw-Hill; Norton; Oliver and Boyd; and Prentice-Hall.

I would also like to acknowledge the excellent work of the staff at Wadsworth. Gary Mcdonald, Pat Dunbar, and Diana Griffin were most helpful during the production phases of the project and made my task much more pleasant than it should have been. The copy editing of Joanne Cuthbertson was superb. A special note of appreciation is extended to my editor, Ken King, who contributed a great deal to this project. Ken's continual support and patience went far beyond the role of the typical editor, and his efforts were always appreciated, both on a personal and professional level.

Several colleagues reviewed the manuscript, either in part or in its entirety, during different stages of development. Professors John M. Knight, Central State University, Oklahoma; Scott E. Maxwell, University of Notre Dame; Karyl Swartz, Herbert Lehman College, CUNY; and David Brinberg, University of Maryland, provided many suggestions that were invaluable to me during the revision process. I would particularly like to acknowledge the input of Professor Alfred Hall of the

College of Wooster. Professor Hall reacted to both the original and a revised manuscript and provided extremely useful feedback. Other reviewers whose comments proved useful included Stephen Edgell, University of Louisville; Teresa Amabile, Brandeis University; Jerry W. Thornton, Angelo State University; Ervin M. Segal, SUNY, Buffalo; Roger Baumgarte, Winthrop College; Arnold Well, University of Massachusetts, Amherst; Scott E. Graham, Allentown College; Brian A. Wandell, Stanford University; and Stephen W. Hinkle, Miami University, Ohio. If there is any merit in this book, it may well be the result of their input. Of course, any shortcomings are my responsibility alone.

Finally, I would like to acknowledge the contributions of Karyl Swartz throughout this project. Her patient ear and willingness to listen to statistic after statistic was but one source of support I valued so much during the writing of this text.

To the Student

This is an introductory statistics text designed for a first course in statistics. I have written the text assuming the student possesses a minimum of mathematical background (simple algebra). For most of you, much of the material in the book will be new. My experience in teaching statistics has led me to conclude that students generally find the logic of statistics easy to grasp. What is difficult is the amount of new material one must assimilate and use. Statistics courses are unique in several respects. Later material relies heavily on a clear understanding of previous material. There is a continual building process and you *must* keep up with the pace your instructor sets. Statistics is not the kind of material you can put off until the night before an exam and then ''cram'' for a test on the next day.

I have developed a number of features in the text that should help you in your study of statistics. First, I have included examples (called Study Exercises) that present a problem and then answer it, based upon the material previously covered. Working through these examples will help you to acquire many important statistical concepts. Second, key terms are **boldfaced**. These terms should be reviewed after reading each chapter. Make sure you understand and can define each one. Third, extensive exercises have been provided. It is strongly recommended that you work through *all* of the exercises as they will reinforce much of what you read. Fourth, I have included Boxes that present examples of interesting research that have used the concepts developed in a chapter. If you read the Boxes carefully, you will not only learn a good deal about social science research, but you will also be able to appreciate more fully the role of statistics in the social sciences. Fifth, I have included a glossary of symbols so you can have a ready reference to the many statistical symbols typically used in statistical texts. Finally, I have included a Method of Presentation section that describes how statistical tests are reported in professional journals and reports. This should help you to understand more fully not only the material presented in this book, but also the material you read in the course of your study of content areas in the social sciences.

Statistical
Preliminaries

Introduction and Mathematical Preliminaries

1.1 / The Study of Statistics

It has become common for courses in statistics to be required of students majoring in the social sciences. Many such students are unclear as to why statistical training is necessary. There are several reasons. Statistics is an integral part of research activity. Important questions and issues are addressed in social science research, and statistics can be a valuable tool in developing answers to these questions. For the student who makes a career of conducting research, statistical analysis should prove to be a useful aid in the acquisition of knowledge.

But the fact of the matter is, many students who take statistical courses will not develop careers that require an active part in research. Although these students may not actually conduct research, they may be required to read, interpret, and use research reports. These reports will usually rely on statistical analyses to draw conclusions and suggest courses of action to take. Knowledge of statistics is therefore important to help one understand and interpret these reports.

Research that uses statistical analysis is clearly having a greater impact on society, both in our everyday lives and in more abstract situations. On television we see commercials that report research "demonstrating" that Brand A is three times as effective as Brand X. In national magazines and newspapers we read the results of surveys of public opinion and attitudes toward politicians. Many magazines include special sections designed to disseminate to the public at large the results of research in the physical and social sciences. As our society becomes more technologically complex, greater demands will be placed on professionals to understand and use results of research designed to answer applied problems. This will generally require a working understanding of statistical methods.

A knowledge of statistical analysis may also help to foster new and creative ways of thinking about problems. Several colleagues have remarked on the new insights they developed when they approached a problem from the perspective of statistical analysis. Statistical "thinking" can be a useful aid in suggesting alternative answers to questions and posing new ones. In addition, statistics helps to develop one's skills in critical thinking, in terms of both inductive and deductive logic. These skills can be applied to any area of inquiry, and hence are extremely useful.

1.2 / Research in the Social Sciences

The major concern with statistics in this book is how they are used in social science research. As such, it will be useful to consider briefly the research process as it commonly occurs in the social sciences.

Most people do not view scientific research as a process but rather as a product. Reference is made to a "body of facts" that is known about some phenomenon. Scientific research is better characterized as an ongoing process consisting of five stages. The first stage is the formulation of a question about some phenomenon or phenomena. Why do people smoke marijuana? Why do some children do better in school than others? Why was Ronald Reagan elected in 1980? Why do some people fail to help another person who is in need of help? The second stage is forming a **hypothesis** concerning the question. A hypothesis is a statement proposing that something is true about a given phenomenon. One might hypothesize that people smoke marijuana because of pressures from their peers to do so. Or one might hypothesize that Ronald Reagan was elected because of his anti-Washington position. The third stage involves designing an experiment or investigation to test the validity of the hypothesis. In such an investigation one makes systematic observations of individuals or groups of individuals in settings that are conducive to testing the hypothesis. The fourth stage is analyzing the data collected in the investigation in order to help the researcher draw the appropriate conclusions. This is generally done with the aid of statistics. The final stage is drawing a conclusion and thinking about the implications of the experiment for future research.

1.3 / Variables

Most social science research is concerned with relationships between **variables.** A variable is a phenomenon that has different values. Sex is a variable with two values, male and female. Weight is a variable, consisting of values such as 101 pounds, 102 pounds, and so on. In contrast, a **constant** is something that does not vary within given constraints. The value for the mathematical symbol π is always 3.1416. It never changes and takes on one and only one value. Hence, it is a constant. If an experiment is conducted only with females, then in the context of the experiment, sex is a constant. It takes on one and only one value: female.*

*Some researchers use the term *level* as a synonym for values. For example, the variable sex is said to have two levels, female and male.

Researchers distinguish between variables in many ways. One such distinction in the social sciences is between an **independent variable** and a **dependent variable.** Suppose an investigator is interested in the relationship between two variables, the effect of information about the sex of a job applicant on hiring decisions by personnel managers. An experiment might be designed in which fifty personnel managers are provided with descriptions of job applicants and asked whether they would hire that applicant. The applicant is described in the same way on several pertinent dimensions to all fifty managers. The only difference is that twenty-five of the managers are told that the applicant is female, and the other twenty-five managers are told that the applicant is male. Each manager then indicates his or her hiring decision. In this experiment the information about the sex of the applicant is the independent variable and the hiring decision is the dependent variable. The hiring decision is termed the *dependent variable* because it is thought to "depend on" the information about the sex of the applicant. In other words, the independent variable is the presumed cause of the dependent variable. A useful tool for identifying an independent and dependent variable is the phrase "The effects of _____ on _____." The variable that fits into the first blank is the independent variable, and the variable that fits into the second blank is the dependent variable (for example, the effects of sugar on the taste of coffee, the effects of child rearing practices on intelligence, and so on).

The term *independent variable* has assumed different meanings in various areas of the social sciences. Some investigators restrict the definition of an independent variable to a variable that is explicitly manipulated in the context of an experiment (such as the information about the sex of the applicant in the last example). We will adopt the more common definition of an independent variable as a presumed cause of the dependent variable. It is not necessary for a variable to be experimentally manipulated in order that it be conceptualized as an independent variable. Rather, if a variable is presumed to cause another variable, then the presumed cause is the independent variable and the consequence of that cause is the dependent variable. Note that just because an investigator presumes that one variable causes another does not mean that the one variable does, in fact, cause the other variable. This is only a presumption made by the researcher for purposes of the investigation. We will return to this issue in Chapter 8 when we analyze the relationship between two variables in detail.

1.4 / Measurement

A major feature of social science research is classification. We frequently classify people or events into different categories. With respect to the variable of sex, there are two categories into which we classify people: male and female. When a teacher assigns grades, there are five categories that are typically used: A, B, C, D, and F.

Measurement involves the assignment of objects (for example, people, events, and so on) to numerical categories according to a set of rules. Obviously, there are many different types of rules one may use and, hence, many different levels or types of measurements.

One level of measurement is **nominal measurement.** Nominal measurement involves using numbers merely as labels. An investigator might classify people according to their religion—Catholic, Protestant, Jewish, and All Others—and use the numbers 1, 2, 3, and 4 to refer to these categories. In this case, the numbers have no special quality about them; they are merely used as labels. In social science research, the basic statistic of interest for variables that involve nominal measurement is usually frequencies (for example, how many people are Democrats, how many are Republicans), proportions, and percentages.

Another level of measurement is **ordinal measurement.** A variable is said to be measured on an ordinal level when the categories can be *ordered* on some continuum or dimension. Suppose we take four individuals who differ in height and assign the number 1 to the tallest individual, the number 2 to the next tallest individual, the number 3 to the next tallest individual, and the number 4 to the shortest individual. In this case, we have measured height on an ordinal level, as it allows us to order the individuals from tallest to shortest. Thus, ordinal measurement allows the researcher to classify individuals into different categories that, in turn, are ordered along a dimension of interest.

Note in the preceding example that height was *not* measured in terms of feet or inches. The tallest individual had a score of 1 on the measurement scale, the next tallest individual had a score of 2, and so on. This set of measures (that is, the rank order from tallest to shortest) represents measures that have ordinal characteristics. As we will subsequently illustrate, height can be measured in other ways whereby the measures have more than just ordinal characteristics.

A third level of measurement is **interval measurement.** Interval measures have all the properties of ordinal measures but allow us to do more than order objects on a dimension. They have the additional property that numerically equal distances on the scale represent equal distances on the dimension being measured. For example, in measuring the weight of an individual, the difference between someone who weighs 125 pounds and someone who weighs 130 pounds is the same as the difference between an individual who weighs 110 pounds and one who weighs 115 pounds. In both instances, the difference is 5 pounds. Thus, interval measures not only provide us with information about the ordering of individuals on a dimension, but also information about the *magnitude of differences* between scale units. Note that this was not true in the previous example using an ordinal measure of height. It was not necessarily true that the difference in height between individuals 1 and 2 was the same as the difference in height between individuals 3 and 4. It was only true that individual 1 was taller than individual 2, who, in turn, was taller than individual 3, who, in turn, was taller than individual 4.

A fourth level of measurement is **ratio measurement.** Ratio measures have all the properties of interval measures (and hence ordinal measures as well), but in addition possess an absolute or natural zero point. This allows the investigator to specify the exact amount of the property being measured. In measuring height, for example, 0 inches implies no height whatsoever. For this measurement scale, there is an absolute zero point. It follows that an individual who is 80 inches tall is twice as tall as an individual who is 40 inches tall. In contrast, without such a zero point, ratio statements such as the preceding are not possible. A measure with an arbitrary zero point does not allow one to make ratio-type statements. Only measures with absolute or natural zero points permit such statements.

Figure 1.1 Three Different Ways of Measuring the Height of Four Buildings

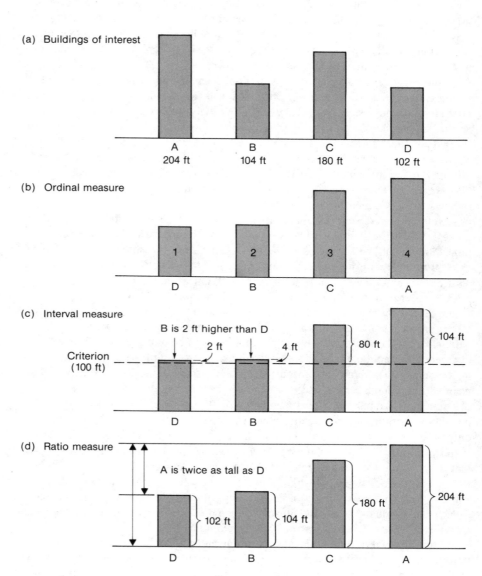

(a) Buildings of interest

(b) Ordinal measure

(c) Interval measure

(d) Ratio measure

We can illustrate the differences between the levels of measurement by examining three different ways of measuring the height of buildings. Assume an architect is interested in studying buildings that are taller than 100 feet in a particular city. Suppose there are four such buildings. Figure 1.1a shows graphically the height of these four buildings and indicates how tall each one is. The three different ways of measuring the height of these buildings are ordinal, interval, and ratio measures. First, we can assign the number 1 to the shortest building, the number 2 to the next shortest building, the number 3 to the next shortest building, and the number 4 to the tallest building (Figure 1.1b). This is ordinal measurement. It allows us to order the buildings on a dimension of height but does not tell us anything about the magnitude of heights. A second method is to measure how many feet taller each one is over the criterion set by the architect (100 feet). In this case, we find that D is 2 feet taller than the criterion, B is 4 feet taller than the criterion, C is 80 feet taller than the criterion, and A is 104 feet taller than the criterion (Figure 1.1c). In contrast to the previous measurement, we can now not only order the buildings on a dimension of height, but we also have information about the relative magnitude of the heights. B is 2 feet taller than D ($4 - 2 = 2$); C is 76 feet taller than B ($80 - 4 = 76$); and so on. We have measured height on an interval scale. Note that on this scale even though building B has a score of 4 (that is, it is 4 feet above the criterion) and building D has a score of 2 (that is, it is 2 feet above the criterion), it is not the case that building B is twice as tall as building D. This is because we do not have a natural or absolute zero point. All measures were taken relative to an arbitrary criterion (100 feet). Finally, we can measure each building from the ground, yielding an absolute zero point. Building D is 102 feet, B is 104 feet, C is 180 feet, and A is 204 feet. We can now state with confidence that building A is twice as tall as building D (Figure 1.1d). We have measured height on a ratio scale.

The four types of measurement just outlined can be thought of as a hierarchy. At the lowest level, nominal measurement allows us to categorize people into different groups. The second level, ordinal measurement, not only allows us to classify people into different groups, but also indicates the relative ordering of the groups on a dimension of interest. Interval measurement, the next level, possesses the same properties as ordinal measurement but, in addition, is sensitive to the magnitude of the differences in the groups on the dimension. However, ratio statements are not possible at this level. It is only at the final level, ratio measurement, that such statements are possible. Ratio measures have all of the properties of nominal, ordinal, and interval measures, but also permit ratio judgments to be made.

When reading a research report, you may encounter references to interval scales, ordinal scales, and so on. Technically, the use of the word scales is somewhat misleading. Nominal, ordinal, interval, and ratio properties are characteristics of a set of measures, not just the scales used to generate those measures. A measure has as its referent not only a particular scale (for example, inches), but an individual on whom the measure is taken, a time at which the measure is taken, and a setting in which the measure is taken. These all must be considered when evaluating the properties of a set of measures. We can illustrate this idea using height as an example. Consider four individuals whose heights are 54, 53, 52, and 51 inches, respectively. We can rank order these individuals from shortest to tallest:

Individual's height in inches (X)	Rank order (Y)
51	1
52	2
53	3
54	4

The rank order measure (Y) has ordinal properties, as does any rank order index. However, for this set of measures (that is, for these four individuals at this point in time), the measures on Y also have interval level properties. The difference between scores for any two individuals with adjacent scores corresponds to the same height difference as any other two individuals with adjacent scores $(4 - 3 = 54 - 53 = 1 = 53 - 52 = 1)$. For this set of measures, Y has interval properties. Note that if a 58-inches-tall individual were added to this set, receiving a rank of 5, Y would no longer exhibit interval properties. It would instead represent a set of measures with only ordinal properties. The point is that the concept of nominal, ordinal, interval, and ratio properties is inherent in measures, not scales.

The determination of whether a variable is measured on a nominal or ordinal level is usually a straightforward matter in the social sciences. This is not necessarily true for some interval and ratio measures. For example, there is controversy as to whether intelligence test scores (such as the Wechsler Adult Intelligence Scale) reflect only an ordinal measure of intelligence or an interval measure of intelligence. Techniques for testing the measurement assumptions of different measures have been developed in psychophysical scaling (for example, Anderson, 1970). The majority of statistical techniques considered in this text assume that the dependent variables are measured on at least an interval level. We will consider this assumption, and the effects of violating it, in more detail in later chapters.

1.5 / Quantitative and Qualitative Variables

Another important distinction is that between a **quantitative variable** and a **qualtative variable.** Any given variable can be classified as either quantitative or qualitative. A quantitative variable is one that takes on an ordered set of values on a dimension. Such a variable is measured on either an ordinal, interval, or ratio level. Individuals can thus be ordered on the dimension in question depending on their score on the scale. A measure of intelligence orders individuals from those of lower intelligence through those of medium intelligence through those of higher intelligence.

A qualitative variable is one that cannot be ordered on some dimension and corresponds to nominal measurement. Examples of qualitative variables include political party identification (Democrat, Republican, Independent), sex (female, male), religion (Catholic, Protestant, Jewish, other), and occupation (plumber, teacher, student, and so on). Qualitative variables reflect no rank order. Rather, they merely distinguish among categories. Thus, a quantitative variable is one that is measured on either an ordinal, interval, or ratio level, whereas a qualitative variable is one that is measured on a nominal level. As we will see later, the distinction between quantitative and qualitative variables is crucial in statistics.

Study Exercise 1.1

For *each* of the following experiments specify the independent variable and the dependent variable. Also identify any variables that are explicitly held *constant* by the experimenter. For both the independent variable and the dependent variable indicate whether the independent variable and the dependent variable represent a quantitative or a qualitative variable.

EXPERIMENT I

Goldberg (1968) was interested in investigating a sex bias among females. A group of 100 female college students was asked to rate an article in terms of its persuasiveness. The article was on the topic of education. The subjects were assigned to one of two groups. One group of 50 women read the article and were told that it was authored by a woman named Joan McKay. The other group of 50 women read the same article but were told it was authored by a man named John McKay. After reading the article, each subject rated the article on a 7 point scale as follows:

not at all
persuasive   1   2   3   4   5   6   7   very
                                            persuasive

The average rating scores were compared for the two groups, that is, the group with a male author vs. the group with a female author. Results indicated that the average rating was much higher when the article was attributed to a male author than to a female author.

Answer.   The independent variable is the sex of the author of the article, male or female. It is a nominal measure and, hence, constitutes a qualitative variable. The dependent variable is the persuasiveness rating. This measure constitutes at least an ordinal measure. It is probably the case that the higher the rating, the more persuasive the article was perceived to be. Because the dependent measure has at least ordinal characteristics, it represents a quantitative variable. Numerous variables have been held constant. One of the most obvious ones is that the study was conducted only with females. Also, the content of the articles was held constant.

EXPERIMENT II

Research on extrasensory perception (ESP) has taken many different directions. Recently, attention has been given to the possibility that hypnosis may be helpful in fostering ESP in people. One standard ESP task involves the use of Zener cards. These are special cards that have only five denominations and look like this:

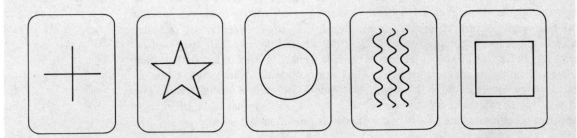

The standard task is to take a deck of 200 cards and have a "sender" shuffle them. The sender looks at the first card, thinks of the denomination of that card, and then the subject guesses what the card is. This process is repeated throughout the entire deck. ESP is measured by the number of correct guesses on the part of the subject (the receiver).

Casler (1964) used this task with 100 female college students. Two conditions were used: subjects who were (a) hypnotized or (b) not hypnotized. Fifty students were assigned to each condition. In the first condition, the women were hypnotized then given the task described above. In the second condition, they were not hypnotized and just completed the task. The average number of correct predictions was computed for the two groups. These averages turned out to be roughly equivalent to each other, and it was concluded that hypnosis does not affect ESP.

Answers. The independent variable is whether or not the subject was hypnotized. This represents a nominal measure and hence, is a qualitative variable. The dependent variable is the number of correct answers on the 200 trials. This represents at least an ordinal measure of ESP and hence it is a quantitative variable. You might be inclined to view this measure as one with ratio characteristics, since there is an absolute zero point (that is, none correct) and you can make ratio type statements (ten correct is twice as many as five correct). Actually, it is unclear whether this is the case. One can conceptualize the number of correct trials as an *index* of ESP ability. Ten correct trials relative to five correct trials may not really reflect *twice* as much ESP ability. Numerous variables have been held constant. Again, the most obvious one is that all subjects were females. Try to specify some others.

1.6 / Discrete and Continuous Variables

The Concept of Discrete and Continuous Variables. Another distinction made in statistics is between **discrete** and **continuous variables.** A discrete variable is one in which there are a finite number of values that can occur between any two points with respect to the variable. An example is the number of cars a person owns. For this variable, there can be only one value between the value of 1 car and 3 cars

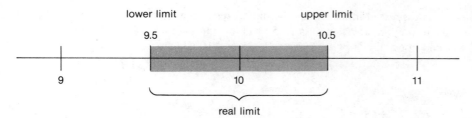

Figure 1.2 Real Limit of 10

(namely, 2 cars). We do not think of someone as owning 1.5 cars or 1.75 cars. In contrast, a continuous variable can theoretically include an infinite number of values between any two points. Time is an example of a continuous variable. Even between the values of 1 and 2 seconds, there are an infinite number of values that could occur (1.5 seconds, 1.52 seconds, and so on).

It should be emphasized that whether a variable is classified as discrete or continuous depends on the nature of the underlying theoretical dimension and not the scale used to measure that dimension. Tests used to measure intelligence, for example, yield scores that are whole numbers (101, 102, and so on). Nevertheless, intelligence is still continuous in nature since ultimately it involves a dimension that permits an infinite number of values to occur, if one had a measuring device sensitive enough to do so.

When reading research reports, you may encounter instances where a statistic for a discrete variable, such as the ideal number of children, is reported in a form that is representative of a continuous variable. For example, you might be told that the average number of children for married couples in the United States is 2.11. In this instance, the inclusion of decimal places is useful since it provides a more sensitive index of the number of children that couples in the United States, considered as a group, tend to have.

Real Limits of a Number. If a variable is continuous in nature, then it follows that the measurements taken on that variable must be approximate in nature. When we say a person reacted to a stimulus in 10 seconds, we don't usually mean exactly 10 seconds, but only approximately 10 seconds because more refined measures are always possible, such as 10.02 seconds or 10.093 seconds. When we say the reaction was in 10 seconds we actually mean it was somewhere between 9.5 seconds and 10.5 seconds since any number less than 9.5 would be rounded to 9 and any number greater than 10.5 would be rounded to 11. Thus, the *real limits* of the number 10 are 9.5 to 10.5. 9.5 is called the *lower limit* and 10.5 is called the *upper limit*. *The real limits of a number are those points falling one-half a unit above that number and one-half a unit below that number.* Figure 1.2 presents the concept of real limits graphically.

The real limits of a number can be stated not only with respect to whole numbers, but for numbers expressed as decimals as well. Consider the number 10.6.

Since it is expressed in tenths and the unit of measurement is .1, one-half a unit is .05. Therefore, the lower limit is 10.55 and the upper limit is 10.65. Similarly, the lower and upper limits of 10.63 are 10.625 and 10.635, respectively.

Study Exercise 1.2

State the real limits of the following numbers, assuming they are measured in the units reported.

(a) 20   (b) 8.4   (c) 12.23   (d) 16.0478

Answers. (a) 19.5 to 20.5
(b) 8.35 to 8.45
(c) 12.225 to 12.235
(d) 16.04775 to 16.04785

1.7 / Mathematical Preliminaries: A Review

The purpose of this section is to review a number of mathematical symbols and concepts that will be used in this book.

Summation Notation. Suppose we have a measure of monthly income for each of five individuals. The scores on this variable are as follows:

Individual	Monthly Income
1	200
2	300
3	200
4	300
5	300

In statistical terminology, the capital letter X is used as a general name for a variable. In this case, X stands for the variable "monthly income." Sometimes the label will be subscripted, X_i. The i represents the individual number (which is arbitrarily assigned). X_1 is the first individual's score on X (monthly income), which in this case is 200. X_5 is the score of the fifth individual on X, which is 300.

On some occasions, an investigator will consider two variables, such as monthly income and number of months worked:

Individual	X Monthly Income	Y Number of Months Worked
1	200	3
2	300	2
3	200	3
4	300	3
5	300	2

If we let X represent monthly income and Y represent the number of months worked, then we can refer to individual scores on X and Y using subscripts: $X_1 = 200$, $Y_1 = 3$, $X_2 = 300$, $Y_2 = 2$, and so on.

Suppose we want to sum the five scores on variable X to determine the total monthly income of the five individuals. In statistics, we have a shorthand way of writing an instruction to sum a set of scores, called *summation notation*. The operation is written as follows:

$$\sum_{i=1}^{5} X_i$$

The summation operation is signaled by the Greek capital letter, sigma (Σ). The number below the sigma means start with individual number 1 and the number above the sigma means add through to individual number 5. In this case:

$$\sum_{i=1}^{5} X_i = X_1 + X_2 + X_3 + X_4 + X_5 = 200 + 300 + 200 + 300 + 300 = 1,300$$

If the notation were written as

$$\sum_{i=2}^{4} X_i$$

it would mean sum the scores of individuals 2 through 4 on variable X. In this case,

$$\sum_{i=2}^{4} X_i = X_2 + X_3 + X_4 = 300 + 200 + 300 = 800$$

Often we let the letter N represent the total number of individuals. In this case

$N = 5$, since there are five individuals. You may thus encounter summation notation as follows:

$$\sum_{i=1}^{N} X_i = X_1 + X_2 + X_3 + X_4 + X_5 \qquad \text{[1.1]}$$

which is the same as having a 5 (or the number that represents the total number of individuals) above the sigma. The same terminology applies to the Y variable.

$$\sum_{i=1}^{N} Y_i = Y_1 + Y_2 + Y_3 + Y_4 + Y_5 = 3 + 2 + 3 + 3 + 2 = 13$$

In addition to the expression in Equation 1.1, you will encounter other forms of summation notation. We will briefly review each. One such term appears as

$$\sum_{i=1}^{N} Y_i^2 \qquad \text{[1.2]}$$

This means that each Y score should first be squared and then summed

$$\sum_{i=1}^{N} Y_i^2 = Y_1^2 + Y_2^2 + Y_3^2 + Y_4^2 + Y_5^2 = 3^2 + 2^2 + 3^2 + 3^2 + 2^2 = 35$$

Or, summation notation may appear as

$$\sum_{i=1}^{N} X_i Y_i \qquad \text{[1.3]}$$

This means that for each pair of scores, each X score should first be multiplied by each Y score, and then these products should be summed

$$\sum_{i=1}^{N} X_i Y_i = X_1 Y_1 + X_2 Y_2 + X_3 Y_3 + X_4 Y_4 + X_5 Y_5$$
$$= (200)(3) + (300)(2) + (200)(3) + (300)(3) + (300)(2)$$
$$= 600 + 600 + 600 + 900 + 600 = 3,300$$

Another summation term you will encounter is

$$\sum_{i=1}^{N} (X_i - Y_i) \qquad \text{[1.4]}$$

This means take the difference between X and Y for each individual, and then sum these differences:

$$\sum_{i=1}^{N} (X_i - Y_i) = (X_1 - Y_1) + (X_2 - Y_2) + (X_3 - Y_3) + (X_4 - Y_4) + (X_5 - Y_5)$$
$$= 197 + 298 + 197 + 297 + 298 = 1,287$$

A fifth summation term is

$$\left(\sum_{i=1}^{N} Y_i \right)^2 \qquad \text{[1.5]}$$

The parentheses signal that the summation operation should be executed first (that is, the five scores should be summed) and then this sum should be squared:

$$(Y_1 + Y_2 + Y_3 + Y_4 + Y_5)^2 = 13^2 = 169$$

Another summation term is

$$c \sum_{i=1}^{N} Y_i \qquad \text{[1.6]}$$

where c represents a constant. Suppose that in this case $c = 20$. Then this term indicates that we should first sum all of the scores in Y, and then multiply the sum by c or, in this case, 20:

$$c \sum_{i=1}^{N} Y_i = 20 \, (3 + 2 + 3 + 3 + 2) = (20)(13) = 260$$

It is very important to understand these six basic forms of summation notation, as we will refer to them constantly throughout this book. Investigators will frequently use a shorthand version of these symbols. The first symbol

$$\sum_{i=1}^{N} X_i$$

may also be written

$$\Sigma X$$

Note that there is no subscript on X, no number beneath the sigma, and no number above the sigma. It is understood that the letter N is above the sigma, the number 1 is below it, and the i is the subscript. Thus,

$$\sum_{i=1}^{N} X_i = \Sigma X$$

$$\sum_{i=1}^{N} X_i^2 = \Sigma X^2$$

$$\sum_{i=1}^{N} X_iY_i = \Sigma XY$$

$$\sum_{i=1}^{N} (X_i - Y_i) = \Sigma(X - Y)$$

$$\left(\sum_{i=1}^{N} Y_i\right)^2 = (\Sigma Y)^2$$

$$c\sum_{i=1}^{N} Y_i = c(\Sigma Y)$$

We will use the shorthand version of these terms wherever possible in this book.

Study Exercise 1.3

Given the following values for X and Y, complete the requested operations.

Individual	X	Y
1	3	3
2	4	3
3	2	3
4	7	5

(a) ΣX (b) $\Sigma(X - Y)^2$ (c) $(\Sigma X)(\Sigma Y)$ (d) $\Sigma X + \Sigma Y$
(e) ΣXY (f) $N\Sigma X$ (g) $(\Sigma X)^2$ (h) ΣX^2

Answers.
(a) $\Sigma X = 3 + 4 + 2 + 7 = 16$
(b) $\Sigma(X - Y)^2 = (3 - 3)^2 + (4 - 3)^2 + (2 - 3)^2$
$+ (7 - 5)^2 = 6$
(c) $(\Sigma X)(\Sigma Y) = (3 + 4 + 2 + 7)(3 + 3 + 3 + 5)$
$= (16)(14) = 224$
(d) $\Sigma X + \Sigma Y = (3 + 4 + 2 + 7) + (3 + 3 + 3$
$+ 5) = 16 + 14 = 30$
(e) $\Sigma XY = (3)(3) + (4)(3) + (2)(3) + (7)(5) = 62$
(f) $N\Sigma X = (4)(16) = 64$
(g) $(\Sigma X)^2 = 16^2 = 256$
(h) $\Sigma X^2 = 3^2 + 4^2 + 2^2 + 7^2 = 78$

Two of the more common notations encountered in statistical formulas that are frequently confused by students are

$$\Sigma X^2 \text{ and } (\Sigma X)^2$$

You should take care to note that these are not the same (compare Problems (g) and (h) in Study Exercise 1.3).

Rounding. It may be necessary to round numbers for purposes of convenience. Thus, $10/3 = 3.333333333. . . .$ Obviously we don't want to write out hundreds of digits, so we round off. There are specific mathematical rules for rounding that will be adopted in this book. Suppose we want to round to hundredths or two decimal places. The rules are as follows (McCall, 1980):

1. If the remaining decimal fraction is less than one-half (.5) a unit, then simply drop the remaining fraction.

In the previous case, the number 3.33333 would become 3.33 since the remaining fraction after the hundredths is less than one-half (.5) a unit. Similarly, 3.3348 would be rounded to 3.33.

2. If the remaining decimal fraction is greater than one-half (.5) a unit, then we increase the preceding digit by one.

The number 3.338 would become 3.34, since the remaining decimal fraction after the hundredths is greater than one-half (.5) a unit. The number 3.335001 would also be rounded to 3.34 according to this rule, since .5001 is greater than one-half (.5) a unit.

3. If the remaining decimal fraction is exactly one-half (.5) a unit, then drop the remaining decimal if the preceding digit is an even number, but add 1 to the preceding digit if that digit is an odd number.

According to this rule, 10.335 would be rounded to 10.34 since the 3 in the hundredths column is an odd number. 10.345 would also be rounded to 10.34 since the 4 in the hundredths column is an even number. The purpose of the latter rule is to avoid a bias in rounding up or down across a large set of numbers, since with the rule, approximately half the time you will round up and half the time you will round down.

The number of decimal places that are used in reporting a statistic in a research report will differ depending upon the nature of the variable being reported. The average annual income of a group of individuals might be rounded to the nearest whole number (for example, $10,030), whereas the average number of seconds it takes a group of rats to run a maze might be reported to two decimal places (for example, 5.32 seconds). The number of decimal places you should report will depend upon how precise you need to be in order to make your point. In general, the present text will report answers to two decimal places.

When computing a statistic such as an average, it may be necessary to do intermediate calculations before you can arrive at a final answer. As a rule of thumb, intermediate calculations should be done using at least one decimal place beyond the number of decimal places you plan to report in your answer. Again, the exact number of decimal places you should use in your calculations will depend on the variable you are studying and the nature of the calculations performed. No hard-and-fast rules can be given to reduce rounding error.

Study Exercise 1.4

Round the following numbers to two decimal places.   *Answers.*   (a) 8.34
                                                                   (b) 7.44
                                                                   (c) 7.56
(a) 8.337   (c) 7.555001   (e) 13.63500   (d) 10.54
(b) 7.443   (d) 10.54500   (e) 13.64

The Concept of Probability. The concept of probability is an essential aspect of statistics. All of us are somewhat familiar with this concept in our everyday life. A weatherperson tells us that the chances of rain tomorrow are 70%. A bettor at the racetrack knows that the odds on a given horse are 3 to 1. The student thinks it is "likely" he or she will obtain an A in a course.

In statistics, the concept of probability has a precise meaning. For a given task, there may be a number of different possible outcomes. If you roll a die, there are six possible outcomes that could occur, 1, 2, 3, 4, 5, or 6. If you draw a card from a standard deck of playing cards, there are 52 different possible outcomes. The **probability** of some outcome, A, can be defined as the ratio

$$p(A) = \frac{\text{number of observations favoring outcome } A}{\text{total number of possible observations}}$$

On a die, what is the probability of rolling a 2? On a given roll, the total number of possible observations is six (that is, 1, 2, 3, 4, 5 or 6). There is only one 2 on a die and hence only one possible observation favoring the event "2." The probability is 1/6 or .17. If one draws a card at random from a standard deck, the probability of drawing an ace is 4/52 or .08. There are 52 possible observations, of which 4 favor an ace.

A given probability must always range from 0 to 1.00. It can never be less than zero or greater than 1.00. The probability of an impossible event is always zero. The probability of rolling an 8 on a single die is zero since the number of observations representing the event ''8'' is zero. It follows that 0/6 = 0. The probability of a completely certain event is always 1.00. For example, if an individual rolls a die, what is the probability that the individual will roll a 1, 2, 3, 4, 5, or a 6? In this case there are six outcomes of which all six will satisfy the event 1, 2, 3, 4, 5, or 6. The probability is therefore 6/6 = 1.00.

A probability is essentially a relative frequency and concerns the likelihood that something will occur. Sometimes probabilities can be logically derived a priori such as in the examples above. Other times, they are estimated empirically based on data an investigator has collected. A good example is that of a baseball player's batting average. If the player has been at bat 400 times and had 100 base hits, then his batting average is 100/400 = .250. All things being equal, the probability that the player will obtain a hit his next time up may be estimated to be .250.

The probability of an event can also be interpreted in terms of a ''long run'' perspective. If we flip a coin ten times, it is unlikely that the coin will come up heads exactly five times and tails the other five times. As we continue to flip the coin, then across a large number of flips (that is, over the long run), the number of heads relative to the total number of flips will approach 1/2, or .50.

1.8 / Populations and Samples

In scientific research, we are often interested in making descriptive statements about a group of individuals or objects. For example, one might state that the average number of children desired by married males in the United States is 2.3. Or, the average weight of males who are six feet tall is 180 pounds. Such statements are made with reference to a **population.** A population is the aggregate of all cases to which one wishes to generalize statements. In the first example, the population is all married males in the United States. In the second example, the population is all males who are six feet tall.

It may be the case that an investigator is unable to make observations on every member of the population about which he or she wishes to make a descriptive statement. In this instance, the investigator will resort to a **sample** of the population. A sample is simply a subset of the population. On the basis of observing the sample, generalizations are made to the population.

When we select a sample for purposes of making a statement about a population on a given dimension (for example, the average number of children), we want to ensure that we are using a **representative sample** of the population. If the population has 60% males and 40% females, we want our sample to reflect this. We don't want a *biased* sample that will lead us to make erroneous statements about the population. In selecting a sample, we want to use procedures that will yield a representative sample. Box 1.1 presents an interesting example of biased sampling as described by Darrell Huff in his excellent book *How to Lie with Statistics* (1954).

Box 1.1 Biased Sampling

"The average Yaleman, Class of '24," *Time* magazine noted once, commenting on something in the *New York Sun*, "makes $25,111 a year." Well, good for him! But wait a minute. What does this impressive figure mean? Is it, as it appears to be, evidence that if you send your boy to Yale you won't have to work in your old age and neither will he? Two things about the figure stand out at first suspicious glance. It is surprisingly precise. It is quite improbably salubrious. . . .

Let us put our finger on a likely source of error, a source that can produce $25,111 as the "average income" of some men whose actual average may well be nearer half that amount. This is the sampling procedure, which is the heart of the greater part of the statistics you meet on all sorts of subjects. Its basis is simple enough, although its refinements in practice have led into all sorts of by-ways, some less than respectable. If you have a barrel of beans, some red and some white, there is only one way to find out exactly how many of each color you have: Count 'em. However, you can find out approximately how many are red in much easier fashion by pulling out a handful of beans and counting just those, figuring that the proportion will be the same all through the barrel. If your sample is large enough and selected properly, it will represent the whole well enough for most purposes. If it is not, it may be far less accurate than an intelligent guess and have nothing to recommend it but a spurious air of scientific precision. It is sad truth that conclusions from such samples, biased or too small or both, lie behind much of what we read or think we know.

The report on the Yale men comes from a sample. We can be pretty sure of that because reason tells us that no one can get hold of all the living members of that class of '24. There are bound to be many whose addresses are unknown twenty-five years later.

And, of those whose addresses are known, many will not reply to a questionnaire, particularly a rather personal one. With some kinds of mail questionnaire, a five or ten per cent response is quite high. This one should have done better than that, but nothing like one hundred per cent.

So we find that the income figure is based on a sample composed of all class members whose addresses are known and who replied to the questionnaire. Is this a representative sample? That is, can this group be assumed to be equal in income to the unrepresented group, those who cannot be reached or who do not reply.

Who are the little lost sheep down in the Yale rolls as "address unknown"? Are they the big-income earners—the Wall Street men, the corporation directors, the manufacturing and utility executives? No; the addresses of the rich will not be hard to come by. Many of the most prosperous members of the class can be found through Who's Who in America and other reference volumes even if they have neglected to keep in touch with the alumni office. It is a good guess that the lost names are those of the men who, twenty-five years or so after becoming Yale bachelors of arts, have not fulfilled any shining promise. They are clerks, mechanics, tramps, unemployed alcoholics, barely surviving writers and artists . . . people of whom it would take half a dozen or more to add up to an income of $25,111. These men do not so often register at class reunions, if only because they cannot afford the trip.

Who are those who chucked the questionnaire into the nearest wastebasket? We cannot be so sure about these, but it is at least a fair guess that many of them are just not making enough money to brag about.

It becomes pretty clear that the sample has omitted two groups most likely to depress the average. The $25,111 figure is beginning to explain itself. If it is a true figure for anything it is one merely for that special group of the class of '24 whose addresses are known and who are willing to stand up and tell how much they earn. Even that requires an assumption that the gentlemen are telling the truth.

One procedure for approximating a representative sample is through the use of **random sampling.** The term *random* has a very precise meaning in scientific discourse. As applied to sampling problems, the essential characteristic of random sampling is that every member of the population has an equal chance of being selected for the sample.

Scientists use random sampling in a variety of ways. In order to obtain a random sample of a population, a survey researcher will make use of a very useful resource known as a random number table. A random number table is a list of numbers generated by a computer that has been programmed to yield a set of truly random numbers. Computers are used to construct such tables since the typical human is not capable of generating random numbers. For example, a person might have a tendency to list mostly even numbers, or those ending only in five and zero.

Appendix B presents a random number table. Suppose an investigator wanted to select a random sample of people from a population. This would involve obtaining a list of all members of the population and then arbitrarily assigning a number to each member. If the population consisted of 500 individuals, a list would be made of the names of these people and numbered from 1 to 500. The investigator would then consult a random number table, such as the one presented in Appendix B. Using the directions provided, the investigator would draw a sample of, say, 50 individuals. The use of random number tables ensures that the selection will be random and not influenced by any unknown selection bias the investigator may have.

It should be emphasized that the use of random procedures does *not* guarantee that a sample will be representative of the population. Random procedures will tend to yield representative samples, but sometimes nonrepresentative samples will result even when random procedures are used. We will discuss the problem of sampling in greater detail in later chapters.

Throughout this text we will refer to various numerical indices based on data from populations and from samples. When the indices are based on data from an entire population, they will be referred to as a **parameter.** When they are based on data from a sample, they will be referred to as a **statistic.**

1.9 / Descriptive and Inferential Statistics

The discipline of statistics has traditionally been divided into two major subfields, descriptive statistics and inferential statistics. The two are highly related to one another, and in some respects, the distinction is arbitrary. **Descriptive statistics** are those used to describe either a population (when measurements have been taken on all members of that population) or a sample. In this case, statistics are used to *describe* a group of individuals (be it a sample or a population) and to convey information about it. **Inferential statistics,** in contrast, involve taking measurements on a sample and then from the observations, inferring something about a population. In this instance, we are attempting to describe a population. However, we do so not by taking measures on all cases in the population, but rather by selecting a sample, observing scores on the variable of interest for that sample, and then *inferring* something with respect to that variable for the entire population.

EXERCISES

For each of the following studies, identify the independent variable and the dependent variable. Indicate whether each is a quantitative variable or a qualitative variable. Justify your answers.

1. Leonard Eron (1963) reported a study in which he examined the possible existence of a relationship between the exposure of young children to violent television shows and the amount of aggression they exhibited toward peers. Eron gathered information concerning the aggressive behavior and television viewing habits of 875 third grade children. By questioning parents about their child's viewing habits, Eron developed a 4-point scale measuring preference on the part of the child for aggressive TV shows. The scale had the following categories; very low preference for aggressive TV shows, low preference for aggressive TV shows, moderately high preference for aggressive TV shows, and high preference for aggressive TV shows. Aggression was measured by peer ratings of each child by at least two peers. These ratings could range from 0 to 32 with low scores indicating lower amounts of aggression.

2. Touhey (1974) was interested in studying the relationship between the type of occupation one held and the prestige people associated with that occupation. In this study, five different occupations were studied: architect, professor, lawyer, physician, and research scientist. A large number of individuals were asked to rate each of these occupations on a 60-point scale measuring perceived prestige. Low scores indicated low levels of prestige and higher scores indicated increasingly higher levels of prestige.

3. Steiner (1972) has discussed a series of experiments that studied the relationship between group size and how quickly a group could solve problems. In one experiment, six different group sizes were created, 2 members, 3 members, 4 members, 5 members, 6 members, and 7 members. Each group was then given a series of problems to solve, and the time until solution was measured for each group and compared.

4. Rubovits and Maehr (1973) were interested in the effects of teacher expectancies on their behavior toward students. Female undergraduates who were enrolled in a teacher training class were asked to prepare a lesson for four seventh and eighth grade students. Just before meeting with the student each teacher was told that two of the students were "gifted" and had high IQs, while the other two were "not gifted" and possessed an average intelligence. In reality, all children were about equal in ability, and these labels were assigned in an arbitrary manner. The teachers were then observed during a forty minute period, while they interacted with the four students. Rubovits and Maehr measured the number of times the teacher interacted with each student. The average number of interactions were then compared for students who were labeled as "gifted" versus those who were labeled as "nongifted."

5. Identify each of the following as being either a qualitative variable or a quantitative variable.

    (a) weight   (c) income   (e) sex

    (b) religion   (d) age

6. Indicate whether each measure listed below is a nominal measure, ordinal measure, interval measure, or ratio measure. Explain the reasons for your choice.

    (a) inches on a yardstick to measure the length of a table.

    (b) social security number

    (c) dollars as a measure of income

    (d) order of finish in a car race

7. What is the difference between ordinal measures and interval measures? Give an example of each.

8. Consider the following data on two variables, X and Y, for each of eight individuals.

Individual	X	Y
1	4	5
2	1	7
3	2	3
4	8	2
5	8	1
6	2	1
7	5	4
8	7	7

Perform the following calculations:

    (a) ΣX (i) ΣX^2

    (b) ΣY (j) ΣY^2

    (c) $\sum_{i=1}^{4} X_i$ (k) $\Sigma X - Y^2$

    (d) $\sum_{i=4}^{8} Y_i$ (l) $\Sigma (X - Y)^2$

    (e) ΣXY (m) $5\Sigma X$

    (f) $(\Sigma X)/N$ (n) $N\Sigma Y$

    (g) $(\Sigma Y)/N$ (o) $(\Sigma X)(\Sigma Y)$

    (h) $\Sigma (X - Y)$ (p) $(\Sigma X) - (\Sigma Y)$

9. Express the following in summation notation, for $N = 5$, on two variables, X and Y.

    (a) $X_1 + X_2 + X_3 + X_4 + X_5$

    (b) $X_1^2 + X_2^2 + X_3^2 + X_4^2 + X_5^2$

    (c) $(X_1 - Y_1) + (X_2 - Y_2) + (X_3 - Y_3) + (X_4 - Y_4) + (X_5 - Y_5)$

    (d) $(X_1^2 + X_2^2 + X_3^2 + X_4^2 + X_5^2)/N$

    (e) $5(X_1 + X_2 + X_3 + X_4 + X_5)$

    (f) $Y_1 + Y_2 + Y_3$

    (g) $X_3 + X_4 + X_5$

    (h) $X_1 Y_1 + X_2 Y_2 + X_3 Y_3 + X_4 Y_4 + X_5 Y_5$

10. Consider the following data for five individuals on two variables X and Y. Also, let the symbol k represent a constant, and let $k = 2$.

Individual	X	Y
1	4	2
2	6	2
3	2	4
4	2	4
5	4	4

    (a) Compute $\Sigma(Xk)$ (c) Compute $\Sigma(Yk)$

    (b) Compute $k(\Sigma X)$ (d) Compute $k(\Sigma Y)$

    (e) Compare your answer in (a) to your answer in (b), and your answer in (c) to your answer in (d). What equation describes this relationship?

    (f) Compute $\Sigma(X/k)$ (h) Compute $\Sigma(Y/k)$

    (g) Compute $(\Sigma X)/k$ (i) Compute $(\Sigma Y)/k$

(j) Compare your answer in (f) to your answer in (g), and your answer in (h) to your answer in (i). What equation describes this relationship?

11. Consider the following data for five individuals on two variables, X and Y. Also, let k be a constant, with $k = 3$.

Individual	X	Y
1	5	1
2	6	3
3	3	4
4	2	1
5	4	1

(a) Compute $\Sigma(X + k)$

(b) Compute $(\Sigma X) + Nk$

(c) Compute $\Sigma(Y + k)$

(d) Compute $(\Sigma Y) + Nk$

(e) Compare your answer in (a) to your answer in (b), and your answer in (c) to your answer in (d). What equation describes this relationship?

(f) Compute $\Sigma(X - k)$

(g) Compute $(\Sigma X) - Nk$

(h) Compute $\Sigma(Y - k)$

(i) Compute $(\Sigma Y) - Nk$

(j) Compare your answer in (f) to your answer in (g), and your answer in (h) to your answer in (i). What equation describes this relationship?

12. Round each of the following numbers to two decimals.

(a) 4.8932

(b) 8.9749

(c) 1.4153

(d) 4.1450

(e) 6.245002

(f) 2.615

(g) 6.3155

(h) .395

(i) .999

(j) 3.666

(k) 12.2538

(l) 9.724001

(m) 1.995

(n) 2.005

13. Compute the probability of each of the following events (assuming a random process).

(a) Drawing an ace, king, or queen from a standard deck of 52 cards.

(b) Throwing a 2 or a 3 on a single die.

(c) For a pair of dice, rolling a combination that totals 7.

(d) Drawing a face card from a standard deck of 52 cards.

14. Suppose you are considering whether to have an operation that is important to your health. A total of 420 operations of this nature have been performed in the past, of which 21 have been successful. Given only this information, what is your best guess as to the probability of success of your operation? What would be your decision? Why?

15. A newspaper conducted a survey in which they asked their readers to indicate their preference for either of two presidential candidates in an election, John Doe or Jane Smith. People were asked to cut out a ballot provided in the paper that day and send it to the newspaper with their preference indicated. One week later the newspaper reported they received 1,000 ballots, of which 800 favored Jane Smith. They stated that the "spirit of the community lies with Jane Smith." A total of 100,000 people live in the community. Is the newspaper's sample probably a representative sample of the community in general? Why or why not?

16. What is the essential characteristic of random sampling?

17. How are descriptive and inferential statistics different?

Frequency and Probability Distributions

2.1 / Frequency Distributions

Suppose an investigator administered a test designed to measure the aggressiveness of fifteen individuals. The scores could range from 0 to 12, with higher scores indicating greater aggressiveness. The scores for the individuals were as follows:

8	8	10
10	7	6
9	6	7
7	8	8
9	7	9

How can we best describe the scores on this test to another individual? One procedure is to list each of the fifteen scores. By examiniation, we can then obtain an intuitive feel about what the scores tend to be like. But suppose that instead of fifteen scores, there were 500 scores. It now becomes impractical to list all of the scores individually. A useful tool for summarizing a large set of data is a **frequency distribution.** A frequency distribution is a table listing the number of individuals who obtained a given score on a variable. We begin by listing the possible scores that were obtained, from highest to lowest, and then counting the number of individuals who had that score:

Score	f
10	2
9	3
8	4
7	4
6	2
	$N = 15$

The letter f is used to represent the word *frequency*. We can see from the above frequency distribution that two people had a score of 10, three people had a score of 9, four people had a score of 8, and so on. Altogether, there were a total of fifteen scores, or $N = 15$.

Considered alone, an index of frequency is not easily interpreted. Suppose you are told the results of a study showing that 200 people in a given town are prejudiced. This tells you little without also knowing, for example, the size of the town, which is 400. With this information, the frequency of 200 takes on more meaning. Half (200/400) of the town is prejudiced! This illustrates a more informative statistic that is used by researchers called a **relative frequency.** A relative frequency is the number of scores of a given type (for example, scores of 10) divided by the total number of scores. In the example on aggressiveness, these would appear as follows:

Score	f	rf	
10	2	2/15 =	.13
9	3	3/15 =	.20
8	4	4/15 =	.27
7	4	4/15 =	.27
6	2	2/15 =	.13
Sum	15	15/15 =	1.00

The letters *rf* are used to represent relative frequency. A relative frequency is simply the **proportion** of times that a score occurred. When this proportion is multiplied by 100, it reflects the **percentage** of times the score occurred. In the example, 13% of the individuals had a score of 10, 20% had a score of 9, and so on.

Study Exercise 2.1

Compute a frequency distribution and a relative frequency distribution for a standard deck of cards, ignoring the suit of the cards.

Answer. There are 52 cards in a deck, and hence $N = 52$. We order the cards from highest to lowest and then compute the relevant frequencies:

Card	f	rf
A	4	.077
K	4	.077
Q	4	.077
J	4	.077
10	4	.077
9	4	.077
8	4	.077
7	4	.077
6	4	.077
5	4	.077
4	4	.077
3	4	.077
2	4	.077

The frequency distribution in this case does not seem very useful since we could have easily stated the nature of the distribution in words: There are thirteen values and four cards representing each. The example does, however, illustrate an important relationship between relative frequency and probability. Recall from Chapter 1 that the probability of an event, *A*, is the number of outcomes of *A* divided by the total number of outcomes. This is exactly what a relative frequency reflects in a distribution: the number of "outcomes" or individuals who obtain a score of *X* divided by the total number of scores. Thus, from the above, the probability of selecting an ace from a well-shuffled deck of cards is .077.

In addition to frequencies and relative frequencies, some investigators report **cumulative frequencies** (represented by *cf*) and **cumulative relative frequencies** (represented by *crf*). For the previous example on test scores, these would appear as follows:

Score	f	rf	cf	crf
10	2	.13	15	1.00
9	3	.20	13	.87
8	4	.27	10	.67
7	4	.27	6	.40
6	2	.13	2	.13

The entries in the cumulative frequency column are obtained by a process of successive addition of the entries in the frequency column. Specifically, for any given score (for example, 7) the cumulative frequency is the f associated with that score (for 7 it is 4) plus the sum of all frequencies below that score. For the score of 7, the cumulative frequency is $4 + 2 = 6$. For the score of 9, it is $3 + 4 + 4 + 2 = 13$. The cumulative relative frequency is computed in the same manner but uses the column of relative frequencies instead of the column of frequencies. For the score of 7, the cumulative relative frequency is $.27 + .13 = .40$. For the score of 9, the cumulative relative frequency is $.20 + .27 + .27 + .13 = .87$.

The advantage of a cumulative frequency is that it allows one to tell at a glance the number of scores that are equal to or less than any given score. One can readily see that ten individuals had a score of 8 or less and that thirteen individuals had a score of 9 or less. By looking at the cumulative relative frequency column, we can see that the proportion of people who had a score of 8 or less was .67 and that the proportion of people who had a score of 9 or less was .87.

When one is concerned with a continuous variable such as aggressiveness, the frequencies, relative frequencies, and cumulative relative frequencies should be thought of in terms of the real limits of the scores. In the above data, although four of the individuals had a score of 8, this is more properly conceptualized as four of the individuals having a score between 7.5 and 8.5. Similarly, the cumulative frequency is conceptualized with respect to the upper real limit of the score. The cumulative frequency in the above data for a score of 8 is 10. Technically, this means that ten individuals had a score of 8.5 or less.

See Method of Presentation on page 29.

2.2 / Frequency Distributions for Quantitative Variables: Grouped Scores

The preceding analysis of frequency distributions examined the case where the test score was quantitative in nature. The scores ranged from 6 to 10 and individuals could be ordered in terms of how high their score was. Also, there were relatively few possible values. In this case, there were only five: 6, 7, 8, 9, and 10. Often we obtain measures on a variable for which there are a large number of values. For example, we might have a sample of 100 people, each with a different income. In constructing a frequency table we don't want to list 100 different values and then indicate a frequency of 1 by each value. This is neither practical nor informative. Rather, we will want to group the data and report it, perhaps as follows:

Income	f	rf	cf	crf
30,001–35,000	14	.14	100	1.00
25,001–30,000	21	.21	80	.86
20,001–25,000	30	.30	65	.65
15,001–20,000	19	.19	35	.35
10,001–15,000	16	.16	16	.16

Method of Presentation

When presenting the results of a frequency analysis, it is rare for an investigator to report the frequencies, relative frequencies, and cumulative relative frequencies. The major reason for this is that journal and book space is costly and consequently the data must be presented as efficiently as possible. Therefore, most research reports focus on relative frequencies and then provide information that will allow the reader to derive, if he or she so desires, the frequencies and cumulative relative frequencies. Actually, sometimes investigators do not report relative frequencies but rather they report the relative frequencies multiplied by 100. As noted earlier, this represents the *percentage* of individuals who obtained a given score. An example of a table that might be presented is as follows:

PERCENTAGE OF INDIVIDUALS WHO APPROVE OR DISAPPROVE OF NUCLEAR POWER PLANTS[1]

Response Category	Percentage
Strongly Approve	24.65
Moderately Approve	19.53
Neither Approve nor Disapprove	8.37
Moderately Disapprove	20.47
Strongly Disapprove	26.98

[1]$N = 215$

This table reports responses to a question asked of 215 college students concerning their approval or disapproval of nuclear power plants. All of the information is present that would allow the reader to determine the frequencies, relative frequencies, and cumulative relative frequencies, if he or she found it necessary to do so. The relative frequencies are derived by dividing each percentage by 100. Thus, 24.65/100 =

.2465, 19.53/100 = .1953, 8.37/100 = .0837, 20.47/100 = .2047, and 26.98/100 = .2698. The footnote to the table indicates the total number of individuals who responded to the question, in this case $N = 215$. The number of people in each category can then be obtained by multiplying each relative frequency by N. Thus, .2465 × 215 = 53, .1953 × 215 = 42, .0837 × 215 = 18, .2047 × 215 = 44, and .2698 × 215 = 58. We now have the information to construct a more detailed table:

Response Category	f	rf	cf	crf
Strongly Approve	53	.2465	215	1.0000
Moderately Approve	42	.1953	162	.7535
Neither Approve nor Disapprove	18	.0837	120	.5582
Moderately Disapprove	44	.2047	102	.4745
Strongly Disapprove	58	.2698	58	.2698

This example does not mean that you, as a reader, need to calculate the frequencies, relative frequencies, and cumulative frequencies for every table you encounter. Usually what is presented will be sufficient for interpretation. However, ideally if a reader wants to construct a more detailed table (for some reason the author hadn't anticipated), the information should be present that will allow the reader to do so. The format suggested here accomplishes this.

The fact that most investigators do not report the actual frequencies and cumulative relative frequencies does not mean that these statistics are of little utility. As we will see in later chapters, they are important concepts in understanding more advanced statistical techniques.

An important consideration in presenting grouped data is the decision as to how to group it together. Three questions are central: (1) How many groups should one use? (2) What should the interval size be for each group? and (3) What should be the lowest value at which the first interval starts? There are no standard rules that govern each of these questions. The nature of the grouping will usually depend upon the particular characteristics of the data. McCall (1980) has suggested several rules of thumb that will be considered here.

Number of Groups. In deciding how many groups to report, one must strike a balance between having too many groups that could overwhelm the reader and make the data incomprehensible, and having too few groups that could sacrifice precision. The problem of too many groups is illustrated in its extreme in our previous example on income, where each income is listed individually and given a frequency of 1. The problem of too few groups may be illustrated in the following table reporting scores of 65 individuals who took a test with possible scores between 0 and 100.

Score	f
50–100	64
0–49	1

This table is not very informative as it provides us with no insights into how the bulk of the individuals performed on the test.

In general, if the number of possible scores is small, one will use fewer groups, whereas if the number of possible scores is large, one will have to use more groups. As a rule of thumb, five to fifteen groups seems to be a useful number.

Size of Interval. Once you have determined how many groups to present, the question of the size of the class interval arises (for example, should the interval be from 30,000–35,000 or from 25,001–35,000?). Suppose we had a set of scores on a test that could range from 0 to 100. The highest score on the test was 99 and the lowest was 45. Suppose further that the decision was made to present the frequency analysis in five groups. A general rule of thumb for determining the interval size is to first subtract the lowest score from the highest score ($99 - 45 = 54$). This difference should then be divided by the number of groups, $54/5 = 10.1$. Since the .1 is awkward, we simply drop it, yielding an interval of 10.

Beginning of Lowest Interval. We now know that we will present five groups with an interval of 10 units per group. The final question is where to begin the lowest interval (45–54 or 41–50, and so on). It is customary to use a number that is nearest to (but not greater than) the lowest score that is evenly divisible by the interval size.

In the example, the lowest score is 45 and the interval size is 10. The closest number to 45 that is evenly divisible by 10 is 40. This should be the starting point for the lowest interval. The distribution, according to these general rules, is:

Score	f	rf
90–99	3	.05
80–89	6	.09
70–79	15	.23
60–69	21	.32
50–59	12	.18
40–49	8	.12

Note, that it was necessary to use six groups instead of five because of the determined interval size and lowest value. Again, the above rules are only guidelines that may be useful in presenting grouped data.

2.3 / Frequency Distributions for Qualitative Variables

When reporting a frequency distribution for a qualitative variable, the most common practice is to list the values of the variable in a column and the corresponding frequencies (or percentages) in columns to the right of this. All of us have encountered such frequency distributions in newspaper polls of candidate preference in political elections. In a small community in Indiana, a poll was taken of the preferences for Ronald Reagan and Jimmy Carter in the 1980 presidential election. The results of the poll were reported in the newspaper as follows:

Candidate	Percentage Favoring
Reagan	62.6
Carter	27.0
Undecided	6.4
Other	4.0

The values of the categorical variable are "Reagan," "Carter," "Undecided," and "Other," and the relevant percentages are presented to the right of each value. The poll was based on a sample of 1,000 individuals.

Given the following 100 scores, construct a frequency distribution to characterize them.

9	29	36	·15	37	12	38	19	37	47
29	14	8	24	13	6	48	19	2	25
39	28	39	49	27	31	9	26	18	1
5	22	16	23	6	46	38	4	49	39
34	15	21	35	11	32	2	19	29	11
43	26	7	44	49	28	29	41	1	41
39	25	24	19	17	5	33	32	27	23
26	44	17	25	31	42	43	12	3	33
22	24	23	45	27	28	3	18	42	22
4	16	21	34	14	35	13	36	21	46

Answer. We will use ten groups to characterize these data. The first step involves defining the interval size. The highest score is 49 and the lowest is 1. The difference between these is $49 - 1 = 48$. This number divided by the number of groups is $48/10 = 4.8$. We round this off to a whole number,

5, which defines the interval size. The next step is to define the beginning of the lowest interval. The lowest number is 1. The closest value that is less than 1 and that is evenly divisible by 5 is 0. We will therefore use ten groups beginning with the number 0, with class intervals of size 5. The analysis is as follows:

Score	f	rf	cf	crf
45–49	8	.08	100	1.00
40–44	8	.08	92	.92
35–39	12	.12	84	.84
30–34	8	.08	72	.72
25–29	15	.15	64	.64
20–24	13	.13	49	.49
15–19	12	.12	36	.36
10–14	8	.08	24	.24
5–9	8	.08	16	.16
0–4	8	.08	8	.08

2.4 / Frequency Graphs

Frequency Graphs for Quantitative Variables. Sometimes rather than presenting tables of a frequency analysis, investigators will report their data graphically, using either a **frequency histogram** or a **frequency polygon.** Consider the frequency analysis presented earlier:

Score	f	rf
10	2	.13
9	3	.20
8	4	.27
7	4	.27
6	2	.13

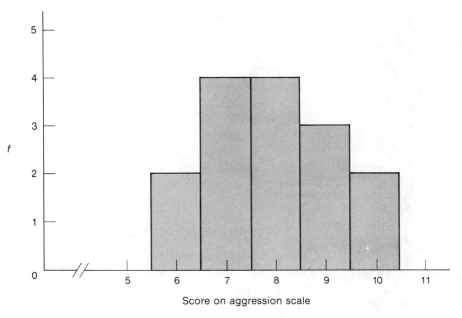

Figure 2.1
Frequency Histogram

These data can also be presented in the form of a frequency histogram that appears in Figure 2.1. The horizontal dimension is called the **abscissa** and the vertical dimension is called the **ordinate.** The abscissa lists the test scores from low to high (as does the first column of the frequency table under the heading ''Score''). In this case, however, the abscissa extends from one unit below the lowest score to one unit above the highest score, from 5 to 11 instead of from 6 to 10. The ordinate represents the frequency with which each score occurred (corresponding to the second column of the frequency table). The bar for a test score of 6 goes up 2 units indicating that two individuals had a score of 6. The bar for a test score of 7 goes up 4 units indicating that four individuals had a score of 7.

The width of the bar for any given score represents the real limits of that score. Consider the bar for a score of 6. The left-most point of the bar represents the real limit 5.5 and the right-most point of the bar represents the real limit 6.5. The midpoint of the bar corresponds to the midpoint of the real limits 5.5 to 6.5, or 6. Notice also that there is a broken line on the abscissa. This is traditionally done when the abscissa ''jumps'' from zero to a larger number and is not drawn to scale. The same principle would hold for the ordinate.

A frequency polygon is similar to a frequency histogram and uses the same ordinate and abscissa. A frequency polygon of the same data appears in Figure 2.2. The major difference from the frequency histogram is that bars are not used, but rather solid dots are indicated directly above the midpoint of the score corresponding to the appropriate frequency. The dots are then connected by a solid line, forming a polygon when the line is closed with the abscissa, hence the name, frequency

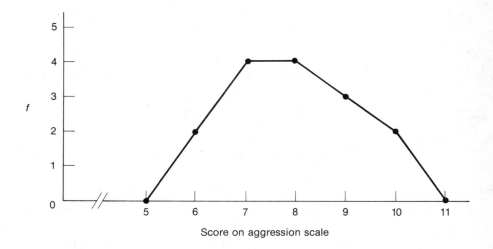

Figure 2.2
Frequency Polygon

polygon. The similarity of the frequency histogram and frequency polygon can be seen in Figure 2.3, where one has been superimposed on the other.

There are no specific rules for when a frequency histogram as opposed to a frequency polygon should be used. Frequency polygons are typically used when the variable being reported is continuous in nature whereas the frequency histogram is used when the variable being reported is discrete in nature. The major reason is that, from a visual perspective, the frequency polygon tends to highlight the "shape" of the entire distribution more than the frequency histogram. The frequency histogram, by comparison, tends to highlight the frequency of occurrence of a specific score rather than the entire distribution. The use of a frequency polygon for a continuous variable is thus more consistent with the notion of emphasizing a continuum.

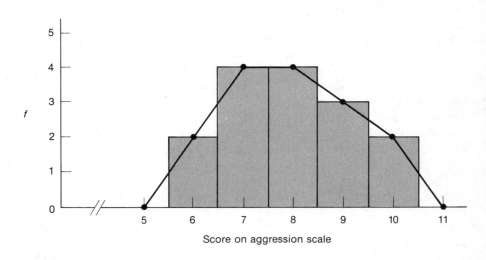

Figure 2.3 Frequency Polygon Superimposed on a Frequency Histogram

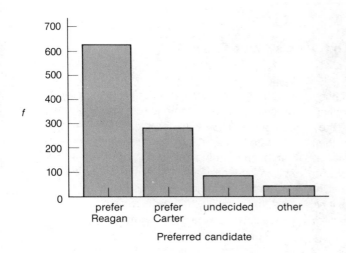

Figure 2.4 Bar Graph for Presidential Poll

Frequency Graphs for Qualitative Variables. Frequency histograms and frequency polygons are used to graph frequency data for quantitative variables. Frequency graphs can also be constructed for qualitative variables. Technically, such graphs are called **bar graphs** and appear as in Figure 2.4. This figure plots the data from the poll of preferences for presidential candidates mentioned in Section 2.3. The values of the variable are listed on the abscissa and the frequencies are listed on the ordinate. The major difference from the frequency histogram is that the bars are drawn such that they do not touch one another. Aside from this feature, the basic principles in constructing a bar graph are the same as those for the frequency histogram.

2.5 / Misleading Graphs

The presentation of data in terms of graphs can be highly informative but can also be misleading. Consider the case where an individual has interviewed 100 people in order to determine how many prefer product A over product B, and vice versa. Suppose that forty-nine preferred A over B and fifty-one preferred B over A. This could be presented graphically as in either of the two frequency histograms presented in Figure 2.5.

If you were to look at graph (a), and not examine carefully the frequencies labeled on the abscissa, then you might conclude that B was substantially preferred over A. In graph (b), however, the differences look much smaller. This is because the distance between the units on the abscissa is smaller in (b) than in (a). When reading graphs, one should always be careful to examine the labels on both the abscissa and the ordinate.

Both the frequency histogram and the frequency polygon represent efficient ways of presenting frequency analyses graphically. Typically, they are used only when necessary and when they nicely illustrate a major trend in the data that might otherwise be difficult to note. It should be mentioned, however, that given an appropriately presented frequency table, the reader can, if he or she wishes, construct a frequency graph from the table.

In presenting frequency graphs it is im-portant that both the abscissa and the ordinate be clearly labeled and marked. The abscissa is sometimes referred to as the "X axis" and the ordinate as the "Y axis." Since frequency graphs can be misleading depending upon how the abscissa and ordinate are formatted, social scientists have adopted a "three-quarter high" rule. This rule states that the ordinate should be presented such that the maximum height (that is, the score with the highest frequency) is equal to three-quarters of the length of the abscissa.

2.6 / Graphs of Relative Frequencies

It is interesting to note the nature of a graph of relative frequencies as compared to that of frequencies. A polygon of the relative frequencies for the example with scores ranging from 6 to 10 in Section 2.4 appears in Figure 2.6. Notice that the shape of the polygon is identical to the shape of the frequency polygon. This should hardly be surprising since all we have done is divide each frequency by the same number, namely N.

Figure 2.5 Example of Misleading Graphs

(a)

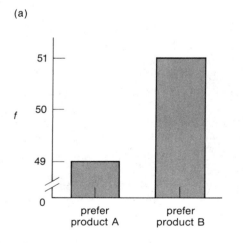

(b)

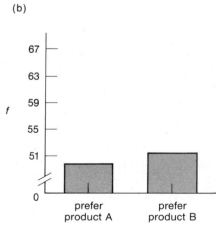

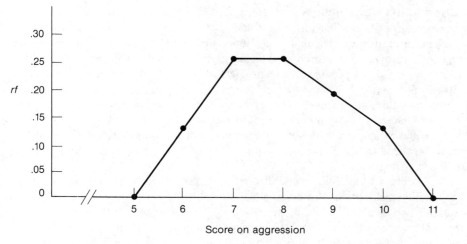

Figure 2.6 Graph of Relative Frequencies

2.7 / Probability Distributions

Probability Distribution for Qualitative and Discrete Variables. In Chapter 1, the probability of an event, *A,* was defined as the number of outcomes that favored *A* divided by the total number of outcomes. In an earlier section of this chapter, the relationship between a relative frequency and probability was noted. Given a population of scores, the probability of randomly selecting a given score from that population equals the relative frequency of that score. Consider the case of a qualitative variable, sex, which has two possible scores, male and female. Suppose we have a population of 200 individuals, with 150 females and 50 males. The relative frequency for females is 150/200 = .75 and for males it is 50/200 = .25. The probability of randomly selecting a female from this population is .75 and the probability of randomly selecting a male is .25. When the potential values on a qualitative or discrete variable are such that a person can have one and only one score (for example, it is impossible for a person to be both male and female—he or she must be either a male *or* a female)*, and when the values considered are **exhaustive** (that is, there are no other possible values that can occur and we have specified all possible values), then the relative frequencies of each score will represent a **probability distribution** with respect to that variable. Figure 2.7 presents a graph of a probability distribution for the population of 200 individuals, with 150 females and 50 males. Since we are, in effect, graphing relative frequencies, the shape of the probability distribution will correspond to the shape of the frequency distribution, as noted in the previous section.

*This is an example of scores that are **mutually exclusive.**

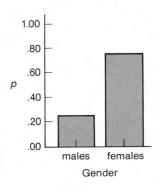

Figure 2.7 Graph of a Probability Distribution for Sex

Continuous Variables. A probability distribution for a continuous variable is conceptualized somewhat differently than that for either a qualitative or discrete variable. Recall from Chapter 1 that a number representing a score on a continuous variable is properly considered from the perspective of the real limits of that number. When we say that a person solved a problem in 30 seconds, we don't mean exactly 30 seconds, but rather somewhere between 29.5 and 30.5 seconds. Because it is always possible, in principle, to have a measuring device that is more accurate than the one we used, we can not meaningfully talk about the probability of obtaining a score equal to an *exact* value for a continuous variable. Rather, probability is better conceptualized as being associated with a range of values such as greater than 29.5 and less than 30.5.

In contrast to qualitative and discrete variables, it is not possible to specify a probability distribution of a continuous variable by listing possible values of the variable and their corresponding probabilities. This is because the number of possible values a continuous variable can have is, in principle, infinite. Statisticians therefore represent a probability distribution of a continuous variable in terms of something called a **probability density function** (PDF).

You can get an intuitive feel for this concept through graphs. Figure 2.8 presents a probability distribution for the continuous variable of intelligence in a population. A continuous probability distribution is always represented as a smooth curve over the abscissa. The abscissa represents the possible values of the continuous variable, with increasingly higher values going from left to right. The ordinate, although not formally demarcated, represents an index of the frequency with which values occur in the population, with higher values going from bottom to top.

The key to understanding a probability density function is conceptualizing the probability as an *area* under the curve, or as it is more formally called, the **density curve.** The total area under the curve represents 1.00, or the probability that a given person will have *some* value on the dimension in question. In Figure 2.8, two points, *a* and *b*, have been marked on the abscissa. These represent the limits of an interval. The shaded portion between *a* and *b* is the area under the curve that corresponds to scores in that interval. The probability of obtaining a value between *a* and *b* may thus be represented by this shaded area under the curve. Using some advanced mathematics, it is possible to compute the size of this area, based upon the case where the total area under the curve equals 1.00. In this way, it is possible to specify the probability of obtaining a set of values falling within some interval for a continuous variable. Specifically, the probability of obtaining a score within a given interval will equal the area of that interval under the density curve. In Figure 2.8, the probability of obtaining a score between points *a* and *b* is .23. We will discuss formal procedures for calculating such areas in Chapter 4.

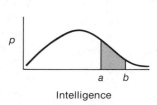

Intelligence

Figure 2.8 Graph of a Probability Distribution for a Continuous Variable

Box 2.1 Frequency Distribution of IQ Scores

The concept of intelligence is a major interest of social scientists. Many attempts have been made to measure this variable, some of which are controversial. One of the most widely used tests for adults is the Wechsler Adult Intelligence Scale (WAIS). This test has been extensively studied and applied to several national samples in the United States. One such application is reported in Wechsler (1958). Based upon 2,052 individuals, the following frequency analysis characterizes the scores on this test:

Score	f	rf	crf	Verbal Description
Above 130	29	.014	1.000	Very superior
120–129	150	.073	.986	Superior
110–119	349	.170	.913	Bright normal
100–109	525	.256	.743	Average
90–99	504	.246	.487	Average
80–89	299	.146	.241	Dull normal
70–79	132	.064	.095	Borderline
below 70	64	.031	.031	Mentally retarded

$N = 2,052$

The verbal descriptions provided on the extreme right are labels often used by clinical psychologists with respect to ranges of scores on the WAIS. Notice that scores labeled "average" (90–109) are the most common. Also, the proportion of individuals with scores above 130 (very superior) is quite small. This is also true of individuals with scores below 70 (mentally retarded). Only 1.4% of the sample had scores above 130 and 3.1% had scores below 70.

In 1926, a social scientist named C. Cox published an extensive study of eminent men in history. He attempted to estimate the IQ scores that these individuals would have achieved had they lived in a day when such a test could be administered. Some of these estimates are as follows:

Francis Galton—English scientist	200
John Stuart Mill—English philosopher	190
Johann Wolfgang von Goethe—German writer, philosopher	185
Gottfried Wilhelm von Leibnitz—German philosopher, mathematician	185
Samuel Taylor Coleridge—English writer, poet	175
John Quincy Adams—American statesman, President	165
David Hume—English philosopher	155
Alfred Tennyson—English poet	155
René Descartes—French philosopher, mathematician	150
Wolfgang Amadeus Mozart—Austrian composer	150
William Wordsworth—English poet, writer	150
Francis Bacon—English philosopher, scientist	145
Charles Dickens—English writer	145
Benjamin Franklin—American inventor, statesman	145
George Frederick Handel—German composer	145
Thomas Jefferson—American statesman, President	145
John Milton—English poet	145
Daniel Webster—American statesman, senator	145

The magnitude of these estimates is impressive, especially in light of the frequency analysis presented above. These scholars were obviously unique individuals.

2.8 / Summary

A frequency distribution is a useful tool for describing a set of scores. It is a table that conveys information about the frequency, relative frequency, cumulative frequency, and cumulative relative frequency of a score. When presenting a frequency distribution for a quantitative variable, it is sometimes useful to group the scores. This requires the consideration of three issues: (1) the number of groups to use, (2) the size of the interval, and (3) the beginning of the lowest interval. Occasionally, frequency analyses will be presented graphically, using either the frequency histogram or the frequency polygon. Frequency distributions are related to probability distributions. For either a qualitative or discrete variable, which has values that are mutually exclusive and exhaustive, a probability distribution is simply the relative frequency for the respective values. For a continuous variable, a probability distribution is best thought of in the context of a probability density, or the area under a density curve.

EXERCISES

An employer kept records of how many days her twenty employees reported in sick during the previous year. For these employees, the scores on this variable were as follows:

8	7	6	4	3
6	3	7	6	6
4	6	6	6	7
6	6	8	7	6

1. Compute a frequency distribution for this set of scores.

2. Compute the relative frequencies, cumulative frequencies, and cumulative relative frequencies for the scores.

3. What percentage of employees was sick for eight days? What percentage of employees was sick for more than six days? What percentage of employees was sick less than five days?

4. Suppose you were to randomly select a score from the above twenty scores. What is the probability this score would be an 8? What is the probability this score would be a 6 or an 8? What is the probability this score would be less than 7?

5. Draw a frequency histogram of the data.

6. Draw a frequency polygon of the data.

7. Draw a frequency polygon of the cumulative relative frequencies.

A principal in a small school measured the intelligence of students in the fifth grade in his school. He used a test where the average score for a national sample was 100. The scores of the students were as follows:

129	110	109	90	89	100	119	120	99	80
100	104	95	107	105	102	98	105	111	94
92	106	111	117	113	93	106	92	105	101
118	103	103	98	96	119	108	118	106	117
128	109	108	127	83	84	122	103	102	88

8. Compute a frequency distribution for these data by grouping the scores into five groups.

9. Compute the relative frequencies, cumulative frequencies, and cumulative relative frequencies for the grouped scores.

10. What percentage of students had a score greater than 109? What percentage of students had a score less than 100?

11. Draw a frequency histogram of the data.

12. Draw a frequency polygon of the data.

13. Draw a frequency polygon of the cumulative relative frequencies. Compare the general shape of this distribution to that in Exercise 7.

Suppose you were commissioned to survey a small community to determine the marital status of all adults over 18. You select a random sample of fifty individuals and ask each individual if they are married (M), divorced (D), widowed (W), or single (S). The data for these individuals are as follows:

M	S	D	W	M	M	S	D	W	M
M	M	S	M	M	M	D	M	M	M
M	W	M	S	S	M	M	M	M	S
M	M	M	M	M	S	S	M	M	M
M	S	D	W	M	M	S	D	W	M

14. Write your report for your employer summarizing these data and using the principles developed in this chapter. Present the data in a form that you think is most informative.

15. Suppose a salesman is trying to sell to someone a house in neighborhood A instead of neighborhood B. The salesman knows that the number of people with incomes above 50,000 dollars in neighborhood A is 30, while in neighborhood B, it is 15.

(a) Draw a bar graph of the data that makes the differences in these two neighborhoods look large.

(b) Draw a bar graph of the data that makes the differences in these two neighborhoods look small.

16. A researcher reported the results of a survey of 1,850 people on their attitudes toward capital punishment. One question asked respondents to indicate if they thought the death penalty should be legal. Responses to this question were scored from 1 (definitely should not be legal) through 5 (definitely should be legal), where 3 = a neutral stance. The results were as follows:

Score	Percentage
5	19
4	31
3	10
2	22
1	18

Convert the above into a frequency analysis, reporting f, rf, cf, and crf.

Measures of Central Tendency and Variability

Describing a set of scores with frequency distributions can be a highly informative way of presenting data about a variable. However, often we are interested in communicating data in a more succinct fashion. Statisticians have developed statistical measures that are useful in characterizing a set of scores for quantitative variables. These will be considered in this chapter.

3.1 / Measures of Central Tendency

When trying to describe a set of scores, one useful piece of information concerns where the scores tend to be, or their **central tendency.** A central tendency refers to the average score, or that score around which other scores tend to cluster. Many indices of central tendency have been proposed and we will consider three of them: the mode, median, and mean.

Mode. The **mode** of a distribution of scores is the most easily computed index of central tendency. It is simply the score that occurs most frequently. For the set of scores 10, 10, 8, 8, 8, 6, the mode is 8, since it occurs most frequently (3 times). If we were to randomly select one score from a set of scores, the value of that score would most likely be equal to the value of the mode as opposed to any other value, since the mode occurs most frequently. In a graph of a distribution of scores, the modal score will have the highest "peak" in the graph.

Although the mode can be a relatively straightforward index of central tendency, it has several disadvantages. The major problem is that there can be more than one modal score. Consider the following scores: 10, 10, 10, 8, 8, 6, 6, 6. Both 10 and 6 are the most frequently occurring scores. In this case, there is ambiguity as to which score is *the* mode since both occur with equal frequency. This set of scores is *bimodal* since it has two modes. When the mode is not equal to one unique value, it loses some of its effectiveness in characterizing the central tendency of a distribution of scores.

Median. Another measure of central tendency is the median. The **median** is that point in the distribution of scores that divides the distribution into two equal parts. In other words, 50% of the scores occur above the median and 50% of the scores occur below the median. In this sense, the median is a measure of central tendency.

There are three different approaches to computing the median. The first concerns the case where there is an even number of scores. Suppose we measured how long it took six students to complete a test. The times for each student are 20, 22, 16, 18, 25, and 27, respectively. To compute the median, we first order the scores from lowest to highest:

16   18   20   22   25   27

The median is the arithmetic average of the two middle scores, or $(20 + 22)/2 = 21$. Figure 3.1a presents this graphically, using a frequency histogram, which also shows the real limits of each score. Note that 50% of the scores occur below 21 and 50% of the scores occur above 21.

The second approach to computing the median applies when there is an odd number of scores. Suppose we measured the test-taking time of seven students and their scores were 6, 8, 14, 12, 16, 6, and 16 minutes, respectively. To compute the median, we order the scores from lowest to highest:

6   6   8   12   14   16   16

The median is simply the middle score, 12. Figure 3.1b presents this graphically. It also highlights the importance of considering the median in the context of the real limits of scores. Examining the values 6, 6, 8, 12, 14, 16, and 16, it is clear that 50% of the scores are *not* less than 12. Only three of the seven scores are less than 12 and three are greater than 12. With an odd number of scores, the problem is what to do with the middle score. The answer lies in the concept of real limits. The individual who obtained the middle score of 12 minutes is properly conceptualized as scoring somewhere between 11.5 and 12.5 minutes. We do not know exactly where in the interval he or she scored, so the most logical thing to do is to divide the interval in half to define the median. This yields the value of 12, or the middle score in the distribution. Note in Figure 3.1b that the point 12 equally divides the distribution in half. There are three and one-half frequency boxes below 12 and three and one-half frequency boxes above 12.

The third approach to computing the median applies when there are duplications of the middle scores, regardless of whether the number of scores is odd or even. This is the more common occurrence in the social sciences. Suppose we measured the test-taking time of ten students and their scores were 6, 8, 6, 8, 6, 8, 9, 10, 9, and 10 minutes. We order the scores from lowest to highest:

6   6   6   8   8   8   9   9   10   10

The average of the two middle scores is 8, if we were to apply our previous approach. However, this is not the median. Only three of the ten scores are less than

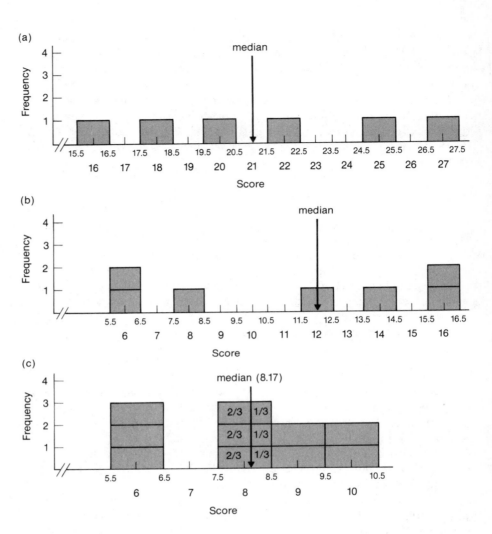

Figure 3.1 Frequency Distributions Illustrating the Median

8! Statisticians have developed a formula for computing the median in this case. The formula is based on interpolation in which the median is assumed to occur within the real limits of the middle score. Consider the example in Figure 3.1c in which there are ten scores and 8 is the middle score. The median occurs somewhere between 7.5 and 8.5, and represents the point where five scores are less than it and five scores are greater than it. There are already three scores less than 8 since three individuals had a score of 6. The median must therefore be defined so that two more scores are less than it. Individuals with a score of 8, technically, have scores between the real limits 7.5 and 8.5. Since three individuals had a score of 8, and we want to define the median so that two of these three individuals score less than it, we can do so by specifying a score 2/3 greater than the lower limit, or 7.5 + 2/3 = 8.17.

This, of course, assumes that the actual scores obtained by the three individuals were equally spaced from one another within the 7.5 to 8.5 limits.

A formula for computing the median can be applied using a frequency analysis. In the last example, a frequency analysis would appear as follows:

Score	f	rf	cf	crf
10	2	.20	10	1.00
9	2	.20	8	.80
8	3	.30	6	.60
6	3	.30	3	.30
	$N = 10$			

The formula is

$$Md = L + \left(\frac{N(.50) - n_L}{n_W}\right)i \qquad \text{[3.1]}$$

where Md represents the median, L is the lower limit of the category containing the median, n_L is the number of scores less than the lower limit of the category containing the median, n_W is the number of scores occurring within that category, and i is the size of the interval containing the median (or more simply, the difference between the upper limit and the lower limit of the category containing the median). We begin by referring to the cumulative relative frequency column. Starting from the bottom, we move up until we find the first number that is greater than or equal to .50. It is .60. This occurs in the category for scores of 8. The lower limit of the category is 7.5, and this represents L. The number of scores *within* the category is 3 and this represents n_W. The number of scores *below* 7.5 is 3. This represents n_L. The interval size, i, is $8.5 - 7.5$, or 1.00. Thus,

$$Md = 7.5 + \left(\frac{(10)(.50) - 3}{3}\right)1.00$$

$$= 7.5 + \left(\frac{5 - 3}{3}\right)1.00$$

$$= 7.5 + \frac{2}{3}$$

$$= 8.17$$

This agrees with what we stated previously. In sum, when there are no ties in the middle scores, the median is (a) equal to the middle score when N is an odd number and (b) equal to the arithematic average of the two middle scores when N is an even number. When there are ties in the middle scores, Equation 3.1 will yield the value of the point that equally divides the scores in half, taking into consideration the real limits of the scores.

The median has an interesting statistical property that underscores its role as a measure of central tendency. Across all individuals, scores will be closer to the value of the median than any other value. Consider the following five scores, which have a median of 3: 1, 2, 3, 7, 8. Suppose we compute the absolute value of how far each score is from the median:

$$|1 - 3| = 2$$
$$|2 - 3| = 1$$
$$|3 - 3| = 0$$
$$|7 - 3| = 4$$
$$|8 - 3| = 5$$

If we sum these absolute deviation scores, we obtain 12. Any value other than the median would yield a sum greater than 12. In this sense, the median is a measure of central tendency since it minimizes the absolute difference between it and the scores in the distribution.

Study Exercise 3.1

A sociologist was interested in determining the median number of years that current employees had been working for a given company. Company records for the 200 employees of the company were used to collect data. The frequency analysis for this variable was as follows:

Number of Years	f	rf	crf
6	20	.10	1.00
5	20	.10	.90
4	30	.15	.80
3	50	.25	.65
2	50	.25	.40
1	10	.05	.15
0	20	.10	.10

Compute the median.

Answer. Using Equation 3.1, we find that

$$Md = 2.5 + \left(\frac{(200)(.5) - 80}{50} \right) 1.00$$

$$= 2.5 + \left(\frac{20}{50} \right) 1.00$$

$$= 2.90$$

This means that the average number of years in the company, as defined by the median, is 2.90.

Mean. The **mean** of a set of scores is probably most familiar to all of us. It is simply the arithmetic average of the scores. It is computed by summing all of the scores and then dividing this sum by the total number of scores. This can be represented by the equation

$$\overline{X} = \frac{\Sigma X}{N} \qquad\qquad [3.2]$$

where $\overline{X}$ is the symbol used to represent a mean of a set of scores on variable X. As an example, consider the following ten scores:

2	5
3	5
3	6
4	8
5	9

The sum of the ten scores is

$$\Sigma X = 2 + 3 + 3 + 4 + 5 + 5 + 5 + 6 + 8 + 9 = 50$$

Hence, the mean is

$$\overline{X} = \frac{\Sigma X}{N} = \frac{50}{10} = 5.00$$

The mean has an important statistical property that will be encountered in later sections of this book; namely, the sum of deviations from the mean will always equal zero. Consider the following five scores, which have a mean of 3: 5, 3, 4, 1, 2. The deviation of each score from the mean is $X - \overline{X}$, or

$$5 - 3 = 2$$
$$3 - 3 = 0$$
$$4 - 3 = 1$$
$$1 - 3 = -2$$
$$2 - 3 = -1$$

If we sum the deviation scores, we get zero. This is true for any set of scores. The sum of deviation scores about the mean will always equal zero. In this sense, the mean is an index of central tendency. It balances the deviation of scores above it with the deviation of scores below it. Note that this is distinct from the median. The mean is the value that minimizes the sum of *signed* deviations, whereas the median is the value that minimizes the sum of *unsigned* deviations.

A researcher developed a test to measure reading ability. The test was intended to be used to help place students in remedial versus advanced reading and English classes. A major concern of this researcher was how long it would take students to complete the test, since it had to be administered during a short testing period. For twelve students in one of his classes, the following times, in minutes, were required to complete the test:

Individual	Time
1	12
2	13
3	11
4	12
5	11
6	13
7	13
8	11
9	12
10	12
11	12
12	12

Compute the mean amount of time it took the twelve students to complete the exam.

Answer. Using Equation 3.1, we find that

$$\Sigma X = 12 + 13 + 11 + 12 + 11 + 13 + 13 + 11 + 12 + 12 + 12 + 12 = 144$$

$$\overline{X} = \frac{\Sigma X}{N} = \frac{144}{12} = 12.00$$

This means that the average amount of time it took to complete the exam, as defined by the mean, was 12.00 minutes.

Comparison of the Mean, Median, and Mode. The mean, median, and mode are all indices of central tendency. The mean is the arithmetic average of scores, the median is the point that divides the distribution into halves, and the mode is the most frequently occurring score. Sometimes the mean, median, and mode of a set of scores all have the same value. But more often than not, they yield different values. Which is the best index of central tendency? Ideally, when trying to characterize a set of scores, it is best to report all three indices. Each represents something slightly different, and the more information we can provide, the better. Most of the inferential statistics used by social scientists make use of the mean, as will be explained in Chapter 6. Consequently, the mean is the most frequently encountered measure of central tendency in research reports.

One way of contrasting the mean, median, and mode is in terms of a "best guess" interpretation of each. Suppose you know only the mean, median, and mode of a large set of scores. Each individual score is randomly selected in succession and after each selection you are to guess the value of the score. What is your best guess? It depends. First, suppose you want the highest probability of being exactly correct. In this case, your best guess is the value of the mode since it is the most frequently occurring score and yields the highest probability of predicting each score exactly. Second, suppose you aren't interested in being exactly correct most often, but rather in making the smallest amount of absolute error, across all scores. In this case, the *sign* of the error is unimportant (that is, whether you overpredict or underpredict the value of the score isn't critical), but the *size* of the error is. Here, your best guess is the value of the median. The median is a measure of central tendency in that it minimizes the absolute (unsigned) error across all scores. Finally, if your goal is to minimize *signed* error, then, across all scores, your best guess is the mean. As noted earlier, this is because, across all scores, the amount of signed error will equal zero.

There is no general rule as to which measure of central tendency is best, since it depends upon what one is trying to accomplish or communicate. For purely descriptive purposes, the median is a very useful measure since it minimizes unsigned error. For purposes of inferring population parameters from samples, the mean is a superior index, as will be explained in Chapter 6.

It may be instructive to consider an example of when the mean, median, and mode can yield a very different characterization of a set of scores. Suppose you are a businessman with seven employees (six others and yourself). The yearly salaries are as follows:

Employee number	Salary ($)
1	3,000
2	3,000
3	4,000
4	5,000
5	6,000
6	7,000
Yourself	175,000

For the set of seven scores, the mean is 29,000, the median is 5,000, and the mode is 3,000. In characterizing the "average" score, you could paint a very different picture of your company depending upon the measure of central tendency used (for example, "Look how generous I am—the average salary is $29,000" as opposed to "Look how small my business is—the average salary is only $5,000").

In general, when there are a few extreme scores in the distribution (in this case, the extreme score is the $175,000 salary), the information conveyed by the mean can become distorted and the median will provide better insights into the central tendency of the data. Note that the median is not really affected by the extreme score. The median would have been the same if your salary was $5,500 or $250,000. In contrast, the mean is affected by the extremity of the score.

Use of the Mean, Median, and Mode. We have presented three measures of central tendency in the context of analyzing a quantitative variable. Some qualifications about the appropriateness of the indices for describing central tendencies are now required. When the quantitative variable is measured on approximately an interval level, all three measures of central tendency are meaningful. Ideally, all measures would be presented in a research report. When the quantitative variable is measured on an ordinal level that departs markedly from interval characteristics, the mean may not be an appropriate index of central tendency and the median is better. This is because the mean relies on calculating a sum, and a sum is only meaningful when the intervals between successive categories are approximately equal. Finally, for qualitative variables, the concept of a mean and median are meaningless since these concepts require ordering objects along a dimension. In this case, the mode or the most frequently occurring category will be the best descriptor of central tendency.

3.2 / Measures of Variability

Measures of central tendency indicate where scores tend to cluster in a distribution. A second important characteristic in analyzing quantitative variables is the extent to which scores are alike or different. Consider the following two sets of scores:

Set I	Set II
5	3
5	4
5	5
5	6
5	7

In both cases, the mean is 5.00. However, the **variability** of the scores, or the extent to which they are similar or dissimilar, is very different. In set I, all of the scores are the same, but in set II, all of the scores are different. Statisticians have developed a number of indices to measure such variability.

Range. One very simple index is the **range,** which is the highest score minus the lowest score. In set I the range is $5 - 5 = 0$ while in set II it is $7 - 3 = 4$. The range is not a very good index of variability since it can be misleading. Consider the extreme case of 900 scores of which 899 are equal to 100 and one of them is equal to 10. The range would be quite large, $100 - 10 = 90$, even though almost every score is identical.

Sum of Squares. A second index, and one we will use extensively in later chapters, is called the **sum of squares** and is abbreviated by SS. The first column of Table 3.1 presents the data we will use to develop this measure. The sum of squares, unlike the range, makes use of every score and *not* just the two most extreme scores. We begin by asking how far scores tend to vary from the typical score. If we let the typical score be represented by the mean, then we are concerned with how much each score deviates from the mean. Column 2 of Table 3.1 presents **deviation scores** in which the mean (in this case, $\overline{X} = 5$) has been subtracted from each original score. If the scores in the distribution tend to be alike, the deviation scores will be close to zero since each will be near the mean. In contrast, if the scores tend to be very different, the deviation scores will tend to be large and nonzero.

We now want to combine the deviation scores in some way to derive a single numerical index of overall variability. We can't sum the deviation scores since, as

TABLE 3.1 COMPUTATION OF SUM OF SQUARES FOR TWO VARIABLES, X AND Y

X	$(X - \overline{X})$	$(X - \overline{X})^2$	Y	$(Y - \overline{Y})$	$(Y - \overline{Y})^2$
2	-3	9	4	-1	1
3	-2	4	5	0	0
3	-2	4	5	0	0
5	0	0	5	0	0
7	$+2$	4	5	0	0
7	$+2$	4	5	0	0
8	$+3$	9	6	$+1$	1
$\Sigma X = 35$	0	SS = 34	$\Sigma Y = 35$	0	SS = 2
$\overline{X} = 5.00$			$\overline{Y} = 5.00$		

noted in Section 3.1, the sum of deviations about the mean will always equal zero. Statisticians have suggested a solution that is very desirable from a statistical perspective. The advantages of this approach cannot be demonstrated now, but it will become apparent in later chapters. The approach involves first squaring each deviation score and then summing them. Column 3 of Table 3.1 does this. Note that the sum of squares for X is 34 and for Y it is 2. This reflects the greater variability in the X scores than the Y scores.

The sum of squares gets its name from the operations performed; it is a shorthand term for the sum of the squared deviations. The general formula for the sum of squares is

$$SS = \Sigma(X - \overline{X})^2 \hspace{2cm} [3.3]$$

Students frequently ask why the mean and not the median is used as the measure of central tendency (the "typical" score) when defining the sum of squares. There are several reasons. One reason is because the sum of the *squared* deviations from the mean will always be less than the sum of the squared deviations from the median. Recall from our earlier discussion that the median minimizes the unsigned error in the "best guess" situation. As it turns out, the mean not only minimizes the signed error but it also minimizes the squared error as well. Other reasons for preferring the mean concern how the two measures of central tendency are used for making inferences about populations based upon sample data. These are discussed in Chapter 6.

Study Exercise 3.3

A researcher was interested in the variability of test scores for five children. On a test where there were 15 points possible, the scores were 10, 8, 6, 4, and 2. Compute the sum of squares for the scores.

Answer. We use Equation 3.3:

X	$(X - \overline{X})$	$(X - \overline{X})^2$
10	4	16
8	2	4
6	0	0
4	-2	4
2	-4	16

$\Sigma X = 30$ $SS = 40$

$\overline{X} = 6.00$

Variance. One problem with the sum of squares as an index of variability is that its size depends not only on the amount of variability among scores but also upon the *number* of scores (N). Consider two sets of scores, where set I = 2, 4, 6, and set II = 4, 4, 4, 4, 4, 6, 6, 6, 6, 6. The sum of squares for set I is 8 and the sum of squares for set II is 10. We can readily see that the scores in set I tend to be more different from each other than the scores in set II. Nevertheless, the sum of squares is larger in set II than in set I. An index that can be used to compare variability among two or more sets of scores should take into account the the number of cases within each set. One possibility is to divide the sum of squares by N, or to compute the average squared deviation score. This is called a **variance** and is defined by the formula

$$s^2 = \frac{SS}{N}$$ [3.4]

The symbol used to represent a variance is a lower case "s" with a square operation. Note that the variance for the data in set I is 8/3 = 2.67 and for set II it is 10/10 = 1.00. The average squared deviation score (the variance) is greater in set I than in set II.

Study Exercise 3.4

In Study Exercise 3.3 the sum of squares for the test scores of five children was 40. What would the variance be?

Answer. The variance is the sum of squares divided by N, or

$$s^2 = \frac{40}{5} = 8.0$$

Standard Deviation. A fourth index of variability among scores is the **standard deviation.** The standard deviation is the positive square root of the variance and is abbreviated by the letter s:

$$s = \sqrt{s^2}$$

In Table 3.1, the standard deviation of X scores is $\sqrt{4.86}$ = 2.20 and of Y scores is $\sqrt{.29}$ = .54.

The standard deviation is probably the most easily interpreted measure of variability among a set of scores. Recall that the variance is the average squared deviation score. Few of us feel comfortable interpreting squared deviation scores, and by taking the square root of the variance, we are, in essence, getting rid of the square and returning to the original units of measurement. The standard deviation thus represents an average deviation from the mean.* On the average, the X scores reported in Table 3.1 deviated 2.20 units from the mean and the Y scores deviated .54 units from the mean. Again, there is less variability in the Y scores than in the X scores.

Students learning about a standard deviation frequently ask what value indicates a large standard deviation. The answer is that it depends on what is being measured. Suppose we measured the number of children that families have in a given country and found a mean of 4.2 and a standard deviation of 3.0. This represents considerable variability since, on the average, scores deviated three "children" from the mean. In contrast, suppose we assess the average income of a neighborhood and observe a mean of $30,000 and a standard deviation of 3. In this case, there is very little variability since scores deviated from the mean by an average of only 3 dollars. When the units are dollars, and the concern is annual income, a standard deviation of 3 is trivial, but when the units are children, a standard deviation of 3 is substantial.

Characteristics of SS, s^2, and s. The sum of squares, variance, and standard deviation are all useful indices of variability. We will make extensive use of all three in this book. It will *always* be the case that SS, s^2, and s, are nonnegative numbers or zero. These statistics can never be negative. This reflects the fact that all are based on squared deviation scores and any number squared must be positive. When the SS equals zero, then by definition, $s^2 = 0$, and $s = 0$. A value of zero on the three statistics means there is no variability in the scores; they are all the same. As the values become greater than zero, more variability among the scores is indicated.

When measures are taken on approximately an interval level, the sum of squares, variance, and standard deviation are proper indices of variability. When measures are taken on an ordinal level that departs markedly from interval characteristics, then these indices may not be appropriate since each is based on a sum of scores. For such measures, an index of variability would be the range, since it does not rely on a sum. For qualitative variables (that is, nominal measures), there is no index of variability comparable to those discussed above, since the categories of scores cannot be ordered (however, see Kirk, 1978, pp. 73–75 for one attempt to measure variability on a qualitative variable).

---

*Specifically, it represents the positive square root of an average squared deviation from the mean.

Method of Presentation

When presenting measures of central tendency and variability, most researchers report means and standard deviations. These are undoubtedly the most frequently encountered descriptive statistics. Occasionally, means will be supplemented with reports of the median, especially when the mean can be a misleading index of central tendency (for example, when there are extreme scores). A common format used in social science journals presents means and standard deviations in a table, such as the following:

MEANS AND STANDARD DEVIATIONS FOR
INTELLIGENCE SCORES OF MALES AND FEMALES

	$\overline{X}$	s	N
Males	101.31	10.62	120
Females	102.48	10.31	115

Sometimes the abbreviation s.d. is used in place of s, and M for $\overline{X}$. It is common to report the size of the samples upon which the statistics are based (N). Note that it is possible to compute the variance and sum of squares from the information provided. The variance for males would be the standard deviation squared, or $10.62^2 = 112.78$. The sum of squares would then be computed by multiplying the variance by N, or $(112.78)(120) = 13,533.60$.

In some instances, you may want to combine two or more groups into one group and calculate a mean for the combined group. It is possible to derive the combined group mean from the individual group means using the following formula (expressed for the case of two groups):

$$\text{Pooled } \overline{X} = \frac{n_1\overline{X}_1 + n_2\overline{X}_2}{n_1 + n_2}$$

where $\overline{X}_1$ is the mean score for group 1, $\overline{X}_2$ is the mean score for group 2, n_1 is the number of individuals in group 1, and n_2 is the number of individuals in group 2. In the preceding example, the mean intelligence scores of males and females combined would be

$$\text{Pooled } \overline{X} = \frac{(120)(101.31) + (115)(102.48)}{120 + 115}$$

$$= 101.89$$

The general formula is

$$\text{Pooled } \overline{X} = \frac{\sum_{j=1}^{k} n_j\overline{X}_j}{\sum_{j=1}^{k} n_j}$$

where k is the number of groups and all other terms are as previously defined. When there are an equal number of subjects in each group, this formula simplifies to

$$\text{Pooled } \overline{X} = \frac{\sum_{j=1}^{k} \overline{X}_j}{k}$$

or, stated verbally, the pooled mean is the mean of the individual group means. It is not possible to compute the combined group standard deviation from the standard deviations of the individual groups.

3.3 / Computational Formulas

The method of computing the sum of squares via Equation 3.3 involves computing a deviation score for each subject. There is an alternative way of computing the index that does not require the computation of deviation scores. Some people find this approach more convenient, so we will present it here. The formula is

$$SS = \Sigma X^2 - \frac{(\Sigma X)^2}{N} \qquad [3.5]$$

Consider the five scores from Study Exercise 3.3. Equation 3.5 requires three terms: N, the sum of X, and the sum of the squared X scores:

X	X^2	
10	100	$N = 5$
8	64	$\Sigma X = 30$
6	36	$\Sigma X^2 = 220$
4	16	
2	4	
$\Sigma = 30$	$\Sigma = 220$	

Then,

$$SS = 220 - \frac{(30)^2}{5}$$

$$= 220 - \frac{900}{5}$$

$$= 40$$

At this point, the variance and standard deviation can be computed as discussed previously:

$$s^2 = \frac{40}{5} = 8.00$$

$$s = \sqrt{8} = 2.83$$

Psychologists have studied how we form impressions of others and how we perceive different traits and characteristics. Anderson (1968) asked a large number of individuals to rate different traits in terms of their desirability or favorableness. The ratings were made on seven-point scales ranging from zero to six, with higher scores indicating that the trait was more desirable. A rating of three was a neutral point. Ratings less than three indicated the trait was undesirable or unfavorable, and ratings greater than three indicated the trait was desirable or favorable. The means and standard deviations for six of the traits were as follows:

Trait	$\overline{X}$	s
Sincere	5.73	.55
Honest	5.55	.68
Narrow-minded	.80	.76
Selfish	.82	.80
Cunning	2.62	1.48
Inexperienced	2.62	.79

The first two traits were rated very positively by the individuals in this study. The means were quite high (5.73 and 5.55) and the standard deviations were relatively small, indicating that most individuals rated the traits using the upper points of the scale. The traits of "narrow-minded" and "selfish" were similarly rated, but in a negative fashion. The traits "cunning" and "inexperienced" both yielded relatively neutral mean scores (2.62 where 3 is neutral). However, notice the large discrepancies in the standard deviations for the two traits. The small standard deviation for "inexperienced" suggests that people consistently tended to rate this trait near the neutral point. The large standard deviation for "cunning" implies something quite different. Apparently, there was considerable variability in these ratings, with many individuals perceiving "cunning" as being a positive trait and many individuals perceiving "cunning" as being a negative trait. When averaged, the mean was near the neutral point. Obviously, the standard deviation helps us to interpret the mean and suggests that the perceptions of cunning were not really all that neutral, but rather exhibited considerable variability across individuals.

3.4 / Skewness and Kurtosis

Thus far, we have considered two ways in which distributions of scores can differ: (1) central tendency and (2) variability. Two other differences are **skewness** and **kurtosis.** Skewness refers to the tendency for scores to cluster on one side of the mean. Figure 3.2 presents three graphs of scores that illustrate this concept. The graph in Figure 3.2a is said to be **positively skewed** because most scores occur below the mean. Such a distribution is characterized by a few extreme scores occurring *above* the mean (hence, a positive skew). The graph in Figure 3.2b represents a distribution that is **negatively skewed** since extreme scores occur below

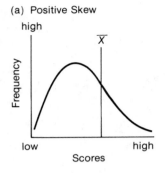

(a) Positive Skew

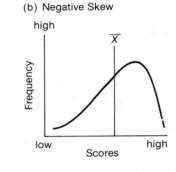

(b) Negative Skew

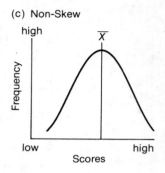

(c) Non-Skew

Figure 3.2 Frequency Graphs of a Positively Skewed Distribution, Negatively Skewed Distribution, and a Distribution That Is Not Skewed

the mean. In Figure 3.3c, the distribution is not skewed, as equal numbers of scores occur above and below the mean.

As you may have already deduced, there is a relationship between the mean and the median with respect to skewness. If the median is less than the mean, then scores will tend to be positively skewed. If the median is greater than the mean, the scores will tend to be negatively skewed.

Kurtosis refers to the flatness or peakedness of one distrition in relation to another. If a distribution is less peaked than another, it is said to be more **platykurtic,** and if it is more peaked than another, it is said to be more **leptokurtic.** It is conventional to label a distribution as being either platykurtic or leptokurtic depending upon whether it is more or less peaked than a particular type of distribution called the *normal distribution*. We will consider this type of distribution extensively in the next chapter. Figure 3.3 presents three graphs of a leptokurtic, platykurtic, and normal distribution.

Statisticians have derived numerical indices of skewness and kurtosis similar to the indices of central tendency and variability already discussed. These are rarely used in the social sciences and hence will not be considered here. Interested readers are referred to Ferguson (1976).

Figure 3.3 Frequency Graphs of a Leptokurtic Distribution, Platykurtic Distribution, and a Normal Distribution

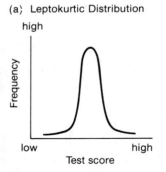

(a) Leptokurtic Distribution

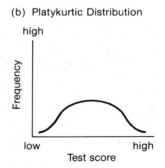

(b) Platykurtic Distribution

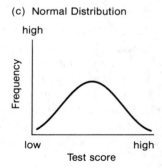

(c) Normal Distribution

3.5 / Sample Notation and Population Notation

The descriptive statistics we will use most frequently are the mean, variance, and standard deviation. Reference will be made to these indices for both samples and populations. It is traditional in statistics to refer to population-derived indices using Greek notation. A sample mean is typically symbolized as $\overline{X}$, whereas a population mean is symbolized as μ (lower case Greek "m"). Of course, the formula for computing the two indices is identical: sum the scores and divide by N. The Greek notation, however, makes it explicit that we are describing a population, whereas the $\overline{X}$ notation makes it explicit that we are describing a sample. The Greek notation for a variance and standard deviation is σ^2 and σ, respectively.

3.6 / Summary

The present chapter has considered several methods for describing a set of scores using summary statistics. Two general characteristics of a distribution of scores were considered: (1) where scores tend to be (measures of central tendency) and (2) how similar scores tend to be (measures of variability). Three different measures of central tendency are the mean, median, and mode. Each measure conveys something different about the central tendency of the scores. Four different measures of variability are the range, sum of squares, variance, and standard deviation. Of these, the standard deviation is probably the most meaningful at a descriptive level. Finally, a set of scores can also be described in terms of its skewness and kurtosis. Overall, these indices can convey a useful description of a set of scores.

EXERCISES

1. An important concern among social scientists is the number of hours children spend watching television. The impact of television on children's attitudes and behavior is considerable. The amount of time children spend watching television tends to differ as a function of age. Presented below are data that are representative of the number of hours children tend to spend watching television during a week for each of two age groups. For each group separately, compute the mean, median, mode, sum of squares, and the standard deviation:

Four Years Old				Twelve Years Old			
10	7	5	6	18	19	22	18
6	7	4	6	21	19	22	20
8	9	5	8	20	17	18	21
7	8	9	7	20	23	20	22

2. How accurate are eyewitness reports of accidents? Social scientists have studied this question in some detail. In one experiment, subjects viewed a film of an accident in which a car ran a stop sign and hit a parked car. The speed of the car was 30 miles per hour. After viewing the film, subjects were asked to estimate the speed of the car. Fifteen subjects gave the following estimates:

15	18	37
40	25	40
35	35	20
30	30	20
28	32	25

Calculate the mean and the standard deviation for these data. How accurate were the estimates considering the mean score for all subjects? How does the standard deviation qualify this?

3. Consider the following five scores: 10, 11, 12, 13, 14. Compute the mean of these scores. Now, generate a new set of five scores by adding a constant of 3 to the old scores. Compute the mean of these scores. Compare the two means. What is the difference between them? Do the same for another set of five scores, based on the original set, but with a constant of 10 subtracted from them. What is the effect on the mean of adding or subtracting a constant to each score in a set of scores?

4. Based on Exercise 3, suppose you measured the average time it took ten children to solve a problem and found it to be 35.3 seconds. You later discovered, however, that your timing device (a watch) was 2 seconds too slow for each child. What would the real mean score be?

5. Consider the following five scores: 10, 20, 30, 40, 50. Calculate the mean of these scores. Now, generate a new set of five scores by multiplying each score by a constant of 2. Calculate the mean of these scores. Compare the two means. In what way are they different? Do the same for another set of five scores, based on the original set, but divide each score by a constant of 10. What is the effect on the mean of multiplying or dividing each score in the set by a constant?

6. Do the principles derived above also apply to the median and the mode? Why or why not?

7. Consider two sets of data, A and B:

A		B	
20	19	53	55
21	20	54	54
20	21	54	53
10	300	55	54

For which set is the median a better descriptor of the central tendency? Why?

8. For the following scores, compute the mean, median, mode, sum of squares, variance, and standard deviation: -3, -3, -2, -2, -2, -1, -1, -1, 0, 0, 0, 0, 1, 1, 1, 2, 2, 2, 3, 3.

9. An organizational psychologist studied how satisfied employees were in two different companies. All employees were given a job satisfaction test in which scores could range from 1 to 7, with higher scores indicating higher satisfaction. The scores were as follows:

Company A	Company B
4	4
6	4
5	4
4	4
3	4
2	4

Compute the mean and standard deviation for each company. Using the result, describe the nature of employee satisfaction in both companies. Why is the standard deviation particularly useful in understanding how satisfaction differs in the two companies?

10. Consider the following five scores: 5, 4, 3, 2, 1. Compute a variance and standard deviation for these scores. Now generate a new set of five scores by adding a constant of 3 to the old scores. Compute a variance and standard deviation for these scores. Compare your results with the first set of scores. Do the same for another set of five scores, but with a constant of 2 subtracted from each score. What is the effect on the variance and standard deviation of adding or subtracting a constant to each score in a set of scores?

11. Suppose you measured the weight of 100 people and found that $\overline{X} = 180.2$ and $s = 10.3$. Then you learned your scale was 1 pound too heavy. What would the correct mean and standard deviation be?

12. Consider the following eight scores: 10, 10, 8, 8, 8, 8, 6, 6. Compute a variance and standard deviation for these scores. Now generate a new set of eight scores by multiplying each score by a constant of 3: 30, 30, 24, 24, 24, 24, 18, 18. Compute a variance and standard deviation for these scores. Compare your results with the first set of scores. Do the same for another set of eight scores, but divide each of the original scores by a constant of 2 first: 5, 5, 4, 4, 4, 4, 3, 3. What is the effect on the variance and standard deviation of multiplying or dividing a set of scores by a constant?

13. Given a set of scores where the mean is 20 and the median is 15, are these scores skewed? If so, how? If the mean was 20 and median was 25, would the scores be skewed? If so, how? If the mean was 20 and the median was 20, would the scores be skewed? If so, how?

14. Generate a set of twenty scores that represent a positively skewed distribution.

15. What does it mean to say that a distribution of scores is leptokurtic?

16. Generate two sets of scores with equal means but unequal standard deviations.

17. Generate two sets of scores with unequal means but equal standard deviations.

18. If the variance of a set of scores is 100, what must the standard deviation be?

19. Without actually calculating it, what must the standard deviation of the following scores be: 6, 6, 6, 6, 6.

Why? What must the variance be? Why? What must the sum of squares be? Why?

20. A consultant you hired measured the satisfaction of your employees and told you the mean score on the scale of satisfaction (ranging from 1 to 7) was 5 and the standard deviation was -2. What would you conclude?

21. Why is the standard deviation more "interpretable" than the variance? That is, what is the advantage of reporting statistics in terms of the standard deviation as opposed to the variance?

22. Under what conditions can the range be a misleading index of variability?

4

Percentiles, Percentile Ranks, Standard Scores, and the Normal Distribution

Suppose I tell you that I have developed a measure of dominance and administered it to a group of seventy-five college students. One of these students, Mary, obtained a score of 50. This information is relatively useless when it is all that is given. Suppose I tell you further that the highest possible score is 100. Now you might infer that Mary is not very dominant. But suppose you are also told that the mean score on the test for the seventy-five college students was 30. Mary is beginning to look a little more dominant. Now I tell you that the standard deviation for this group was 5. This is even more enlightening. The mean score was 30 and, on the average, scores tended to deviate from the mean by about 5 units. Mary's score is 20 units *above* the mean (50 − 30), or 4 standard deviations above the mean, indicating that relative to the rest of the students, her score is very high.

The point to be made is that, in general, observations are meaningful only in relation to other observations. When you are told that a person is 7 feet tall, this is meaningful because you have an intuitive grasp of the distribution of height in humans. You know, for example, that most people are between 5 and 6 feet tall and that someone who is 7 feet tall is unusual. In essence, you are intuitively using information about the mean and standard deviation of height, as described above. In the present chapter, we will consider several methods for understanding the meaning of a score within a distribution of scores.

4.1 / Percentiles and Percentile Ranks

One approach that can be used to express the relative standing of a score in a distribution uses percentage as a basis. The percentage of scores in the distribution that occurs at or below a given value, *X,* is the **percentile rank** of that value. Consider a national survey of 2,000 adults in the United States who responded to the question, "Approximately how many hours per week do you watch television?" If 63.5% of the responses are at or below a score of 30 hours, then the percentile rank of 30 is 63.5. One way to convey the relative position of a score is, thus, to compute its percentile rank. On the other hand, we may want to know the reverse; namely, what score is it that 75% of the respondents score at or below. This score is referred to as the 75th **percentile.** We will now consider how to compute a percentile and a percentile rank.

Computation of a Score Corresponding to a Given Percentile Value.

Suppose we administered a test designed to measure intelligence to a group of 200 children. A frequency analysis of the scores obtained by the children is presented in Table 4.1. The question of interest is to specify the score that corresponds to a given percentile, P. If P is 70, then we want to specify the score that defines the 70th percentile. Actually, we have already considered a special case of the formula for answering this question in the previous chapter on the median. Recall that the median is that score in the distribution of which 50% of the scores are greater and 50% of the scores are less. The median corresponds to the 50th percentile. Thus, the procedures that were used in calculating the median are the same as those for calculating a percentile. The formula for the median, as presented in Equation 3.1, is

$$Md = L + \left(\frac{(N)(.50) - n_L}{n_W}\right) i$$

where L is the lower limit of the category containing the median, n_L is the number of individuals with scores less than L, n_W is the number of individuals with scores within the category containing the median, and i is the size of the interval between the lower and upper limits in the category containing the median. The .50 in the

TABLE 4.1 FREQUENCY ANALYSIS OF INTELLIGENCE SCORES FOR 200 CHILDREN

Score	f	rf	cf	crf
105	9	.045	200	1.000
104	16	.080	191	.955
103	20	.100	175	.875
102	26	.130	155	.775
101	29	.145	129	.645
100	30	.150	100	.500
99	25	.125	70	.350
98	21	.105	45	.225
97	14	.070	24	.120
96	10	.050	10	.050
	$N = 200$			

numerator refers to the percentile of interest, but it is expressed in the format of a proportion. For the median, the percentile of interest is the 50th (.50). If we were interested in the 70th percentile, the formula would appear as follows:

$$X_{70} = L + \left(\frac{(N)(.70) - n_L}{n_W}\right) i$$

where X_{70} represents the score defining the 70th percentile, L is the lower limit of the category containing the 70th percentile, n_L is the number of individuals with scores less than this lower limit, n_W is the number of individuals with scores within the category containing the 70th percentile, N is the total number of individuals, and i is the size of the interval of the category containing the 70th percentile. The method of computation is then analogous to the steps used in computing the median. We start at the bottom of the column of cumulative relative frequencies (see Table 4.1), and move up the column until we find the first number that is greater than or equal to .70 (the percentile of interest, expressed in proportion format). In this case, it is .775 and occurs at the score 102. The lower limit (L) of 102 is 101.5. The number of individuals (n_L) whose score is less than L is 129. The number of individuals (n_W) who had a score of 102 (or, more precisely, scored between 101.5 and 102.5) is 26. Thus,

$$X_{70} = 101.5 + \left(\frac{(200)(.70) - 129}{26}\right) 1.0$$

$$= 101.5 + \left(\frac{140 - 129}{26}\right)$$

$$= 101.5 + .42$$

$$= 101.92$$

In this instance, a score of 101.92 reflects the 70th percentile. Note that this score did not actually occur in the set of scores listed in Table 4.1. This is because the real limits of the numbers were considered, as discussed in Chapter 3. If expressed in terms of the original scale, the 70th percentile would be reported as 102 since 101.92 is rounded to 102.

The general formula for computing a score defining a given percentile is

$$X_P = L + \left(\frac{(N)(P) - n_L}{n_W}\right) i \qquad \text{[4.1]}$$

where X_P represents the score defining the Pth percentile, L is the lower limit of the category containing the Pth percentile, P is the Pth percentile expressed in the format of a proportion, N is the total number of individuals or cases, n_L is the number of individuals with scores less than L, n_W is the number of individuals with scores within the category containing the Pth percentile, and i is the size of the interval of the category containing the percentile of interest.

Study Exercise 4.1

A researcher was interested in determining the size of babies at birth. A group of 1,000 newborn infants was studied and their lengths and weights at birth were measured. The distribution of scores for length (measured to the nearest inch) was as follows:

Length	f	rf	cf	crf
24	27	.027	1000	1.000
23	48	.048	973	.973
22	77	.077	925	.925
21	125	.125	848	.848
20	226	.226	723	.723
19	225	.225	497	.497
18	124	.124	272	.272
17	73	.073	148	.148
16	52	.052	75	.075
15	23	.023	23	.023

Compute the score defining the 90th percentile.

Answer.   Using Equation 4.1, we find that

$$X_{90} = 21.5 + \left(\frac{(1{,}000)(.90) - 848}{77} \right) 1.0$$

$$= 21.5 + \left(\frac{52}{77} \right)$$

$$= 22.18$$

Thus, rounding off, 90% of newborn infants were 22 inches long or less.

Computation of the Percentile Rank of a Given Score. Just as Equation 4.1 can be used to compute the score corresponding to a given percentile, it is possible to specify a formula that will permit the computation of the percentile rank for any given score. In the set of scores presented in Table 4.1, what is the percentile rank corresponding to the score of 101? The formula for computing this is

$$PR_X = \left(\frac{(.5)(n_W) + n_L}{N} \right) 100 \qquad \text{[4.2]}$$

where PR_X represents the percentile rank of the score X, n_W is the number of individuals whose score was equal to X, n_L is the number of individuals whose score was less than X, and N is the total number of subjects. For the score of 101 from the data presented in Table 4.1,

$$PR_X = \left(\frac{(.5)(29) + 100}{200} \right) 100$$

$$= \left(\frac{114.5}{200} \right) 100 = 57.25$$

In this instance, a score of 101 reflects the 57.25 percentile rank. Rounded to whole numbers, the answer would be expressed as 57.

See Method of Presentation on page 67 and Box 4.1 on page 68.

Study Exercise 4.2

For the data presented in Study Exercise 4.1, compute the percentile rank corresponding to a length of 20.

Answer. Using Equation 4.2, we obtain the following.

$$PR = \left(\frac{(.5)(226) + 497}{1,000} \right) 100$$

$$= \left(\frac{610}{1,000} \right) 100$$

$$= 61.0$$

Newborn infants who are 20 inches long correspond to the 61st percentile rank (that is, 61% of newborn infants are 20 inches long or less).

4.2 / Standard Scores

A percentile rank represents one index of the relative position of a score in a set of scores. However, a percentile reflects only an *ordinal* measure of relative standing. To say that a score occurred at the 80th percentile is simply to state that 80% of the individuals scored at or below that score. But *how much* lower did these other

Method of Presentation

The most common presentation of percentile ranks is in manuals of educational and psychological tests that social scientists consult to interpret scores. These manuals generally list the scores that a person could obtain on the test and the corresponding percentile rank of the score. Percentile ranks must always be interpreted relative to the group upon which the scores are based. Someone who scores at the 95th percentile on the Graduate Record Exam (an aptitude test given to applicants for graduate school) probably has a higher general aptitude than someone who scores in the 95th percentile on the Scholastic Aptitude Test (an aptitude test given to applicants for undergraduate schools). This is because only the brighter students who have done well in college tend to take the GRE exam, whereas the SAT reflects a more broad-based population. An example of part of a table presenting test scores and their percentile rank equivalents for two different groups is as follows:

RAW SCORE EQUIVALENT

Percentile	Males	Females
99	140	152
98	138	150
97	136	148
96	134	146

In this table, there are different norms for males and females. A score of 140 would be at the 99th percentile for males, whereas a score of 152 would be at the 99th percentile for females. Notice that in this presentation, not all test scores are listed. Instead, only scores that correspond to particular percentiles are given. Nevertheless, the relative location of a score of 139 for a male, for example, would be apparent from these data.

individuals score? Consider the following scores on two tests, X and Y, in which there are 150 points possible:

X	Y
100	100
99	42
98	43
95	40
96	40
95	39

Box 4.1 General Aptitude of Different Majors

The Miller Analogies Test is a psychological test designed to measure general aptitude of applicants for graduate and professional schools. The test consists of a series of items in the form of analogies. An example item might appear as follows (where you are to fill in the blank with the appropriate word from those given):

A *book* is to *trees* as a *skirt* is to _____ .

(a) shoes   (b) sheep   (c) dress   (d) women

The test has been administered to a wide array of young professionals and a list of percentile ranks are included in the test manual. There are 100 points possible on the test. A list of the scores defining selected percentiles for different groups of graduate students follows:

Percentile Rank	English	Law School	Social Work
99	87	84	81
90	80	73	67
80	74	63	61
70	68	58	58
60	65	53	54
50	59	49	50
40	53	45	46
30	46	40	41
20	41	35	37
10	35	30	27
1	7	18	9

In this table, a score of 53 has a percentile rank of 40 for graduate students in English (literature and language) whereas the identical score has a percentile rank of 60 for law school students. It may be interesting for you to contrast your own major with that of others in terms of what score defines different percentiles.

RAW SCORE EQUIVALENT

Percentile Rank	Physical Sciences	Medical Science	Social Science
99	93	92	90
90	88	78	82
80	82	74	76
70	78	67	69
60	74	60	64
50	68	57	61
40	63	53	56
30	58	47	51
20	51	43	46
10	43	34	39
1	28	24	18

In both cases, a score of 100 would fall at the same percentile rank. However, the score of 100 on the Y test is certainly more distinctive than the score of 100 on the X test. Another index of relative standing that would reflect these differences is a **standard score.**

The Concept of a Standard Score. Recall the example at the beginning of this chapter concerning the dominance measure. Mary obtained a score of 50 out of 100. When told that the average score was 30, a score of 50 took on more meaning. Mary was 20 units above the mean. When told that the standard deviation was 5, the significance of Mary's score was even clearer. By comparing Mary's score to the mean and standard deviation, considerable insight into its relative position was gained. A standard score does just this. It converts a raw score into a form that takes into consideration the score relative to the mean and standard deviation of a group of scores.

Table 4.2 presents two sets of scores that will be used for illustration, X and Y. Consider the X variable. Each score for an individual can be converted into a standard score by the following formula:

$$\text{standard score for } X = \frac{X - \overline{X}}{s} \qquad [4.3]$$

The standard score is the difference between the original score and the mean, divided by the standard deviation. The numerator of Equation 4.3 reflects the number of units the score is above or below the mean. When the numerator is

TABLE 4.2 RAW SCORES AND STANDARD SCORES FOR TWO VARIABLES, X AND Y

Individual	Raw Score on X	Standard Score on X	Raw Score on Y	Standard Score on Y
1	1	−1.42	1	−2.13
2	3	0	3	0
3	5	1.42	3	0
4	2	−.71	3	0
5	3	0	5	2.13
6	4	.71	3	0
7	1	−1.42	3	0
8	3	0	3	0
9	5	1.42	3	0
10	3	0	3	0

$$\overline{X} = 3.00 \qquad\qquad \overline{Y} = 3.00$$
$$SS_X = 18.00 \qquad\qquad SS_Y = 8.00$$
$$s_X^2 = 2.00 \qquad\qquad s_Y^2 = .89$$
$$s_X = 1.41 \qquad\qquad s_Y = .94$$

divided by the standard deviation, the result expresses the number of standard deviations the score is above or below the mean. Mary was 20 units above the mean. The standard deviation was 5 and hence Mary was 4 standard deviations above the mean (20/5 = 4). Her standard score was 4. John had a score of 25, which is 5 units below the mean. His score was one standard deviation below the mean and his standard score was −1 (−5/5 = −1). *A standard score is the number of standard deviation units that a score occurs above or below the mean.* It summarizes the individual's relative standing, taking into consideration the mean and standard deviation of the distribution.

In Table 4.2, compare the standard score on X for the first individual with the standard score for the same individual on Y. Note that although the raw scores are the same (both are 1), the standard scores are different, −1.42 as compared to −2.13. The negative signs indicate that in both cases the raw score was below the mean of its respective distribution. For the X variable, a score of 1 is only 1.42 standard deviation units below the mean of its distribution while for the Y variable, it is 2.13 deviation units below the mean of its distribution. A score of 1 is more distinctive in the context of Y scores than it is in the context of X scores. The standard score reflects this.

Properties of Standard Scores. Standard scores have several important properties. When a standard score is positive, it indicates that the original score is greater than the mean and when it is negative, it indicates that the original score is less than the mean. When a standard score is 0, the original score is equal to the mean. If we sum the standard scores for variable X (that is, sum column 3 of Table 4.2), the result will be zero. The same result will occur if we sum the standard scores for Y. *The sum of a set of standard scores will always be zero.* This is because standard scores reflect deviation scores and the sum of deviation scores about the mean is zero. It follows that if the sum of a set of standard scores is zero, the mean of a set of standard scores is also zero.

It is also the case that if we were to compute a standard deviation of the column of standard scores (column 3 in Table 4.2), this standard deviation would equal 1.0. The standard deviation of a set of standard scores is always 1.0. A proof of these properties of standard scores is presented in Appendix 4.1.

See Study Exercise 4.3 on page 71.

4.3 / Standard Scores and the Normal Distribution

A standard score yields considerable information about the relative position of a score in a distribution. Such scores are even more meaningful when they occur in what is called a **normal distribution.** If we were to construct a frequency polygon

Given a set of scores with $\overline{X} = 20$ and $s = 2$, compute a standard score for $X = 17$.

$$= \frac{-3}{2}$$

$$= -1.5$$

Answer. Using Equation 4.3,

This means that a score of 17 is 1.5 standard deviations below the mean.

$$\text{Standard score} = \frac{17 - 20}{2}$$

reflecting the height of all adult men, the distribution would have approximately the following shape:

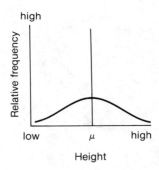

Specifically, the set of scores is approximately *normally distributed,* or possesses the characteristics of a *normal distribution.* The normal distribution is one type of distribution that has been studied extensively by mathematicians. It was given the name *normal* in the early nineteenth century by a French mathematician named Quetelet. Quetelet believed that the shape of the distribution characterized a large number of phenomena, such as height, weight, intelligence, and numerous psychological variables.

The normal distribution is actually a very specific type of distribution that has been precisely defined mathematically. The formula describing it is:

$$Y = \frac{1}{\sqrt{2\pi\,\sigma^2}}\, e^{-(X-\mu)^2/2\sigma^2} \qquad [4.4]$$

where Y is the height of the curve for particular values of X, μ is the mean of the distribution, σ^2 is the variance of the distribution, π is 3.1416, and e is the base of naperian logarithms. If μ and σ are known, different values of X may be substituted in the equation and the corresponding values of Y obtained. If paired values of X and Y are plotted graphically, they will form a normal curve, as illustrated above. Note that everything is constant in this equation except μ and σ and the individual scores, X, that define these parameters. Thus, there is a different normal distribution for every possible μ and σ. Figure 4.1 presents some examples of normal distributions with different values for μ and σ.

All normal distributions have a number of common properties. All are symmetrical about the mean and all are characterized by a "bellshape" (see Figure 4.1). In all cases, the mean equals the median, which in turn equals the mode. When we know that a set of scores is normally distributed, we can invoke some useful statistical properties to better understand the distribution. One useful feature is that the proportion of scores occurring above or below a given standard score is the same

Figure 4.1 Examples of Normal Distributions with Different Means and Variances (Adapted from Johnson and Liebart, 1977)

(a) Equal Means and Different Variances

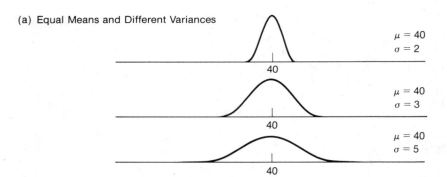

(b) Different Means and Equal Variances

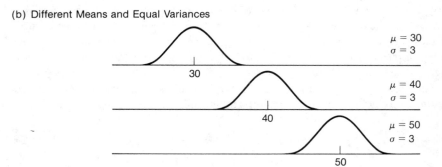

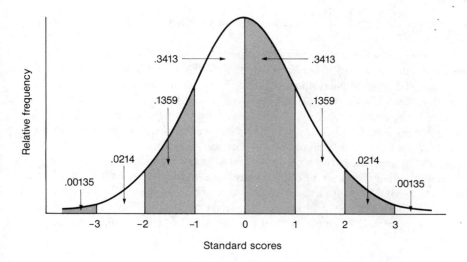

Figure 4.2 Proportion of Scores Occurring Between Selected Standard Scores in a Normal Distribution

in all such distributions. It is always the case, for example, that 50% of the scores occur above the mean in a normal distribution. It is also the case that 15.67% of the scores will be 1 or more standard deviation units above the mean. Thus, knowing that a set of scores is normally distributed allows us to make probability statements with respect to those scores.

Figure 4.2 presents a normal distribution and the proportion of scores that occur between different standard scores. The proportion of scores between 0 and 1 standard scores is .3413, while the proportion of scores between 1 and 2 standard scores is .1359. The proportion of scores greater than or equal to a standard score of 3 is .00135. Note that these proportions are symmetrical about the mean. The proportion of scores occurring between 0 and +1 standard scores is the same as the proportion of scores between 0 and −1 standard scores. Appendix C presents a table that summarizes the proportion of scores in a normal distribution that are greater than selected standard scores in a normal distribution (for example, greater than or equal to 1.00). A standard score in a normal distribution is generally referred to as a *z* **score.*** Let us explore how this table, coupled with our knowledge of the normal

---

*Some texts refer to any standard score as a *z* score. However, in traditional statistics, a *z* score is used to represent a standard score *in a normal distribution*. This distinction will be maintained in the present text.

distribution, can give us insights into scores on a variable that are normally distributed.

For the sake of illustration, try to estimate how many hours a week you spend watching television. This might be easiest to do if you think of each night of the week individually, Monday through Sunday, and estimate how many hours you typically watch television on those days. Then sum these across the seven days, yielding an estimate of how much you tend to watch television per week. For the author of this text, the estimate is 14 hours per week. How does this (or your own estimate) compare with that of others? Let us compare our scores with results suggested by national surveys of adults in the United States.

Suppose that the mean number of hours of television watched by adults in the United States is 25.4, with a standard deviation of 6.1. Suppose also that the scores in this distribution closely approximate a normal distribution. How unique is a score of 14 hours per week? We can convert this to a standard score using Equation 4.3:

$$z = \frac{X - \mu}{\sigma}$$

$$= \frac{14 - 25.4}{6.1}$$

$$= -1.87$$

Watching television 14 hours per week is 1.87 standard deviations *below* the national average. Using Appendix C, we find that the proportion of scores less than or equal to a standard score of -1.87 is .031. Thus, 14 hours defines the 3.1st percentile. Stated another way, less than 3.1 percent of adults in the United States watch 14 hours or less of television per week. Approximately 96.9% (100% − 3.1%) of adults in the United States watch 14 hours or more of television per week. Relative to most American adults, the amount of time the author spends watching television is quite low. Do a similar analysis for your own estimated viewing behavior.

Suppose we want to know what percentage of Americans watch between 30 and 40 hours of television per week. We can use our knowledge of the normal distribution and Appendix C to estimate this. First, convert the scores of 30 and 40 into z scores:

$$z = \frac{30 - 25.4}{6.1} = .75$$

and

$$z = \frac{40 - 25.4}{6.1} = 2.39$$

Examining Appendix C, we find that .7734 or 77.34% of scores are less than or

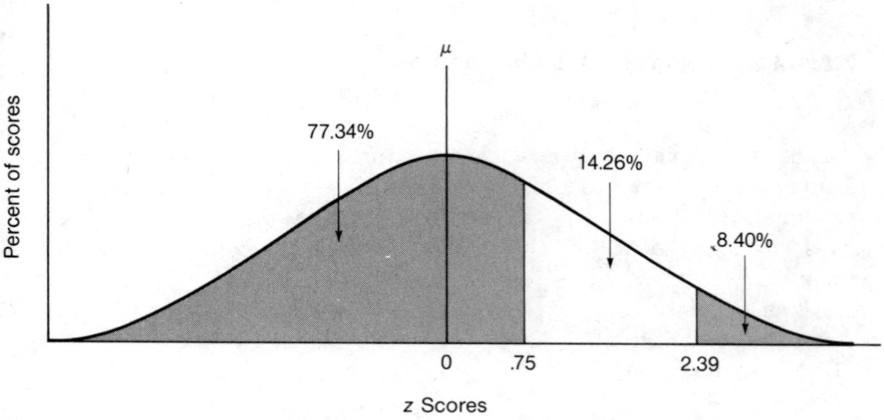

Figure 4.3 Percentage of Scores Between z Scores of .75 and 2.39

equal to a z score of .75 and that .0084 or 8.4% of the scores are greater than a z score of 2.39. This is illustrated in Figure 4.3. The percentage of scores occurring *between* these scores is 100% minus the 8.40% above a score of 2.39 and minus the 77.34% below a score of .75, or $100\% - 8.40\% - 77.34\% = 14.26\%$. Thus, roughly 19% of adult Americans are estimated to watch between 30 and 40 hours of television per week.

Study Exercise 4.4

Given a set of scores that are normally distributed, with a mean of 100 and a standard deviation of 10, what proportion of scores is greater than or equal to 120? Less than or equal to 90?

Answer. We begin by converting the raw score into a z score.

$$z = \frac{120 - 100}{10} = 2.0$$

Using Appendix C, we find that the proportion of scores greater than a score of 2.0 is .0228. Thus, the proportion of scores greater than 120 is .0228. For the question of the proportion of scores less than or equal to 90, we obtain

$$z = \frac{90 - 100}{10} = -1.0$$

Using Appendix C, we find that the proportion of scores less than a z score of -1.0 is .1587. Thus, the proportion of scores less than 90 is .1587.

Most psychological and educational test manuals report standard score equivalents of raw scores in order to aid the interpretation of any given score. Standard scores, however, have many negative values and are expressed in decimals, a property which some people find confusing. For this reason, standard scores are frequently transformed to T scores, using the following equation

$$T = 50 + 10 \text{ (standard score)}$$

A **T score** is directly analogous to a standard score but instead of $\overline{X} = 0$ and $s = 1$, T scores have $\overline{X} = 50$ and $s = 10$. A standard score of $-.5$ would be transformed to a T score as follows:

$$T = 50 + 10(-.5)$$
$$= 50 + (-5)$$
$$= 45$$

An example of a test manual report of raw scores and their T-score equivalents (with $\overline{X} = 50$ and $s = 10$) might appear as follows:

Raw Score	T Score
100	70
99	68
98	66
97	64

In this example, a raw score of 100 corresponds to a T score of 70. Thus, a score of 100 is 2 standard deviation units above the mean since a T score of 70 (with $\overline{X} = 50$ and $s = 10$) corresponds to a standard score of 2.0.

Sometimes T scores are transformed further so that they are also normally distributed. When this is done, the report or manual should state this explicitly. The mathematics of the transformation are complex and can be found in Gulliksen (1960).

4.4 / Empirical and Theoretical Distributions

An important distinction in statistics is that between an **empirical distribution** and a **theoretical distribution.** An empirical distribution is based upon actual measurements collected in the real world. Figure 4.4a presents a frequency histogram for an empirical distribution of intelligence test scores of students in a high school. A theoretical distribution is not constructed by taking actual measurements but is derived by making assumptions and representing these assumptions mathematically. The normal distribution is an example of a theoretical distribution. Figure 4.4b presents a theoretical normal distribution of intelligence scores. The distribution is theoretical because we have not used actual frequencies in constructing the

(a) Empirical Distribution of Intelligence Scores

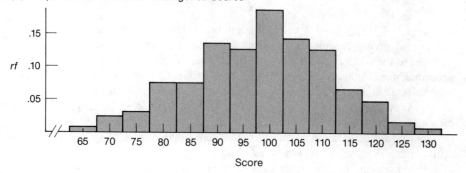

Figure 4.4 Empirical and Theoretical Distributions

(b) Theoretical Distribution of Intelligence Scores

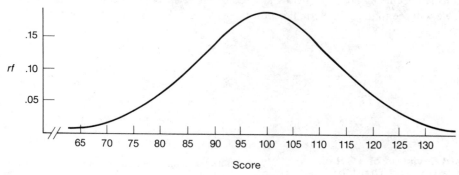

polygon, but rather used the mathematical assumptions noted earlier in this chapter (see Equation 4.4).

Notice that there is a close correspondence between the shapes of the empirical and theoretical distributions in Figure 4.4. Theoretical distributions are frequently used to represent reality when practical circumstances make it impossible to construct an empirical distribution. A distribution of the intelligence of all adults in the United States would be impossible to construct empirically. However, one might be able to closely approximate this by reference to the normal distribution. If one is willing to make the assumption that intelligence approximates a normal distribution, then our knowledge of this theoretical distribution can be used to help us gain insights into the nature of intelligence in the real world. In fact, this is exactly what we did in the prvious example on television viewing behavior. As we will see in later chapters, inferential statistics makes extensive use of theoretical distributions.

4.5 / Recapitulation and Summary

Chapters 2, 3, and 4 have developed a number of techniques for describing a set of scores and giving meaning to a score within a distribution of scores. Chapter 2 introduced the concept of frequency distributions, which represents a very effective procedure for communicating how a set of scores is distributed. Chapter 3 considered selected summary statistics focusing upon measures of central tendency, variability, skewness, and kurtosis. Both of these chapters address the issue of communicating the nature of a set of scores. Chapter 4, in contrast, focused in large part on identifying the location of an individual score within a set of scores by considering percentiles, percentile ranks, and standard scores. Finally, knowing that a set of scores is normally distributed also conveys information about the nature of the scores. Normal distributions are bell-shaped and symmetric. The proportion of scores occurring above or below any score can be specified using Appendix C. As we will see in later chapters, this property is a central aspect of inferential statistics.

APPENDIX 4.1

Proof That the Standard Deviation of a Set of Standard Scores Equals One and That the Mean of a Set of Standard Scores Equals Zero

Let z represent a standard score such that

$$z = \frac{X - \overline{X}}{s}$$

Then

$$\sum z = \frac{\sum(X - \overline{X})}{s}$$

$$\sum z^2 = \frac{\sum(X - \overline{X})^2}{s^2}$$

From Chapter 3,

$$s^2 = \frac{\sum(X - \overline{X})^2}{N}$$

We substitute the right-hand side of this expression for the denominator:

$$\sum z^2 = \frac{\sum(X - \overline{X})^2}{\dfrac{\sum(X - \overline{X})^2}{N}}$$

$$= N$$

Since the mean of a set of standard scores is zero, its variance equals

$$s^2 = \frac{\sum(z - 0)^2}{N}$$

$$= \frac{\sum z^2}{N}$$

As shown above,

$$\sum z^2 = N$$

Hence,

$$s^2 = \frac{\Sigma z^2}{N}$$

$$= \frac{N}{N}$$

$$= 1.00$$

The square root of s^2 is the standard deviation:

$$s = \sqrt{s^2}$$

$$= \sqrt{1.00}$$

$$= 1.00$$

To prove that the mean of z equals 0:

$$\Sigma z = \Sigma \left(\frac{X - \overline{X}}{s} \right)$$

$$= \frac{1}{s} \Sigma (X - \overline{X})$$

$$= \frac{1}{s} (\Sigma X - \Sigma \overline{X})$$

$$= \frac{1}{s} (\Sigma X - N \overline{X})$$

$$= \frac{1}{s} \left(\Sigma X - N \cdot \frac{\Sigma X}{N} \right)$$

$$= \frac{1}{s} (\Sigma X - \Sigma X)$$

$$= 0$$

Then,

$$\overline{z} = \frac{\Sigma z}{N}$$

$$= \frac{0}{N}$$

$$= 0$$

EXERCISES

1. Lie detectors or polygraphs are used to help determine whether a person has knowledge of a crime. These devices are based upon autonomic changes in the nervous system. The assumption is that lying will be reflected in physiological changes that are not under the voluntary control of the individual. By measuring the individual's physiological response when asked certain questions, inferences are made about the veridicality of his or her answers. Lie detection tests typically ask a series of neutral questions (for example, What is your name, Where do you work) among which the critical questions are embedded. Consider the case where an individual has been asked twenty questions of which one is the critical item. The mean score on a galvanic skin response for the twenty questions was 49.4 with a standard deviation of 3. For the critical question, the score was 61.4. Convert this into a standard score and draw a conclusion.

2. A major form of identification in criminal investigations is fingerprints. Fingerprints vary on many different dimensions, one of which is called the ridgecount. Suppose you know that the ridgecounts of people in the United States follow a normal distribution with a mean of 165 and a standard deviation of 10. Suppose further that a set of fingerprints was found at the scene of a crime and it was determined that the ridgecount was at least 200 (the exact value being in question because of smudging). Finally, suppose that out of ten suspects, one of them has a ridgecount of 225. What would you conclude and why?

3. A researcher administered a questionnaire to 500 people that was designed to measure knowledge of positions that two presidential candidates held on major issues. Scores on the test could range from 0 to 10, with higher scores indicating more knowledge. The following frequency analysis resulted.

Score	f	rf	cf	crf
8	6	.012	500	1.000
7	44	.088	494	.988
6	51	.102	450	.900
5	49	.098	399	.798
4	200	.400	350	.700
3	98	.196	150	.300
2	52	.104	52	.104

$$N = 500$$

Given the above, compute the score, defining the following percentiles:

(a) 20th   (c) 60th   (e) 99th

(b) 40th   (d) 80th   (f) 50th

4. Given the above set of data, compute the percentile ranks corresponding to a score of:

(a) 7   (b) 5   (c) 3   (d) 2

5. Convert each of the following standard scores into a T score with a mean of 50 and a standard deviation of 10.

(a) −2.0   (c) 1.5   (e) 0

(b) 2.0   (d) −1.5   (f) 4.0

6. Given a set of scores that is normally distributed with a mean of 20 and a standard deviation of 5, what is the proportion of scores that is:

(a) greater than 25   (d) less than 17

(b) less than 15   (e) greater than 20

(c) between 15 and 28

7. John told his parents that he received a 90 out of 100 on an English exam and a 60 out of 100 on a math exam. He did not tell them that the English exam yielded $\overline{X} = 70$, $s = 20$ and that the math exam yielded $\overline{X} = 40$, $s = 3$. On which exam was John's performance truly better?

8. What are the major characteristics of a normal distribution?

9. Given a group of scores with $\overline{X} = 100$ and $s = 10$, what score would correspond to a standard score of:

(a) 2   (c) 0   (e) 1.5

(b) −2   (d) −1.5   (f) 3.0

10. Convert the following scores into T scores with a mean of 50 and a standard deviation of 10.

(a) −1.0   (c) −2.0

(b) 1.0   (d) 1.3

11. Given a distribution with a mean of 20 and a standard deviation of 3, compute the standard score equivalent of the following scores:

(a) 21   (c) 18   (e) 15

(b) 23   (d) 20   (f) 24.3

12. Using the following scores, compute the score defining the 50th percentile: 100, 100, 99, 99, 99, 98, 98, 97, 97, 96, 95, 94.

13. Using the following scores, convert a score of 50 to a standard score: 60, 55, 50, 50, 55, 60.

14. If John obtained a score of 80 on a test, which one of the following distributions would allow for the most favorable interpretation of that score (assuming that higher values are more favorable)?

(a) $\overline{X} = 60$, $s = 5$ (c) $\overline{X} = 60$, $s = 1$

(b) $\overline{X} = 60$, $s = 10$ (d) $\overline{X} = 60$, $s = 20$

15. Given a normal distribution with a mean of 100 and a standard deviation of 5, what score would

(a) 20% of the cases be greater than

(b) 5% of the cases be greater than

(c) 15% of the cases be less than

(d) 2.5% of the cases be greater than

(e) 2.5% of the cases be less than

16. Are normal distributions skewed?

17. What is the value of the mean and standard deviation of any set of standard scores? Why?

Probability 5

The concept of probability forms the foundation of inferential and several descriptive statistical methods. The purpose of the present chapter is to provide a background in elementary probability theory that will serve as the basis for understanding a wide array of statistical problems.

When we flip a coin, two possible outcomes can result; we can obtain a head or we can obtain a tail. In probability theory, the act of flipping the coin is called a **trial** and each unique outcome is called an **event.** There are several different conceptualizations of probability. The first is the classical viewpoint and is based on logical analysis. The probability of an event, A, is formally defined as the number of observations favoring event A divided by the total number of possible observations:

$$p(A) = \frac{\text{number of observations favoring event } A}{\text{total number of possible observations}} \qquad [5.1]$$

In the coin example, there are two possible observations. On any given trial there is one observation favoring a head and one observation favoring a tail. The probability of obtaining a head is 1/2 or .50. Suppose you randomly select an individual from a population consisting of sixty males and forty females. What is the probability you will select a male? There are sixty observations favoring the event "male," and a total of one hundred observations. According to Equation 5.1, the probability of selecting a male is 60/100 = .60.

A second interpretation of probability focuses upon the long run. According to this interpretation, if we flip a coin a large number of times, then over the long run, we expect the proportion of heads to approach .5. Over an infinite number of trials, the proportion of heads relative to the total number of observations is .5.

Suppose an investigator administered to all married males in a town a set of scales measuring two variables: (1) satisfaction with one's marriage, and (2) satisfaction with one's job. The first variable had two values, men who were satisfied with their marriage versus men who were dissatisfied with their marriage. The second variable also had two values, men who were satisfied with their job versus men who were dissatisfied with their job. By combining these two variables, it is possible to classify the men into one of four groups: (1) those who are satisfied with both their marriage and their job, (2) those who are satisfied with their job but not their marriage, (3) those who are satisfied with their marriage but not their job, and (4) those who are dissatisfied with both their marriage and their job. The results of this study can be displayed in a two-way **frequency table** (also called a **contingency table**), as in Table 5.1. A total of 410 males were interviewed. The rows indicate whether the person was satisfied with his job and the columns indicate whether the person was satisfied with his marriage. The entries within each box or "cell" refer to the number of males who were in that group. For example, there were 156 males who were satisfied with both their job and their marriage, 52 who were satisfied with their marriage but not their job, 54 who were satisfied with their job but not their marriage and 148 who were dissatisfied with both their job and their marriage. The numbers just outside the boxes are the sum of the frequencies of a given row or column. These are called **marginal frequencies.** In this case, we can see that a total of 210 males were satisfied with their jobs, while 200 males were dissatisfied with their jobs. Similarly, 208 males were satisfied with their marriage, whereas 202 males were dissatisfied with their marriage. We will now use these data to develop some key concepts in probability theory.

TABLE 5.1 TWO-WAY FREQUENCY TABLE FOR MARITAL AND JOB SATISFACTION

	Satisfied with Marriage	Dissatisfied with Marriage	Totals
Satisfied with job	156	54	210
Dissatisfied with job	52	148	200
Totals	208	202	410

5.1 / Probabilities of Simple Events

In the language of probability theory, the variable of job satisfaction has two possible outcomes: (1) being satisfied with one's job or (2) being dissatisfied with one's job. These outcomes are said to be **mutually exclusive** in that it is impossible for both outcomes to occur for a given individual: If a person is classified as being satisfied with his job, he can't also be classified as being dissatisfied with his job. (In probability theory, a set of outcomes is said to be mutually exclusive if, on a given trial, the occurrence of one outcome precludes the occurrence of the remaining outcomes.) The variable of marital satisfaction also has two possible outcomes: (1) being satisfied with one's marriage or (2) being dissatisfied with one's marriage. These outcomes are also mutually exclusive.

If we were to select a married male at random from the community in the preceding example, what would the probability be that he is satisfied with his job? According to Equation 5.1, it would be the number of males who were satisfied with their jobs (which equals 210) divided by the total number of males (which equals 410):

$$p \text{ (satisfied with job)} = \frac{210}{410} = .512$$

The probability of a male who is dissatisfied with his job would be the number of males who were dissatisfied with their jobs (200) divided by the total number of males (410):

$$p \text{ (dissatisfied with job)} = \frac{200}{410} = .488$$

Note that the sum of these two probabilities is equal to 1.00 (.512 + .488 = 1.00). Given a set of outcomes that are mutually exclusive and *exhaustive* (that is, which exhaust the possibility of potential outcomes that could occur), the sum of the probabilities of the outcomes will always equal 1.00. Such a set of outcomes and their associated probabilities is known as a **probability distribution.**

Study Exercise 5.1

For the data in Table 5.1, what is the probability of randomly selecting a male who is satisfied with his marriage? A male who is dissatisfied with his marriage?

Answer. Using Equation 5.1,

$$p \text{ (satisfied with marriage)} = \frac{208}{410} = .507$$

$$p \text{ (dissatisfied with marriage)} = \frac{202}{410} = .493$$

Thus, for these data, the likelihood that someone is satisfied with his marriage is roughly equal to the likelihood that he is dissatisfied with his marriage, everything else being equal.

5.2 / Conditional Probabilities

A useful concept for analyzing the data in Table 5.1 is that of a **conditional probability.** A conditional probability refers to the probability that an event will occur given that some other event occurs. In the context of Table 5.1, this might be expressed as follows: Given that a person is satisfied with his job, what is the probability that the person is satisfied with his marriage? A total of 210 males were satisfied with their jobs. Of these 210 males, 156 were satisfied with their marriage. The probability that a male is satisfied with his marriage *given* that he is satisfied with his job is thus

$$p \text{ (satisfied with marriage|satisfied with job)} = \frac{156}{210} = .743$$

The general symbolic form for a conditional probability is "$p(A|B)$," which is read "the probability of event A, given event B." The conditional probability was computed in the above example by considering only those individuals who exhibited outcome B and then deriving the proportion of these individuals who also exhibited outcome A.

Conditional probabilities are useful for understanding the relationship between events. This can be illustrated by considering the concept of **independence.** An event A is said to be independent of event B if $p(A) = p(A|B)$. In other words, the occurrence of event B is unrelated to the occurrence of event A. If this is true, then A is said to be independent of B. Consider the two events "being satisfied with one's marriage" and "being satisfied with one's job." The probability of the first event was computed as being .507. We also found that the conditional probability of being satisfied with one's marriage given one was satisfied with his job was .743. It is clear that being satisfied with one's marriage is not independent of being satisfied with one's job for the married males in this town. The probability of A, in this case, *is* related to the occurrence of B in that it is substantially raised given event B (from .507 to .743).*

See Study Exercise 5.2 on page 85.

5.3 / Joint Probabilities

Another type of probability an investigator might be interested in is a **joint probability.** This refers to the probability of observing *both* event A and event B. In the data in Table 5.1, we might want to determine the probability that an individual is satisfied with his job *and* satisfied with his marriage. In this case, there are 156 such individuals. Since there was a total of 410 individuals in the study, and if we let A

---

*If two events are independent, it is not only the case that $p(A) = p(A|B)$, but the reverse is also true; that is, $p(B) = p(B|A)$. Note that even though two events are related (nonindependent), this does not necessarily mean they are causally related (see Chapter 8).

For the data in Table 5.1, compute the conditional probability of being dissatisfied with one's marriage given being dissatisfied with one's job.

Answer. A total of 200 males were dissatisfied with their jobs, of which 148 were dissatisfied with their marriage. Thus,

$$p \text{ (dissatisfied with marriage} | \text{dissatisfied with career)}$$

$$= \frac{148}{200} = .740$$

= satisfied with job and B = satisfied with marriage, then

$$p(A,B) = \frac{156}{410} = .380$$

where $p(A,B)$ stands for the probability of both event A *and* event B occurring.

For the data in Table 5.1, compute the joint probability of both being dissatisfied with one's job *and* being dissatisfied with one's marriage.

Answer. There were 148 individuals who were dissatisfied with both their marriages and their jobs. Out of a total of 410 individuals, this would yield a joint probability of 148/410 = .361.

5.4 / Recapitulation

We have discussesd three types of probabilities one could potentially compute: a probability of a simple event, a conditional probability, and a joint probability. Each of these can provide information that is useful to an investigator. The probability of a simple event tells us the likelihood that a given outcome will result. A conditional probability tells us the likelihood that a given outcome will result given some other event. Finally, a joint probability tells us the likelihood of two events occurring together. Let us explore the insights such concepts can give.

Suppose that you are a counselor and are discussing with a couple their possible decision to have a child. You are exploring the advantages and disadvantages of this decision. You know that the incidence of a certain type of disease at birth is relatively low among couples. In fact, the probability of this occurring is estimated to be .001. However, it also turns out that the offspring of couples who are in racial group X are particularly susceptible to the disease. The conditional probability that a baby will have the disease given that its parents are in racial group X is estimated to be .020. In this instance, the occurrence of the birth defect is not independent of the race of the couple and your advice to the couple might differ depending upon their race.

Let us consider another example. As you may have suspected, there exist certain mathematical relationships between joint probabilities, conditional probabilities, and probabilities of simple events. One common relationship is characterized by the following equation:

$$p(A,B) = p(A) P(B|A)$$ [5.2]

In our original example we were concerned with males' satisfaction with their marriages and jobs. Let A = being satisfied with one's job and B = being satisfied with one's marriage. Suppose we were particularly interested in males who were satisfied with both their jobs and their marriages. Equation 5.2 states that the probability that someone is satisfied with *both* his job and his marriage is related to the probability that he is satisfied with his job ($p(A)$), weighted by the conditional probability that being satisfied with his job implies that he is satisfied with his marriage ($p(B|A)$). Suppose as an investigator you observed that $p(A,B)$ was low (people in your study were generally not satisfied with both their job and their marriage), that $p(A)$ was low (people were generally not satisfied with their job), but that $p(B|A)$ was high (people who were satisfied with their jobs were, in general, satisfied with their marriages). Such a patterning of data might suggest that if you could somehow increase the satisfaction that a person has with his job, you might also increase to a substantial degree the person's satisfaction with both his job *and* his marriage. Note that this would *not* be the case if the conditional probability of $p(B|A)$ was low, for example, equal to .01. Equation 5.2 can thus give insights into the nature of a joint probability.

5.5 / Adding Probabilities

Another type of probability focuses on the likelihood of observing event *A or B*. In Table 5.1, we might be interested in specifying the probability that a person is either satisfied with his job *or* satisfied with his marriage. One way to determine this would be as follows: First determine the number of people who are satisfied with their jobs (210) and the number who are satisfied with their marriages (208). Sum these two totals (210 + 208 = 418). This does *not* represent the number of people who are satisfied with either their jobs or marriage, however, because we have counted people who are satisfied with both their jobs and marriages twice. So we subtract these people out (418 − 156 = 262) and divide this by the total number of people. This yields 262/410 = .639.

The above process illustrates another relationship among probabilities:

$$p(A \text{ or } B) = p(A) + p(B) - p(A,B) \qquad \textbf{[5.3]}$$

$$= .507 + .512 - .490$$

$$= .639$$

Study Exercise 5.4

For the data in Table 5.1, compute the probability of being satisfied with one's job or being dissatisfied with one's marriage.

Answer. There were 148 individuals who were dissatisfied with both their marriages and their jobs. Out of a total of 410 individuals, this would yield a joint probability of 148/410 = .361.

$$p(A \text{ or } B) = \frac{210 + 202 - 54}{410} = .873$$

5.6 / Relationships Among Probabilities

It is possible to derive a number of formulas that relate different probabilities to one another. An extended discussion of these is beyond the scope of this book. However, for the sake of completeness, we will list the formulas most frequently encountered in the social science literature.

1. In Equation 5.2, it was noted that $p(A,B) = p(A)p(B|A)$. If A and B are independent, then $p(B) = p(B|A)$. It follows by simple substitution that *when A and*

B are independent,

$$p(A,B) = p(A)p(B)$$
$$\text{and}$$
$$p(B,A) = p(B)p(A)$$

[5.4]

2. In Equation 5.3, it was noted that $p(A \text{ or } B) = p(A) + p(B) - p(A,B)$. If A and B are mutually exclusive, then $p(A,B)$ must equal zero. It follows that *when A and B are mutually exclusive,*

$$p(A \text{ or } B) = p(A) + p(B)$$

[5.5]

3. It is possible to express a conditional probability in terms of the following formula, often referred to as *Bayes' theorem:*

$$p(A|B) = \frac{p(A,B)}{p(B)}$$

[5.6]

Equations 5.2 through 5.6 have wide applicability to a number of problems. As an example, suppose a student applied to two medical schools, A and B, and suppose further that he assessed the probability he would get into A as being .20 and B as being .30. What would be the probability that the student would be admitted to either school A or school B? Let $A =$ being admitted to school A and $B =$ being admitted to school B. Equation 5.3 states that

$$p(A \text{ or } B) = p(A) + p(B) - p(A,B)$$

Assuming A and B are independent events, we can use Equation 5.4 to compute $p(A,B)$,

$$p(A,B) = p(A)p(B)$$

$$= (.20)(.30)$$

$$= .06$$

Then, using Equation 5.3, we find that

$$p(A \text{ or } B) = (.20) + (.30) - (.06)$$

$$= .44$$

The probability that the student would be admitted to at least one of the two schools is .44. Let us now consider some additional implications of probability theory in the context of sampling and counting rules.

Box 5.1 Opinions and Probability Theory

A major concern of social psychologists has been the relationships between beliefs that an individual holds. How are a person's beliefs structured? If we change a belief, will this produce a change in other beliefs? If so, can we predict these changes?

One approach to the study of relationships between beliefs an individual holds has been the application of mathematical probability theory. In this approach, beliefs are characterized as subjective probabilities (that is, probability judgments made by an individual). The belief that "nuclear energy is dangerous" is characterized by the individual's subjective probability that this statement is true. Some individuals may believe that nuclear energy is *definitely* dangerous, whereas others might believe that nuclear energy is *probably* dangerous. Still others might believe that it is not dangerous at all. In accord with mathematical probability theory, a subjective probability may range from .00 to 1.00. A subjective probability of 1.00 means that the individual is completely certain that a given statement is true whereas a subjective probability of .00 means that the individual is completely certain that a given statement is *not* true. Numbers between .00 and 1.00 represent increasing degrees of certainty that the belief statement is true.

If beliefs are characterized as subjective probabilities, some psychologists have suggested that it may be possible to predict and understand these subjective probabilities using mathematical probability theory. For example, consider the two statements

A: Nuclear power plants are dangerous

B|A: If nuclear power plants are dangerous, then nuclear power plants should be shut down.

It is possible to measure an individual's subjective probability concerning each of these statements. If subjective probabilities are organized in accord with probability theory, then knowledge of these two subjective probabilities should allow us to predict the individual's belief in the statement.

A,B: Nuclear power plants are dangerous and they should be shut down.

Recall from Equation 5.2 that

$$p(A,B) = p(A)\, p(B|A)$$

Thus, the belief in A,B should be predictable from the product of the beliefs in A and B|A. If the individual believes that the probability nuclear power plants are dangerous is .90, and if he also believes with a probability of .95 that nuclear power plants should be shut down given that they are dangerous, then his belief in statement A,B should be (.95)(.90) = .855.

Research on the relationship between subjective and objective probabilities has provided some interesting insights into belief organization and change. Although there is not always a strong correspondence between them, the degree of agreement has been striking in many respects. Interested readers are referred to Wyer and Goldberg (1970) for a discussion of this research.

5.7 / Sampling With and Without Replacement

The preceding discussion of probability theory has important implications for researchers who draw random samples from populations, such as in survey research. When sampling individual cases from a population, two different situations arise. First, a given case can be randomly selected, the measurement of interest taken, and then the case can be returned to the population. Then, with the population fully intact, the random selection procedure can be performed again. This process is called **sampling with replacement** since each case is "replaced" back into the population before another case is randomly selected. In contrast, **sampling without replacement** involves selecting a case at random, and then without replacing that case, selecting another case at random, and so on. Whether sampling is done with or without replacement can affect the probability of observing some event.

Consider the data in Table 5.1 and suppose we are interested in determining the probability of randomly selecting first someone who is satisfied with his job and then someone who is dissatisfied with his job. In this case, event A is "selecting someone who is satisfied with his job as the first case" and event B is "selecting someone who is dissatisfied with his job as the second case." We can determine the joint probability of these events using Equation 5.2:

$$p(A,B) = p(A)\,p(B|A)$$

Let us first consider the case of sampling with replacement. Before any case is selected, the probability of selecting someone who is satisfied with his job as the first case ($p(A)$) is $210/410 = .512$. After this has been done, the case is returned to the population. This means that the outcome of the first random selection in no way affects the outcome of the second random selection, that is, the two events are independent:

$$p(B) = p(B|A)$$

Since $p(B) = .488$ (or $200/410$), $p(B|A)$ must also equal .488 and hence, using Equation 5.2,

$$p(A,B) = (.512)(.488) = .2499$$

Now consider the case of sampling without replacement. It is no longer the case that the outcome of the first selection is independent of the second selection. The probability that we will select someone who is dissatisfied with his job given that we selected someone who is satisfied with his job (and then not replaced him) is different. It is now the case that $p(B|A) = 200/409 = .489$. Hence,

$$p(A,B) = (.512)(.489) = .2504$$

a value which is slightly greater than the previous one.

If the number of cases in the population is large relative to the size of the sample (for example, a ratio of twenty cases in the population for every one case in the sample), sampling with or without replacement will not affect probabilities appreciably. However, when the number of cases is small, the different sampling procedures can produce very different results. This should be kept in mind when interpreting the results of probability studies using sampling procedures.

5.8 / Counting Rules

We have defined a probability as the number of observations favoring some event A divided by the total number of observations. For some complex events, it would be very tedious to enumerate all possible outcomes in order to compute a probability. For example, in a study of social interaction, if we wanted to state the number of ways we could seat five different people in each of five positions at a table, the task of enumerating each sequence would be quite laborious. Fortunately, there are some counting formulas that permit such computations to be easily performed.

Suppose we have a set of three objects A, B, and C. Suppose we want to specify for this set all possible subsets of two of the objects. There are two different conditions under which we can attempt this task. One condition is where the ordering of the two objects mattered, such that the combination "AB" is different from the combination "BA," since the two objects occur in a different order. The other condition is where ordering does not matter, and "AB" and "BA" would be instances of the same event. The list of all possible subsets of two under each condition would be as follows:

Ordering Doesn't Matter	Ordering Matters	
AB	AB	BA
AC	AC	CA
BC	BC	CB

A **permutation** of a set of objects or events is an *ordered* sequence, whereas a **combination** of a set of objects is a sequence in which the internal ordering of events is irrelevant. The general notation for specifying the number of permutations (P) of n things taken r at a time is

$$_nP_r$$

In the above example, $n = 3$ since there are three objects, A, B, and C; and

$r = 2$ since we are concerned with how we could order these three objects "taking two at a time." The general notation for specifying the number of *combinations* (C) of n things taken r at a time is

$$_nC_r$$

Before specifying three major counting rules, we must introduce one other concept, that of a **factorial** of a number. The factorial of a number, n, is formally defined as

$$n! = n(n-1)(n-2)(n-3) \ldots \ldots (1) \qquad \qquad [5.7]$$

where $n!$ stands for "n factorial." Consider the following examples:

$$6! = (6)(5)(4)(3)(2)(1) = 720$$
$$5! = (5)(4)(3)(2)(1) = 120$$
$$3! = (3)(2)(1) = 6$$

In factorial notation, the expression 0! equals 1.0. The first counting rule concerns computing the number of *permutations* of n things taken r at a time. The following formula can be used:

$$_nP_r = \frac{n!}{(n-r)!} \qquad \qquad [5.8]$$

In our previous example, we had three objects $(A, B,$ and $C)$ and we wanted to specify the number of ordered sequences of two objects that could be derived from these. Using Equation 5.8,

$$_3P_2 = \frac{3!}{(3-2)!} = \frac{(3)(2)(1)}{(1)} = 6$$

There are six possible permutations of three objects taken two at a time (AB, AC, BA, BC, CA, CB).

The second rule is used to compute the number of *combinations* of n things taken r at a time:

$$_nC_r = \frac{n!}{(n-r)!r!} \qquad \qquad [5.9]$$

In the previous example with three objects taken two at a time, we would find

$$_3C_2 = \frac{3!}{(3-2)!2!} = \frac{(3)(2)(1)}{(1)(2)(1)} = \frac{6}{2} = 3$$

There are three possible combinations of three objects taken two at a time (*AB, AC, BC*). Appendix D presents factorials for numbers 0 through 20, making calculations such as those listed above relatively simple.

A third useful counting rule can be specified as follows: If any one of K mutually exclusive and exhaustive events can occur on each of n trials, then the number of different sequences of n events that can occur is $(K_1)(K_2) \ldots (K_n)$. Suppose you toss a coin on the first trial (two possible outcomes) and roll a die on the second trial (six possible outcomes). Then the total number of different sequences that could result would be $(2)(6) = 12$ (for example, a head and a 1, a head and a 2, and so on).

Although these three counting rules may seem abstract, they can be used to good effect in a number of instances. Consider the following example on crime and statistics by Zeisel and Kalvin (1972), which makes use of the third counting rule:

> This simple example comes from a Swedish trial on a charge of overtime parking. A policeman had noted the position of the valves of the front and rear tires on one side of the parked car, in the manner pilots note directions: one valve pointed, say, to one o'clock, the other to six o'clock, in both cases to the closest "hour." After the allowed time had run out, the car was still there, with the two valves still pointing toward one and six o'clock. In court, however, the accused denied any violation. He had left the parking place in time, he claimed, but had returned to it later, and the valves just happened to come to rest in the same positions as before. The court had an expert compute the probability of such a coincidence by chance. The answer was that the probability is 1 in 144 (12×12), because there are twelve positions for each of two wheels. In acquitting the defendant, the judge remarked that if all four wheels had been checked and found to point in the same directions as before, then the coincidence claim would have been rejected as too improbable and the defendant convicted: four wheels with twelve positions each can combine in 20,736 ($= 12 \times 12 \times 12 \times 12$) different ways so the probability of a chance repetition of the original position would be only 1 in 20,736. Actually, these formulae probably understate the probability of a chance coincidence because they are based on the assumption that all four wheels rotate independently of each other, which, of course, they do not. On an idealized straight road all rotate together, in principle. It is only in the curves that the outside wheels turn more rapidly than the inside wheels, but even then the front and rear wheels on each side will presumably rotate about the same amount.

For those readers who attend horse races, the first counting rule could prove to be illuminating. A common bet at horse races is the "trifecta," or predicting which horses will come in first, second, and third in a given race. In a ten horse race, how many ways are there that one could order three of them? Using our first counting rule

$$_{10}P_3 = \frac{10!}{(10-3)!} = \frac{(10)(9)(8)(7)(6)(5)(4)(3)(2)(1)}{(7)(6)(5)(4)(3)(2)(1)}$$

$$= (10)(9)(8) = 720$$

Only 1 of these 720 orders will be correct. Thus, in the absence of any information about the horses, the probability of accurately predicting a trifecta in a ten horse race is 1/720 or .0014. Unless you have some reliable inside information, betting a trifecta is probably not advisable.

The Binomial Expression. Another useful probability counting rule is the **binomial expression.** Consider an experiment on extrasensory perception (ESP) where an individual who claims to have psychic powers is required to predict the outcome of each of ten tosses of a coin. The individual could, in principle, correctly predict none out of ten, one out of ten, two out of ten, and so on through ten out of ten. If we assume that the person has no psychic powers, then the probability he will correctly guess the outcome of a given toss is .50. Another way of saying this is that the probability of a "success," p, is .50 and the probability of a "failure," q, is $1 - p = .50$. Under the assumption that $p = .50$ and $q = .50$, what is the probability that the person would correctly guess each of the above scores (zero out of ten, one out of ten, and so on)? The binomial expression is a formula that allows us to address this question. It can formally be stated as follows: In a sequence of n independent trials that each have only two outcomes (arbitrarily called a "success" and a "failure"), with the probability p of success and the probability q of failure, then the probability of r successes in n trials is

$$\left(\frac{n!}{r!(n-r)!}\right) p^r q^{n-r} \qquad [5.10]$$

As an example, let us compute the probability of obtaining ten out of ten correct predictions. In this instance, the number of trials (that is, coin tosses) is 10 (n) and the number of successes is also 10 (r). With $p = .50$ and $q = 1 - p = .50$, we find that

$$p(10 \text{ correct}) = \left(\frac{10!}{10!(10-10)!}\right)(.50^{10})(.50^{10-10})$$

$$= .001$$

The probability of nine successful predictions is

$$p(9 \text{ correct}) = \left(\frac{10!}{9!(10-9)!}\right)(.50^9)(.50^{10-9})$$

$$= .010$$

Table 5.2 presents the probabilities for each possible score under the assumption that $p = q = .50$. Before undertaking the experiment, we want to specify what outcomes are consistent with the conclusion that the individual has psychic powers. We might arbitrarily state that if the results of the experiment could have occurred by chance with only a small probability, say .05 or less, then this will be judged to be consistent with the conclusion that the person has psychic abilities. We note from Table 5.2 that the probability of correctly guessing nine or more outcomes is $.001 + .010 = .011$ and that the probability of correctly guessing eight or more outcomes is $.044 + .010 + .001 = .055$. Using the .05 criterion, we decide that correctly guessing nine or more outcomes will cause us to conclude that the results are consistent with an individual having psychic powers.* Any fewer correct predictions will be attributed to chance guessing. Try flipping a coin ten times and see if you can meet this criterion.

TABLE 5.2 PROBABILITY OF DIFFERENT OUTCOMES IN ESP EXPERIMENT

Number of Correct Predictions	Probability
10	.001
9	.010
8	.044
7	.117
6	.205
5	.246
4	.205
3	.117
2	.044
1	.010
0	.001

In experiments such as the above, it is usually the case that more than ten trials are used. Suppose instead that 500 trials had been used and 280 correct predictions were made. What is the possibility of this outcome, assuming $p = .50$? The application of the binomial expression would be an arithmetical nightmare in this instance. Fortunately, the normal distribution (Chapter 4) can be used to obtain a very close approximation to the relevant probabilities. We will now consider such an application.

*There may, however be alternative explanations of such an outcome. All we really know is that *something* other than random chance is probably operating and this "something" may be psychic ability. However, it could also be the result of something else. This issue is considered in Chapter 8.

The Binomial and Normal Distributions. For any given probability distribution it is possible to compute a mean and standard deviation of the distribution, but in a fashion that is somewhat different from the procedures in Chapter 3. The formula for a mean in a probability distribution is

$$\mu = p(X) \cdot X \qquad\qquad [5.11]$$

where $p(X)$ is the probability of randomly selecting a score equal to the value of X and X is the score in question. Consider the following five scores, 1, 3, 3, 3, 5. Since 1 occurs once, 3 occurs three times, and 5 occurs once, the probability of the scores 1, 3, and 5 are 1/5, 3/5, and 1/5, respectively. Applying Equation 5.11 should yield the mean. We find

X	$p(X)$	$p(X) \cdot X$
5	$\frac{1}{5}$	1.0
3	$\frac{3}{5}$	1.8
1	$\frac{1}{5}$	.2
		$\Sigma = 3.0 = \mu$

Note that the computed mean, 3.0, is the same as what we would have obtained had we applied the traditional formula developed in Chapter 3, $\Sigma X/N = 15/5 = 3.0$.

Given Equation 5.11, it is possible to compute the mean of our original ESP experiment with ten trials and $p = .50$. This has been done in Table 5.3. The mean is 5.0, indicating that on the average, we would expect five correct guesses. Although Equation 5.11 is a general formula for a mean in a probability distribution, in the case of the binomial distribution, there is a simple computational formula that can be used instead:

$$\mu = np \qquad\qquad [5.12]$$

where n is the number of trials and p is the probability of success. In the ten trial ESP experiment,

$$\mu = (10)(.50) = 5.0$$

The standard deviation of a binomial distribution can be computed by

$$\sigma = \sqrt{npq} \qquad\qquad [5.13]$$

where $q = 1 - p$. In the ESP experiment:

$$\sigma = \sqrt{(10)(.5)(.5)}$$

$$\sigma = 1.58$$

Figure 5.1 presents a bar graph of the probability distribution in Table 5.2. A normal distribution has been superimposed over the exact probabilities of each outcome. Note the similarity in shapes of the distributions. It turns out that the binomial and normal distributions are closely related and we can use this fact to good advantage.

Let us return to the example where the individual correctly guessed 280 out of 500 tosses of a coin. Using Equations 5.12 and 5.13, the mean and standard deviation of the distribution is

$$\mu = np$$

$$= (500)(.50)$$

$$= 250$$

and

$$\sigma = \sqrt{npq}$$

$$= \sqrt{(500)(.50)(.50)}$$

$$= 11.18$$

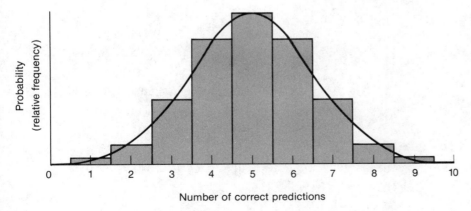

Figure 5.1 Relationship Between the Binomial Distribution and Normal Distribution for $n = 10$, $p = .50$

TABLE 5.3 COMPUTATION OF MEAN FOR DATA IN TABLE 5.2

X	p(X)	p(X) · X
10	.001	.010
9	.010	.090
8	.044	.352
7	.117	.819
6	.205	1.230
5	.246	1.230
4	.205	.820
3	.117	.351
2	.044	.088
1	.010	.010
0	.001	.000
		$\Sigma p(X) \cdot X = 5.00$

TABLE 5.4 COMPARISON OF BINOMIAL PROBABILITIES AND CORRESPONDING NORMAL APPROXIMATIONS

Number of Correct Predictions	Binomial Probability	Normal Approximation
10	.001	.002
9	.010	.011
8	.044	.044
7	.117	.114
6	.205	.205
5	.246	.248
4	.205	.205
3	.117	.114
2	.044	.044
1	.010	.011
0	.001	.002

Since the distribution of scores approximates a normal distribution, we can convert a score of 280 into a z score using the standard score formula from Chapter 3:

$$z = \frac{X - \mu}{\sigma}$$

$$= \frac{280 - 250}{11.8}$$

$$= 2.54$$

Using Appendix C, we find that the probability of obtaining a z score of 2.54 or greater is .0055. This is less than our criterion value of .05. Thus, 280 correct guesses would be consistent with (but does not necessarily confirm) the hypothesis that the person possesses psychic powers.

The accuracy of the approximation between the binomial distribution and the normal distribution depends on the values of n and p. As n increases, the approximation is closer. As p becomes closer to .50, the approximation is closer. Table 5.4 illustrates the correspondence between the two distributions for our original ESP example with ten outcomes.

5.9 / Summary

In this chapter we have considered some basic principles of elementary probability theory. This has included the concept of probabilities of simple events, conditional probabilities, joint probabilities, adding probabilities, and sampling with and without replacement. These concepts are used extensively in several areas of the social sciences and can be used to gain considerable insights into the relationship between events. We will return to further uses of probability theory in later chapters.

Study Exercise 5.5

A person who claims to have psychic abilities tries to predict the outcome of a roll of a die on each of 100 trials. He correctly predicts 21 rolls. Using a small probability value of .05 as a criterion, what would you conclude?

Answer. For this problem, we need to use the normal approximation to the binomial. Since there are six numbers on a die, the probability of success, p, on any one trial is $1/6 = .167$. The probability of failure is $q = 1 - p = 1 - .167 = .833$. The mean of the relevant probability distribution is

$$\mu = np = 16.7$$

The standard deviation is

$$\sigma = \sqrt{npq} = \sqrt{(100)(.167)(.833)} = 3.73$$

A score of 21 is translated into a z score as

$$z = \frac{X - \mu}{\sigma} = \frac{21 - 16.7}{3.73} = 1.15$$

From Appendix C, we find that the probability of obtaining a z score of 1.15 or greater is .1251. Since this is not less than the value of .05, we conclude that the person's performance does not allow us to say that he has psychic powers.

EXERCISES

The Equal Rights Amendment (ERA) has been the subject of considerable controversy in the United States. The proposed amendment to the constitution concerns equal rights for both sexes and consists of three simple statements: "Equality of rights under the law shall not be denied or abridged by the United States or by any State on account of sex. The Congress shall have power to enforce, by appropriate legislation, the provisions of this article. This amendment shall take effect two years after the date of ratification." A social scientist was interested in the extent to which people in a southern city were in favor of or opposed to the ERA and, accordingly, interviewed a random sample of 350 people from this city. The number of males and females supporting or opposing the ERA is summarized in the following two-way frequency table:

	Male	Female	Totals
Favors ERA	60	120	180
Opposes ERA	115	55	170
Totals	175	175	350

1. Using the above data, compute the probability that an individual living in this city favors the ERA. Compute the probability that an individual living in this city opposes the ERA. Interpret the values obtained.

2. What is the probability that an individual favors the ERA given that the individual is a male? What is the probability that an individual favors the ERA given that the individual is a female? What is the probability the individual opposes the ERA given that the individual is a male? What is the probability the individual opposes the ERA given that the individual is a female?

3. Are being a male and favoring the ERA statistically independent in the above data? Why or why not? Be specific.

4. What is the probability that an individual living in the city will be a male who favors the ERA? What is the probability that an individual living in the city will be a female who opposes the ERA? What is the probability that an individual living in the city will be either a male who favors the ERA *or* a female who opposes the ERA?

5. How many joint probabilities are there in the above table? Name them.

6. Recent statistics indicate that the percentage of non-white adults living in the United States is 11%. For non-whites, the percentage of those unemployed is 18%, while for whites, the percentage of those unemployed is 6%. Using these data, fill in the following table with the appropriate probabilities in each of the cells and in the marginals:

	White	Non-white
Employed		
Unemployed		

A social scientist interviewed a random sample of 500 people in order to study the relationship between a person's political party preference and his attitudes toward legal abortions. The following frequency table was observed:

	Favors Legal Abortions	Opposes Legal Abortions	Totals
Democrat	160	40	200
Republican	40	160	200
Independent	75	25	100
Totals	275	225	500

7. Using the above data, what is the probability that an individual in this sample favors legal abortions? What is the probability that an individual opposes legal abortions? What is the probability an individual will identify with the Democratic party? The Republican party? An Independent?

8. What is the probability that an individual favors legal abortions, given that the individual is a Democrat? What is the probability that an individual will identify with the Democratic party, given he or she favors legal abortions? Why are these probabilities different?

9. Are being a Democrat and opposing legal abortions statistically independent? Why or why not? Be specific.

10. What is the probability that an individual in the sample will be a Republican who favors legal abortions? What is the probability that an individual in the sample will be a Republican who opposes legal abortions?

11. How many joint events are there in the above table? Name them.

12. If events A and B are independent and the probability of A is .40 and the probability of B is .30, what is the probability of observing events A and B?

13. Compute the following combinations:

(a) $_5C_3$ (b) $_6C_5$ (c) $_4C_2$ (d) $_5C_5$

14. Compute the following permutations:

(a) $_5P_3$ (b) $_6P_5$ (c) $_4P_2$ (d) $_5P_5$ (e) $_4P_4$

15. Compute 5!. Compare your answer with the answer to 14d. Compute 4!. Compare your answer with the answer to 14e. What generalization can you make from this?

16. Suppose you are doing an experiment that involves showing five different slides to a group of subjects. You are concerned that the order in which you present the slides could affect the outcome of the experiment. In how many ordered sequences can the slides be presented? How might you deal with the problem of order effects in this instance?

17. A researcher was interested in the effects of four different independent variables, A, B, C, and D, on a dependent variable. For practical reasons, she could study only two of the variables. How many different combinations of two variables are there that she could potentially study? Suppose she could study three variables. How many different combinations of three variables are there that she could potentially study?

18. How many different distinct outcomes can occur on three rolls of a die?

19. A group of students was given a ten item true-false test. If a student simply made random responses, what is the probability he or she would

(a) get 9 correct

(b) get 8 correct

(c) get 7 or more correct

(d) get 1 correct

(e) get 2 correct

20. Calculate the mean and standard deviation of a binomial distribution with $n = 150$ and $p = .6$.

21. Calculate the mean and standard deviation of a binomial distribution with $n = 156$ and $q = .2$.

22. What factors influence the extent to which a binomial distribution approximates a normal distribution?

23. Twenty percent of individuals who seek psychotherapy will exhibit a return to normal personality irrespective of whether or not they receive psychotherapy (a phenomenon called spontaneous recovery). If a researcher studies one hundred people who seek psychotherapy, what is the probability that at least thirty of them will exhibit spontaneous recovery?

24. Sixty-five percent of people eligible to vote in the 1976 presidential election did not do so. If we obtain a random sample of ten voters and X is the number of people who didn't vote, compute the probability distribution for possible values of X. What are the mean and standard deviation of this distribution?

Estimation and Sampling Distributions

6

Suppose you have been given the task of describing the income of all married women in the United States. It would be a formidable, if not impossible, job to contact each of these women in order to determine her income. Instead, you might decide to select a random sample and then try to make generalizations about the population from the sample. The problem of making inferences about populations from sample data is fundamental to statistical techniques used in the social sciences. This chapter is concerned with issues in estimating population parameters from sample statistics.

6.1 / Finite versus Infinite Populations

Estimating population parameters can be accomplished with reference to either small, finite populations, or with populations that are infinite in size (or that are so large that for all practical purposes they can be considered infinite). Most applications of statistics in the social sciences are concerned with very large or infinite-sized populations. Social science research is typically conducted with the goal of explaining the behavior of large numbers of individuals, not only those individuals of the present but also, at times, those of the past and the future. As such, the statistics in most social science disciplines are applicable to extremely large, if not infinite, populations.

Given that social science research is concerned with large or infinite-sized populations, it is impossible for investigators to select truly random samples from such large populations. Recall from Chapter 1 that a random sample requires listing all members of the population and then using random number tables to select a sample. Because this is not possible for large, infinite-sized populations, social scientists typically select a set of individuals to study, and then *assume* that these individuals are randomly drawn from some population. Exactly what population the individuals are assumed to represent depends, in part, on the characteristics of the subjects in the study. If a learning experiment was conducted on 200 college students at a Midwestern university, the investigator might want to generalize his or her results to people in general. In this case, the population is conceptualized as consisting of "all people" and the college students are said to represent a random sample from this population. Obviously this may be a very questionable conceptualization. Perhaps the population should be conceptualized in terms of "all college students." Or maybe in terms of "all college students at Midwestern universities." Or maybe in even more specific terms. Social scientists sometimes disagree about what the appropriate population is for purposes of generalization based upon an investigation of a small set of individuals. When interpreting research, you should always keep in mind who the results should generalize to (or in other words, exactly what the relevant population is).

For the sake of illustration, several of the examples in this chapter will use relatively small, finite populations. Sampling from an infinite population (as social science research is conceptualized as doing) is analogous to sampling from a small, finite population *with replacement* (that is, where each sample member is returned to the population before selecting the next sample member). Accordingly, we will occasionally develop principles for estimating population parameters from examples using finite populations. In such cases, sampling is done with replacement so as to mimic the case of infinite populations. For a discussion of statistical applications to finite populations, see Hays (1963).

6.2 / Estimation of the Population Mean

Consider the case of one hundred families in a small town. An investigator wants to describe the inhabitants in terms of the average number of children in each family. Because of practical limitations, the investigator is unable to interview all one hundred families, so she instead resorts to a sample. For the sake of convenience, let us say that the sample size will be 10. In this case, the *population* is the one hundred families in the town and the *sample* is the ten families who are selected to be interviewed. Table 6.1 lists the number of children in each family of the population. In reality, this is unknown to the investigator, but we will refer to it for illustrative purposes. The true population mean, μ, is 3.5 with a variance, σ^2, equal to 2.3. Again, the investigator is unaware of the values of either μ or σ^2.

Let us randomly select ten families from this set of one hundred and let these represent the ten families interviewed by the investigator. The scores for these families are 3, 4, 4, 5, 2, 4, 1, 5, 4, 1, with a mean equal to 3.3. Note that this value is *not* equal to the true population mean. The fact that a sample statistic may not equal the value of its corresponding population parameter is said to be the result of **sampling error.** The amount of sampling error is defined as the difference between the value of the sample statistic ($\overline{X}$) and the value of the population parameter (μ). In the above case, it is $3.3 - 3.5 = -.2$. In practice, an investigator does not know the value of the population parameter, in which case it is impossible to compute the exact amount of sampling error that occurs.

TABLE 6.1 NUMBER OF CHILDREN IN FAMILIES FOR A HYPOTHETICAL POPULATION

3	6	5	4	4	1	3	4	4	5
4	4	6	4	3	3	5	3	4	4
4	3	4	3	5	3	4	3	5	2
2	3	2	5	4	2	3	5	2	3
0	2	4	3	2	4	2	2	1	3
4	3	3	2	4	5	4	4	2	4
3	5	4	5	2	3	5	3	4	3
2	4	3	7	5	1	4	4	5	4
5	2	2	3	7	4	7	3	3	1
2	0	7	4	3	6	5	2	3	1

$$\mu = 3.5$$
$$\sigma^2 = 2.3$$

In the absence of any other information, the sample mean that one observes is probably one's "best guess" about the value of the population mean. One reason is because of an important statistical property that the mean possesses. This property can be illustrated with reference to the family size example above. Suppose in addition to the first sample, we select another random sample of size 10 from the population of one hundred families. The scores for this sample are 1, 3, 3, 4, 3, 7, 5, 1, 3, 4, with a mean of 3.4. Suppose that this process is repeated over and over until we have computed the mean score for *every possible random sample of size 10*. Table 6.2 illustrates fifteen of the many sample means that would be observed and the amount of sampling error that occurs in each instance (that is, $\overline{X} - \mu$). If we were to compute the average (mean) amount of sampling error that resulted across all possible samples of size 10, we would find that the average would equal zero. In other words, some of the sample means overestimate the true population mean while others underestimate it. Across all samples of a given size, however, these overestimations cancel the underestimations and the *average* amount of error is zero. This property of the sample mean makes it a useful estimate of the population mean. On the average, the amount of sampling error we observe across all possible samples will be zero. In statistical terms, the sample mean is said to be an **unbiased estimator** of the population mean. An unbiased estimator of a population parameter is one whose average (mean) over all possible random samples of a given size equals the value of the parameter.

TABLE 6.2 SELECTED MEAN SCORES FOR SAMPLES OF SIZE 10

Sample Scores	$\overline{X}$	$\overline{X} - \mu$
3, 4, 4, 5, 2, 4, 1, 1, 4, 3	3.1	− .4
1, 3, 3, 4, 3, 7, 5, 1, 3, 4	3.4	− .1
3, 5, 4, 3, 2, 4, 5, 4, 3, 2	3.5	.0
0, 2, 7, 2, 5, 2, 5, 5, 4, 3	3.5	.0
6, 3, 4, 4, 4, 2, 3, 3, 4, 3	3.6	.1
3, 3, 4, 4, 5, 3, 2, 2, 3, 3	3.2	− .3
7, 4, 7, 3, 3, 3, 5, 4, 5, 2	4.3	.8
5, 4, 5, 2, 2, 3, 4, 3, 5, 3	3.6	.1
6, 7, 4, 4, 4, 5, 4, 4, 2, 2	4.2	.7
2, 2, 3, 3, 1, 2, 3, 2, 3, 4	2.5	−1.0
3, 4, 3, 4, 3, 4, 3, 4, 3, 4	3.5	.0
1, 3, 3, 2, 4, 3, 4, 3, 2, 0	2.5	−1.0
4, 6, 4, 3, 3, 4, 4, 5, 4, 1	3.8	.3
2, 2, 2, 2, 5, 5, 5, 3, 5, 4	3.5	.0
6, 7, 7, 4, 3, 0, 5, 5, 4, 2	4.3	.8

6.3 / Estimation of the Population Variance

In the previous section, we selected a sample of size 10 from the population of one hundred families and computed a mean of 3.3. The variance of the ten scores in the sample was 2.0. This also is different from the true population variance ($\sigma^2 = 2.3$) and, again, reflects sampling error. Unlike the sample mean, however, the sample variance is *not* our best guess of the true population variance. Statisticians have determined that the sample variance is a *biased* estimator of the population variance and tends to underestimate the population variance across all possible samples of a given size.

In Chapter 3, the formula for a variance was defined as the sum of squared deviation scores divided by N, or

$$s^2 = \frac{SS}{N} \qquad [6.1]$$

As noted above, this is not our best estimate of the population variance since it is biased. Statisticians have shown that an unbiased estimator can be obtained from sample data by making a minor modification in the denominator of Equation 6.1:

$$\hat{s}^2 = \frac{SS}{N - 1} \qquad [6.2]$$

where $\hat{s}^2$ is the symbol used to represent a sample estimate of the population variance, or as it is more commonly called, a **variance estimate.** Equations 6.1 and 6.2 are identical except that $N - 1$ is used in the denominator instead of N. This "correction" involving the subtraction of 1 from N serves to make the variance estimate larger than the sample variance and hence, corrects for the conservative bias of the statistic. In an example with ten scores of 3, 4, 4, 5, 2, 4, 1, 5, 4, 1 with a mean of 3.3, we find

$$s^2 = \frac{20.10}{10} = 2.01 \qquad\qquad \hat{s}^2 = \frac{20.10}{9} = 2.23$$

Our best guess as to the value of the population variance, based on these data, would be 2.23. Again we recognize that there is sampling error involved and this estimate may be wrong.

The sample standard deviation is defined as the square root of the variance, $\sqrt{s^2}$. By the same token, the estimate of the population standard deviation is the square root of $\hat{s}^2$, or

$$\hat{s} = \sqrt{\hat{s}^2} \qquad [6.3]$$

In the above example, $s = \sqrt{2.01} = 1.42$ and $\hat{s} = \sqrt{2.23} = 1.49$. Our best guess about the standard deviation of the population is 1.49. The symbol $\hat{s}$ is referred to as the **standard deviation estimate.**[*]

The preceding relationships are summarized in the following table:

Statistical Term	Sample Value	Sample Estimate of Population Value	Population Value
Mean	$\bar{X} = \dfrac{\Sigma X}{N}$	$\bar{X} = \dfrac{\Sigma X}{N}$	$\mu = \dfrac{\Sigma X}{N}$
Variance	$s^2 = \dfrac{SS}{N}$	$\hat{s}^2 = \dfrac{SS}{N - 1}$	$\sigma^2 = \dfrac{SS}{N}$
Standard Deviation	$s = \sqrt{s^2}$	$\hat{s} = \sqrt{\hat{s}^2}$	$\sigma = \sqrt{\sigma^2}$

*Technically, $\hat{s}$ is not an unbiased estimate of σ. A minor correction factor is necessary to make $\hat{s}$ an unbiased estimate of σ (see Hays, 1963, p. 208). However, if the sample size is greater than 10, the amount of bias tends to be small. In addition, the statistics to be used in later chapters circumvent the problem introduced by this bias, so we will use $\hat{s}$ as the sample estimate of the population standard deviation.

Some texts do not distinguish between the sample variance, s^2, and the variance estimate, $\hat{s}^2$, as has been done in the present context. Rather, the unbiased estimate, $\hat{s}^2$, is introduced as *the* variance of a sample. For some applications, this may lead to confusion and, hence, the more traditional distinctions are maintained here.

Study Exercise 6.1

A sample of ten individuals took an intelligence test that could range in scores from 0 to 150. This sample yielded a mean score of 100.00 and a sum of squares equal to 50.00. Compute the standard deviation for the sample and estimate what the standard deviation is in the population.

Answer. The sample variance is the sum of squares divided by N, whereas the variance estimate is the sum of squares divided by $N - 1$. Thus,

$$s^2 = \frac{50}{10} = 5.00 \qquad \hat{s}^2 = \frac{50}{9} = 5.56$$

The respective standard deviations are the square roots of the variance:

$$s = \sqrt{5.0} = 2.24 \qquad \hat{s} = \sqrt{5.56} = 2.36$$

The sample standard deviation is 2.24 and our estimate of the standard deviation in the population is 2.36.

6.4 / Degrees of Freedom

An important concept in statistical estimation is that of degrees of freedom. Suppose you are told that a friend received a score of 95 on a statistics exam. Suppose you are also told that there were 100 points possible and that your friend missed 5 points. You have been given three distinct pieces of information:

1. Your friend received a score of 95 on a test.

2. The test had 100 points possible.

3. Your friend missed 5 points.

If you had been told any two of the above pieces of information, you could have deduced the third (a score of 95 out of 100 means 5 points were missed). Thus, you have actually been given two independent pieces of information and one piece of information that is dependent on (follows from) the other two. Which two of the three are called "independent" and which one of the three is called "dependent" is arbitrary. The fact of the matter is that only two pieces of information are independent. In statistics, the phrase **degrees of freedom** is used to indicate the number of pieces of information that are "free of each other" in the sense that they cannot be deduced from one another. In the above example, there are two degrees of freedom, since there are two pieces of independent information.

Statistical indices, such as the sum of squares and the variance, have a certain number of degrees of freedom since they are based on certain pieces of information (scores on the variable) that are "free of each other." Let's consider the sum of squares. When computing a sum of squares it is necessary to derive deviation scores about the mean. Consider the following four scores: 8, 10, 10, 12. The mean of these scores is 10, and if we were to compute the deviation scores of the first three scores, they would be -2, 0, and 0. Recall from Chapter 3 that the sum of deviation scores about the mean will always equal zero. If you are told that there are four scores in a distribution, and the deviation scores for three of these are -2, 0, and 0, you know what the last deviation score must be. In this case, it must be 2. The last deviation score to be computed is not *free to vary* but is determined by the previous deviation scores. Thus, there are three degrees of freedom. In general, a sum of squares will have $N - 1$ degrees of freedom associated with it.* In our example, there were four scores and the number of degrees of freedom was $N - 1 = 3$.

The above leads us to a basic principle in estimation techniques. *As the number of degrees of freedom of an estimate increases, the probability that the estimate is accurate also increases.* Of concern here is the distinction between the number of degrees of freedom involved as opposed to the total number of pieces of information. When an estimate is based on a large number of independent pieces of information, then it will tend to be relatively accurate. However, if the information tends to be dependent on each other, there is greater room for error. Technically, the accuracy of a variance estimate is not a function of sample size (N), but rather is a function of the degrees of freedom (number of independent pieces of information) used in calculating it ($N - 1$).

We will encounter the concept of degrees of freedom in later chapters. As an aside, we should note that the variance estimate, $\hat{s}^2$, is formally defined in statistics as the sum of squares divided by the degrees of freedom associated with it, or

$$\hat{s}^2 = \frac{SS}{N - 1} = \frac{SS}{df} \qquad \text{[6.4]}$$

where df = degrees of freedom. Equation 6.4 derives from the fact that the df associated with a sum of squares is $N - 1$.

*We will encounter exceptions to this in later chapters.

To summarize thus far, a sample statistic may differ from the value of its corresponding population parameter because of sampling error. The mean and the variance estimate are both unbiased estimators of the population mean and population variance, respectively. An unbiased estimator is one whose average (mean) across all possible random samples of a given size is equal to the value of the population parameter. In general, as the degrees of freedom of a sample statistic increase, the more accurate that statistic will be in estimating the corresponding population parameter.

6.5 / Sampling Distribution of the Mean and Sampling Error

Given that sampling error operates, it would be desirable to obtain an estimate of the degree to which sampling error is likely to affect our results. Such an estimate is possible via a theoretical concept known as a sampling distribution. We will consider the case of estimating μ from $\overline{X}$ and the corresponding use of the sampling distribution of the mean.

Suppose we are trying to estimate the population mean of the number of children in one hundred families listed in Table 6.1 from a sample size of 10. We randomly select a sample of ten families and observe a mean score of 3.1. Although it is not done in practice, suppose we select another random sample of size 10 from the one hundred families and compute the mean for this sample. The scores are 3, 4, 4, 5, 2, 4, 1, 5, 4, 1 and the mean is 3.3. Suppose we repeat this process over and over until we have computed the mean scores for all possible samples of size 10. This set of "scores," consisting of the mean scores for all possible samples of size 10, constitutes a sampling distribution of the mean. A **sampling distribution of the mean** is a theoretical distribution consisting of the mean scores of all possible samples of a given size from a population.

In practice, we would never compute a sampling distribution of means. However, there are features of such distributions that statisticians make use of. As noted, a sampling distribution of the mean consists of a set of all possible mean scores based upon samples of a given size (in this case, 10). If we wanted, we could compute a mean and standard deviation of these "scores," just as we could any other distribution. If we were to compute a mean for all of these scores, it would equal 3.5, the true population mean. *The mean of a sampling distribution of means is always equal to the population mean.* The reason for this can be stated intuitively. When we select our samples of size 10, some of the sample means will overestimate the true population mean and others will underestimate it. However, when we average all of these, the underestimations will cancel the overestimations, with the result being the true population mean. This characteristic of a sampling distribution of means is very important for the statistical concepts to be developed later.

Suppose in addition to the mean of our sampling distribution we also calculated the standard deviation of the sampling distribution. What would this reflect? Recall that a standard deviation reflects an average deviation from the mean. If the mean in this case is the true population mean, μ, then this standard deviation would reflect the average deviation of our sample means from the population mean. The standard deviation of a sampling distribution of the mean is called the **standard error of the mean.** It reflects how much confidence we can have in a sample mean as an estimate of the population mean. If the standard error of the mean is small, then all of the sample means based on that N tend to yield the same result and all are close to the population mean. If the standard error of the mean is large, then the sample means based on the N tend to yield very different results, some of which may or may not be close to the population mean.

If you are told in the example with children that the standard error of the mean for samples of size 10 was 1.3, then this means that on the average, samples of size 10 differed 1.3 units (in this case, "children") from the true population mean. This represents a fair amount of error and one would have to be concerned with the soundness of the particular sample mean observed. If, on the other hand, the standard error of the mean for samples of size 10 was only .001, then this means that on the average, sample means deviated only .001 unit (or "children") from the true population mean. You would now have more confidence in your sample mean as an estimate of the population mean. The utility of the concept of the standard error of the mean is that statisticians have developed a method for estimating it based upon sample data. This helps us to interpret the sample mean relative to the population mean.

Computation of the Standard Error of the Mean. It is a statistical fact that the standard deviation of a sampling distribution of the mean is equal to

$$\sigma_{\bar{X}} = \frac{\sigma}{\sqrt{N}} \qquad [6.5]$$

where $\sigma_{\bar{X}}$ is the symbol used to represent the standard error of the mean, σ is the standard deviation of scores in the population, and N is the sample size. It is beyond the scope of this book to prove Equation 6.5 (see Hays, 1963), but an intuitive feel can be given for why it works. Equation 6.5 suggests that two factors influence the size of the standard error of the mean. The first is the sample size. When large samples are used, the standard error will tend to be smaller. If in the example on size of families we had used a sample size of 99 instead of 10, it is clear that the sampling distribution would consist of sample means that are, for all intents and purposes, very similar to one another. However, with smaller sample sizes, greater variability is possible. The second factor that influences the size of the standard error is the variability of scores in the population, or σ. When the population variability is small, the standard error of the mean will tend to be small, other things being equal. To take an extreme example, if all of the scores in the population were identical (that is, the variance $= 0$), then every sample would yield a mean that reflects the population mean. For example, if all scores were equal to 4, then the average of any random sample would also have to be 4, as would the population mean. However, when there is considerable variability in the population, it is more likely that the sample mean could be influenced by the overinclusion of extreme scores. This will tend to yield greater variability in the sample mean and, consequently, a larger standard error of the mean.

In practice, we are rarely in a position where the population standard deviation, σ, is known. Rather, we typically have data based upon one sample. We therefore estimate the standard error of the mean by substituting $\hat{s}$ into Equation 6.5 for σ:

$$\hat{s}_{\bar{X}} = \frac{\hat{s}}{\sqrt{N}} \qquad [6.6]$$

where $\hat{s}_{\bar{X}}$ is the estimate of the standard error of the mean and all other terms are as previously defined. Let's consider a numerical example.

Suppose we have two populations, A and B. A sample of ten individuals is interviewed from population A and a sample of ten individuals is interviewed from population B. Both samples were randomly chosen. For each member in the sample, a measure of the person's age is obtained. These data are reported in Table 6.3. We begin by calculating the mean score for each sample. For sample A it is 20.0 and for sample B it is 30.0. We next calculate the population estimate of the standard deviation ($\hat{s}$) in each sample, which turns out to be .82 for sample A and 4.14 for sample B. The respective standard errors of the mean are then calculated by dividing these standard deviations by the square root of N. The standard error for sample A is .26 and for Sample B it is 1.31.

On the basis of this, we would have more confidence in using the mean of sample A as an estimate of the mean for population A than we would in using the mean of sample B as an estimate of the mean for population B. Since both samples are identical in size, $N = 10$, the reason for the difference in the standard errors of the mean is probably due to differential variability of scores in the populations. This is reflected in the differences in the size of standard deviations. For sample A, $\hat{s} = .82$ while for sample B, $\hat{s} = 4.14$.

A number of points about the sampling distribution of the mean can now be summarized:

1. The mean of a sampling distribution of means will equal the population mean.

2. The standard deviation of a sampling distribution of means is called the standard error of the mean.

3. The standard error of the mean can never be a negative number. This is because it is a standard deviation. When the standard error of the mean is small (tends toward zero), one can have increased confidence in the sample mean as an estimate of the population mean.

4. Two factors that influence the size of the standard error of the mean are the sample size and the variability of scores within the population.

TABLE 6.3 NUMERICAL EXAMPLE FOR CALCULATING A STANDARD ERROR OF THE MEAN

Population A			Population B		
X	$(X - \overline{X})$	$(X - \overline{X})^2$	X	$(X - \overline{X})$	$(X - \overline{X})^2$
19	-1	1	26	-4	16
21	1	1	30	0	0
20	0	0	34	$+4$	16
19	-1	1	35	$+5$	25
20	0	0	25	-5	25
21	1	1	30	0	0
21	1	1	24	-6	36
20	0	0	30	0	0
20	0	0	30	0	0
19	-1	1	36	$+6$	36
		SS = 6			SS = 154

$\Sigma X = 200$
$\overline{X} = 20.0$
SS = 6
$\hat{s}^2 = SS/(N - 1) = 6/9 = .67$
$\hat{s} = \sqrt{\hat{s}^2} = .82$
$\hat{s}_{\overline{X}} = \hat{s}/\sqrt{N} = .82/\sqrt{10} = .26$

$\Sigma X = 300$
$\overline{X} = 30.0$
SS = 154
$\hat{s}^2 = SS/(N - 1) = 154/9 = 17.11$
$\hat{s} = \sqrt{\hat{s}^2} = 4.14$
$\hat{s}_{\overline{X}} = \hat{s}/\sqrt{N} = 4.14/\sqrt{10} = 1.31$

5. There is a different sampling distribution for every sample size. A sampling distribution of the mean for random samples of size 10 is different from a sampling distribution of the mean for random samples of size 90.

6. The standard error of the mean can be estimated by dividing the population estimate of the standard deviation by the square root of the sample size.

An Example of an Empirical Sampling Distribution of the Mean. We can demonstrate some of the above points by considering a contrived example with a small population, using sampling with replacement. Consider a population consisting of four scores, 2, 4, 6, and 8. We can construct an empirical sampling distribution of the mean for samples of size 2 by listing all possible random samples of size

2 and their corresponding means. They are

Sample	$\overline{X}$
2, 2	2.0
2, 4	3.0
2, 6	4.0
2, 8	5.0
4, 2	3.0
4, 4	4.0
4, 6	5.0
4, 8	6.0
6, 2	4.0
6, 4	5.0
6, 6	6.0
6, 8	7.0
8, 2	5.0
8, 4	6.0
8, 6	7.0
8, 8	8.0

The true population mean, μ, is $(2 + 4 + 6 + 8)/4 = 5.00$. If we compute the mean of the sixteen mean scores above, we find it equals $80/16 = 5.00$. Thus, the mean of the sampling distribution equals μ. The standard deviation of the sixteen scores is 1.58. This is the standard error of the mean and indicates how far, on the average, sample means deviated from μ. This same value would result if we computed $\sigma/\sqrt{N}$, as indicated by Equation 6.5. In fact, the standard deviation (σ) of the four scores, 2, 4, 6, and 8, is 2.24, and $2.24/\sqrt{2} = 1.58$.

Suppose instead of a sample size of 2, we construct a sampling distribution for random samples of size 3. Table 6.4 presents the mean scores for all possible random samples of size 3. The mean of these scores is, as before, equal to $320/64 = 5.0$, the value of μ. The standard error of the mean is $\sigma/\sqrt{N} = 2.24/\sqrt{3} = 1.29$. This equals the standard deviation of the sixty-four mean scores in Table 6.4. Note that the standard error of the mean for samples of size 2 $(\sigma_{\overline{x}} = 1.58)$ is larger than that for samples of size 3 $(\sigma_{\overline{x}} = 1.29)$. This is because, on the average, larger sample sizes will lead to more accurate mean estimates, everything else being equal.

6.6 / Central Limit Theorem

In previous sections we stated that it is possible to treat a sampling distribution of means like any other distribution of scores and discussed the implications of computing the mean and standard deviation of such a distribution. It is also possible to construct a frequency graph of a sampling distribution in order to better visualize the shape of the distribution of sample means. Figure 6.1 presents three different populations and corresponding frequency graphs of empirical sampling distributions of the mean for each of two sample sizes, $N = 10$ and $N = 30$. Note that in all cases, the sampling distribution is bell-shaped and symmetrical. The most frequently occurring sample mean is the true population mean with very deviant sample means being infrequent. The frequency of highly deviant sample means is much less when

TABLE 6.4 ALL POSSIBLE RANDOM SAMPLES AND SAMPLE MEANS FOR SAMPLES OF SIZE 3 (BASED ON MINIUM, 1970)

Population: 2, 4, 6, 8

Sample	Mean	Sample	Mean	Sample	Mean
2, 2, 2	2.00	4, 4, 6	4.67	6, 8, 2	5.33
2, 2, 4	2.67	4, 4, 8	5.33	6, 8, 4	6.00
2, 2, 6	3.33	4, 6, 2	4.00	6, 8, 6	6.67
2, 2, 8	4.00	4, 6, 4	4.67	6, 8, 8	7.33
2, 4, 2	2.67	4, 6, 6	5.33	8, 2, 2	4.00
2, 4, 4	3.33	4, 6, 8	6.00	8, 2, 4	4.67
2, 4, 6	4.00	4, 8, 2	4.67	8, 2, 6	5.33
2, 4, 8	4.67	4, 8, 4	5.33	8, 2, 8	6.00
2, 6, 2	3.33	4, 8, 6	6.00	8, 4, 2	4.67
2, 6, 4	4.00	4, 8, 8	6.67	8, 4, 4	5.33
2, 6, 6	4.67	6, 2, 2	3.33	8, 4, 6	6.00
2, 6, 8	5.33	6, 2, 4	4.00	8, 4, 8	6.67
2, 8, 2	4.00	6, 2, 6	4.67	8, 6, 2	5.33
2, 8, 4	4.67	6, 2, 8	5.33	8, 6, 4	6.00
2, 8, 6	5.33	6, 4, 2	4.00	8, 6, 6	6.67
2, 8, 8	6.00	6, 4, 4	4.67	8, 6, 8	7.33
4, 2, 2	2.67	6, 4, 6	5.33	8, 8, 2	6.00
4, 2, 4	3.33	6, 4, 8	6.00	8, 8, 4	6.67
4, 2, 6	4.00	6, 6, 2	4.67	8, 8, 6	7.33
4, 2, 8	4.67	6, 6, 4	5.33	8, 8, 8	8.00
4, 4, 2	3.33	6, 6, 6	6.00		
4, 4, 4	4.00	6, 6, 8	6.67		

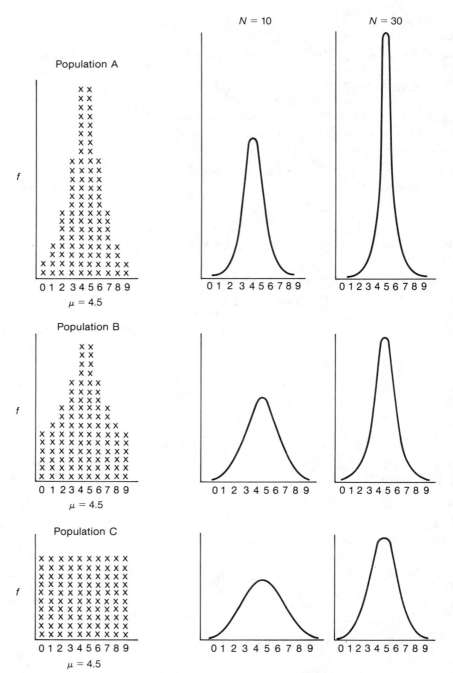

Figure 6.1 Sampling Distribution For Three Populations and Two Sample Sizes (From Johnson and Liebert, 1977)

the sample size is larger (compare the graphs for $N = 10$ and $N = 30$) and when the population variability is lower (compare the graphs for population A with population B, within sample sizes). This is consistent with principles developed earlier in the chapter. Notice also that even though all scores in population C occur with equal frequency, the shape of the sampling distribution of means is still bell-shaped, with highly deviant sample means (from the true population mean) being less frequent.

This bell-shaped distribution should look familiar since we discussed a similarly shaped family of distributions in Chapter 4, the normal distribution. An important concept in statistical theory known as the **central limit theorem** states that *the shape of a sampling distribution of the mean tends to approximate a normal distribution as the sample size becomes larger*. When the sample size is greater than 30, the approximation is usually quite close. For sample sizes less than 30, the approximation is less exact, though in many instances, the fit will be reasonable. The relationship of the sampling distribution of the mean to the normal distribution is very important in statistical theory and we will refer to this relationship extensively in later chapters.

Box 6.1 Polls and Random Samples

Most of us have encountered the concept of sampling error in the context of political polls published in newspapers. A consideration of sampling procedures used by major polling agencies and how sampling error affects their results may be useful. Recall from Chapter 1 that a random sample requires a list of every member of the population. Such a list is often not feasible, or even possible, to construct. For example, no master list of names of all people in the United States currently exists, and even if it did, such a list would become dated very quickly. Because of this, many polling agencies use a variant of random sampling, called *area sampling*, when doing national polls. In this method, the United States is first divided into a large number of homogeneous geographic regions. A subset of these regions is then randomly selected using a random number table. Each of these regions is then subdivided into a large number of geo-

graphic areas (based on census tracts or districts). A subset of these areas is then randomly selected. Within each area a list of the city or town streets is compiled. A subset of streets is then randomly selected. Finally, on each of the chosen streets, the house addresses are listed, and a subset of these is, in turn, randomly selected. The interviewer then approaches that house and randomly selects an individual within it. This completes the process. The overall region is selected at random. The tract is selected at random. The streets are selected at random. The houses are selected at random. And the individuals are selected at random. The end result is that every individual has an equal chance of being included in the sample.

Although this is true in theory, it does not really hold in practice. The chosen dwelling might be vacant. Or the individual within the dwelling may not agree to participate in the poll.

No nationwide survey has ever reached every person designated in a random sample and, accordingly, special "correction" factors must be adopted. A detailed consideration of these is presented by Gallup (1976). The process of area sampling is an extremely expensive one and is usually only carried out by major polling firms. Many of the smaller companies, newspapers, and magazines obtain samples in ways that would call their representativeness into question. The result of public opinion polls must always be interpreted in light of their sampling procedures.

One of the most famous polling errors in presidential elections was the prediction made by the *Literary Digest* in 1936. The polling procedure used by this organization had accurately predicted previous elections. It consisted of mailing out millions of postcard ballots to individuals who were listed in the phone book or who were on lists of automobile owners. This system was effective as long as the more affluent (people with phones and cars) were as equally likely to vote Democratic or Republican as the less affluent (people without phones and cars). With the advent of the New Deal, however, a shift in the American electorate occurred with the more affluent tending to vote Republican and the less affluent tending to vote Democratic. The result was that the *Literary Digest* sample was over represented with Republican voters, leading to a prediction that the Republican candidate (Landon) would defeat the Democratic candidate (Roosevelt) by a margin of 57% to 43%. Of course, Landon did not win and was, in fact, defeated by Roosevelt by a margin of 62.5% to 37.5%.

Even when random procedures are used, a sample many not be broadly representative of its population. Perhaps by chance, an unrepresentative sample will occur. What factors influence whether a random sample will truly reflect the population? One factor noted in this chapter is the size of the sample. In general, the larger the sample size, the better the estimate will be of a population statistic. The Gallup agency reports margins of errors associated with its polls and these are directly related to the size of the sample. For example, for a typical analysis of political opinions of people in the U. S. (for example, the percentage of people favoring a candidate), the following margin of errors, in percentage points, occurs:

Sample Size	Error
1500	2.5
750	4.0
400	6.0
200	8.0
100	11.0

For samples based on $N = 1500$, the reported percentage will be accurate within approximately 2.5 percentage points. For samples based on $N = 100$ the margin of error is much larger (11 percentage points).

6.7 / Types of Sampling Distributions

So far we have only considered the sampling distribution of the mean although it is possible to conceptualize sampling distributions of other statistics, such as the median, mode, or a variance. In the case of the median, for example, the sampling distribution reflects the median of all possible random samples of size N selected from a population. Given the same population, the sampling distribution of the mean will show less variability than the sampling distribution of the median and the sampling distribution of the mode. It is for this reason that the mean is usually preferred by statisticians as a descriptor of central tendency, since its standard error is lower.

6.8 / Confidence Intervals

Suppose we are trying to estimate a population mean for students at a large university on an intelligence test. We do so by selecting a random sample of one hundred students and administering the test to them. Suppose the mean score was 107. If we wanted to estimate with one value what the true population mean is, our best guess would be 107. But we would also not be very confident that the true population mean is exactly 107, because of sampling error.

Another approach to specifying the population mean would be to state a range of values that we are relatively confident the population mean is within (for example, between 100 and 110). The larger the range of values we specify, the more likely it is that the true population mean will be contained within it. Statisticians have developed a procedure for specifying such a range of values using our knowledge of the normal distribution and probability theory.

The interval to be constructed is called a **confidence interval.** The values that describe the boundaries of the interval are called the **confidence limits.** The degree of confidence we have that the population mean is contained within the confidence interval is stated in terms of a probability value or a percentage. The most common one is the "95% confidence interval."

Confidence intervals are conceptualized with respect to a sampling distribution. In the present example, we are concerned with a sampling distribution of the mean based on random samples of size 100. Suppose that the standard deviation of the sampling distribution, $\sigma_{\bar{X}}$, is known to equal 2.0. We also know from the central limit theorem that the sampling distribution is approximately normally distributed. Finally, suppose for the sake of illustration that the true population, μ, is equal to 105. Given that the sampling distribution is approximately normally distributed, we know that roughly 67% of all sample means will fall between one standard deviation below μ and one standard deviation above μ. Since the standard deviation of the sampling distribution of the mean is $\sigma_{\bar{X}}$ and equals 2.0, this means that roughly 67% of all sample means based upon $N = 100$ will fall between 103 (105 − 2) and 107 (105 + 2). Between what two numbers will 95% of the sample means occur? Using Appendix C, we find that 95% of the scores in a normal distribution occur between −1.96 standard deviations and 1.96 standard deviations of the mean. This means that 95% of all sample means based upon $N = 100$ will be between (− 1.96) (2.0) = − 3.92 and (1.96) (2.0) = 3.92 units of μ or between 101.08 (105 − 3.92) and

108.92 (105 + 3.92). The 95% confidence limits are, in general notation, defined as

$$\mu - (z_{.05})(\sigma_{\overline{X}}) \text{ to } \mu + (z_{.05})(\sigma_{\overline{X}}) \qquad [6.7]$$

where $z_{.05}$ is the z score defining the 95% confidence limits (in this case, 1.96) and all other terms are as previously defined. In our present example, the 95% confidence interval is between 101.08 and 108.92.

There is, however, a complication. We were able to describe the true 95% confidence interval because we knew the value of the population mean, μ. In practice, we do not know this and, in fact, μ is what we are trying to estimate based on data from one sample! Consequently, confidence intervals are computed using the sample mean in place of the population mean in Equation 6.7:

$$\overline{X} - (z_{.05})(\sigma_{\overline{X}}) \text{ to } \overline{X} + (z_{.05})(\sigma_{\overline{X}}) \qquad [6.8]$$

In our example, the sample mean we observed was 107. Using Equation 6.8, the 95% confidence intervals would be estimated as

$$107 - (1.96)(2.0) \text{ to } 107 + (1.96)(2.0)$$

$$= 107 - 3.92 \text{ to } 107 + 3.92$$

$$= 103.08 \text{ to } 110.92$$

The use of a single sample mean in the construction of a confidence interval raises some important interpretational distinctions. Suppose we draw all possible random samples of size 100 from a population and each sample mean, compute the confidence intervals based upon Equation 6.8. What we would find is that 95% of these confidence intervals would contain the value of μ, and 5% of the confidence intervals would not. *A 95% confidence claim reflects a long-term performance of an extended number of confidence intervals across all possible random samples of a given size.* In practice, only one confidence interval is constructed and that one interval either contains the population mean or does not contain the population mean. We never know for sure whether a particular confidence interval contains μ. However, with a 95% confidence interval, we can be reasonably certain that the observed confidence interval does include μ (since 95% of the confidence intervals constructed from samples of size N would include μ).

In most applications that use confidence intervals, the 95% confidence interval is the one most frequently computed. Some investigators will construct 99% confidence intervals in which $z_{.01} = 2.58$. The confidence level selected should depend upon how important it is not to have a "false" interval (that is, one that does not contain μ). The lower the confidence percentage (for example, 80% as opposed to 95%), the more likely it is that false intervals will be observed.

In general, the higher the degree of confidence (for example, 99% as opposed to 90%), the larger the size of the confidence interval. In our intelligence example, the 95% interval was 103.08 to 110.92, whereas the 99% interval would be 101.84 to 112.60. Since the size of the interval is also influenced by $\sigma_{\bar{X}}$ (see Equation 6.8), it follows that the sample size, N, will influence the size of the interval. As N increases, $\sigma_{\bar{X}}$ decreases, meaning that as N increases, the size of the confidence interval will decrease.

There is one final complication that we cannot consider until the next chapter. The above example was developed for the case where $\sigma_{\bar{X}}$ is known. We are more frequently in the position that $\sigma_{\bar{X}}$ is not known but rather must be estimated using sample data. In these cases, the conceptual foundations of a confidence interval are slightly different. This will be developed in Chapter 7. It should also be noted that confidence intervals can be constructed for many statistics and not just the mean.

Study Exercise 6.2

An investigator administers a reading test to a sample of eighty students and observes a mean of 83.0. The standard deviation for the population is known to be 17.35. Calculate the 95% and 99% confidence intervals.

Answer. Since the population standard deviation, σ, is known, it is possible to compute the standard error of the mean, $\sigma_{\bar{X}}$, using Equation 6.5:

$$\sigma_{\bar{X}} = \frac{\sigma}{\sqrt{N}} = \frac{17.35}{\sqrt{80}} = 1.94$$

The 95% confidence interval uses a z value of 1.96, as indicated in Appendix C. Using Equation 6.8, we find that the 95% confidence interval is

$$\bar{X} - (z_{.05})(\sigma_{\bar{X}}) \text{ to } \bar{X} + (z_{.05})(\sigma_{\bar{X}})$$

$$= 83.0 - (1.96)(1.94) \text{ to } 83.0 + (1.96)(1.94)$$

$$= 79.20 \text{ to } 86.80$$

The 99% confidence interval uses a z value of 2.58 and is represented by

$$\bar{X} - (z_{.01})(\sigma_{\bar{X}}) \text{ to } \bar{X} + (z_{.01})(\sigma_{\bar{X}})$$

$$= 83.0 - (2.58)(1.94) \text{ to } \bar{X} + (2.38)(1.94)$$

$$= 77.99 \text{ to } 88.01$$

We have 95% and 99% confidence that the above ranges, respectively, contain the value of the population mean.

6.9 / Summary

Whenever social scientists study samples, they must deal with the problem of sampling error, or the fact that a sample statistic may differ from the value of its corresponding population parameter. An unbiased estimator is one whose mean over all possible random samples equals the value of the population parameter being estimated. A key concept for understanding the nature of sampling error is that of sampling distributions. A sampling distribution of the mean refers to a distribution of mean scores based on all possible random samples of a given size. The mean of a sampling distribution of the mean will always equal the true population mean. The standard deviation of a sampling distribution of the mean, called a standard error of the mean, indicates the average deviation that random samples of size N deviate from the true population mean. The size of the standard error of the mean is influenced by two factors, the sample size, N, and the variability of scores in the population, σ. When the sample size is large ($N > 30$), the central limit theorem states that the shape of a sampling distribution of the mean will tend towards normality. These properties are crucial for inferential statistics and will be referred to extensively in later chapters. When estimating the value of a population parameter, it is possible to state a range of values within which the population value occurs with some probability. Such a range is called a confidence interval.

EXERCISES

1. How is a sampling distribution of the mean different from a frequency distribution as discussed in Chapter 2?

2. Define the standard error of the mean. What information does it convey?

3. Given a population with three scores, 2, 4, and 6, specify all possible samples of size 2 one could obtain from this sample (hint: there are six of them). Compute the mean score for each sample. Next compute the mean score across the six means. How does this result compare with the population mean score, μ? What principles does this illustrate?

4. Distinguish a standard error of the mean from a standard deviation of a set of raw scores.

5. A random sample of size 30 is drawn from each of two populations. Population A has a $\mu = 10$ and $\sigma = 5$. Population B has a $\mu = 8$ and $\sigma = 7$. Which sample mean will probably be a better predictor of its true population mean? Why?

6. Two populations each have a $\mu = 10$ and a $\sigma = 6$. A random sample of size 20 is taken from population A and a random sample of size 40 is taken from population B. Which sample mean will probably be a better predictor of its true population mean? Why?

7. If a sampling distribution of the mean has an associated standard error of the mean equal to zero, what does this imply about sample means drawn from the relevant population?

8. If a sampling distribution of the mean has an associated standard error of the mean equal to zero, what does this imply about the variability of scores in the population (σ)?

9. Match the term with the appropriate symbol.

(a) standard error of the mean   1. $\hat{s}_{\bar{X}}$

(b) standard deviation in a sample   2. s

(c) standard score from a normal distribution   3. $\sigma_{\bar{X}}$

(d) standard deviation in a population   4. z

(e) variance estimate   5. σ

(f) estimated standard error of the mean   6. $\hat{s}^2$

(g) standard deviation estimate   7. $\hat{s}$

10. What two factors influence the size of the standard error of the mean?

11. What is the central limit theorem?

12. Why is the mean usually preferred to the median as a measure of central tendency in inferential statistics?

13. For the following sample data, compute the mean and the estimated standard error of the mean: 2, 3, 3, 4, 4, 4, 5, 5, 5, 5, 6, 6, 6, 7, 7, 8.

14. For the following sample data, compute the mean and the estimated standard error of the mean: 4, 4, 4, 4, 4, 4, 5, 5, 5, 5, 6, 6, 6, 6, 6, 6.

15. Compare the estimated standard error of the mean in Exercise 13 with that in Exercise 14. Which sample is probably a better estimate of its true population mean? Why?

16. A researcher administered a test of mathematical ability to a sample of 169 students. The mean score for the sample was 74.4. The standard deviation for the population, σ, was known to be 13.0. Compute the 90%, 95%, and 99% confidence intervals for these data.

17. In Exercise 16, suppose that σ was equal to 7.0 and not 13.0. Compute the 90%, 95%, and 99% confidence intervals in this case. What is the effect of decreasing the size of σ on the size of the intervals? Why?

18. In Exercise 16, suppose the sample size was 100 instead of 169 and that $\sigma = 13.0$. Compute the 90%, 95%, and 99% confidence intervals in this case. What is the effect of increasing N on the size of the intervals? Why?

7 Hypothesis Testing: Inferences About a Single Mean

In the present chapter we consider basic principles of hypothesis testing. This material is central to all of inferential statistics and should be studied carefully. We will rely on several concepts developed in previous chapters and you should ensure that you are familiar with these concepts before covering the present material:

standard score

z score

normal distribution

probability of an event

sampling distribution of the mean

central limit theorem

You may wish to return to previous chapters and review these concepts.

7.1 / A Simple Analogy for Principles of Hypothesis Testing

Some of the basic steps of hypothesis testing in social science research can be characterized by a simple analogy. Suppose you were given a coin and asked to determine whether it was a fair coin, or biased towards heads or tails. You might respond to this task by conducting a simple experiment. If you assume that the coin is fair, then you would expect that flipping the coin a large number of times would result in heads about one-half the time, since the probability of a head is .50. So, you flip the coin 100 times and count the number of heads that occur. Suppose the tosses came out heads 52 times. This does not correspond *exactly* to what you would expect based upon a probability of .50, but it certainly is close (52 out of 100 as compared with 50 out of 100). Because the result is not very discrepant from the expected result, and because you know that you could have deviated somewhat from this expected result just by chance alone, you might conclude that the coin is not biased. But suppose the result of 100 flips had yielded 65 heads? Or 95 heads? At some point, the discrepancy from the expected result becomes too great to attribute it to chance and, in this case, you would reject your original assumption that the coin is fair.

In social science research, the process of hypothesis testing is very similar to the above. The investigator begins by stating a hypothesis that is assumed to be true (the coin is fair). Based upon this assumption, an expected result is specified (we should obtain 50 heads out of 100 flips). The data are collected and the results are compared with the expected result (52 heads versus an expectation of 50 heads). If the results are so discrepant from the expected results that the difference cannot be attributed to chance, then the original hypothesis is rejected. Otherwise, it is not rejected.

7.2 / Statistical Inference and the Normal Distribution

Suppose an investigator is interested in the intelligence of the type of student who attends Victor University. Specifically, she wants to know if the mean intelligence score of such students is above or below the typical intelligence score of college students in general. Suppose that past research concerning a particular intelligence test has shown that a score of 100 would represent the intelligence level of the typical college student. The question of interest, then, is whether the average score of the population of students who attend Victor University is different from 100.

Competing Hypotheses. We can state this issue more precisely in terms of two competing hypotheses or possibilities regarding the population mean. First, the population mean of students who attend Victor University could, in fact, equal 100, yielding the hypothesis that

$$H_0: \mu = 100$$

where "H_0:" is read as "The hypothesis that," and μ represents the true population mean of the students who attend Victor University. Second, the true population mean of the students may not equal 100, but rather may be higher or lower than 100. We can state this in terms of an **alternative hypothesis:**

$$H_1: \mu \neq 100$$

The task at hand is to choose between these two competing hypotheses.

The investigator decides to do this by collecting some data. A random sample of fifty students currently attending Victor University is administered the intelligence test. The mean score for the sample is 105. These data would seem to support the second hypothesis stated above, since the sample mean is not equal to 100, but

105. However, we know from Chapter 6 that a sample mean may not be an accurate descriptor of the population mean because of sampling error. Maybe the true population mean is 100 and we observed a sample mean of 105 because of sampling error. We need to test the viability of this possibility. Our knowledge of sampling distributions and the normal distribution will help us in this regard.

Analysis of Sampling Distributions. In the coin flipping example, we tested whether the coin was fair by stating a hypothesis that we assumed was true and then deriving an expected result from an experiment. Technically, such a hypothesis is called a **null hypothesis.** We will do the same for the problem with intelligence scores. Assume that the true population mean for Victor University students is equal to 100. If we select a random sample of fifty students from the population, we would expect the mean score for the sample to be near 100. We would not expect it to be exactly 100 since sampling error may be operating (just as we would not expect flipping a fair coin 100 times to yield exactly 50 heads and 50 tails). How much sampling error can we reasonably expect to have and still assume that μ equals 100? If the sample mean was equal to 101, could this be due to sampling error? What about 110?

We can be quite specific on this matter by making reference to a sampling distribution of means based upon all possible random samples of size 50. Recall that the mean score of a sampling distribution of the mean equals the true population mean. We have assumed, for purposes of our statistical test, that this equals 100. The standard deviation of the sampling distribution of the mean is the standard error of the mean ($\sigma_{\bar{X}}$). It represents how much, on the average, sample means based on $N = 50$ deviate from the true population mean. Suppose we knew the value of the standard error of the mean to be 2.5*. This implies that, on the average, sample means based on $N = 50$ deviate 2.5 units from the true population mean.

If the true population mean is 100 and the standard error of the mean is 2.5, how much sampling error can we reasonably expect in our data? Recall from Chapter 6 that a sampling distribution of means based upon relatively large N is approximately normally distributed. If the mean of this sampling distribution is 100 and the standard deviation (as indexed by the standard error) is 2.5, we know that roughly 67% of all scores in the distribution occur between one standard error above and one standard error below the mean (see Chapter 4), or between 97.5 (100 − 2.5) and 102.5 (100 + 2.5). Between what two scores would 95% of the sample mean occur, if $\mu = 100$? If we consult the z table in Appendix C, we find that 95% of the scores in a normal distribution will fall between 1.96 standard deviations above the mean and 1.96 standard deviations below the mean. If 95% of the sample means fall between ±1.96 standard errors of the mean, the probability that we would observe

---

*In practice, we usually won't know this, but we can estimate it. We will consider the case where estimates of the standard error are used in a later section of this chapter.

a sample mean *outside* of this range is less than .05 (since less than 5% of the sample means are outside of this range). This is a highly unlikely event. Our sample mean *was* outside of this range. It was 2.0 standard errors above the mean. Given this, we might reasonably conclude that the observed sample mean cannot be attributed to sampling error and that the assumption that $\mu = 100$ is untenable. We would therefore reject the null hypothesis.

We can formally summarize the steps we went through as follows:

1. Translate the research question into two competing hypotheses, a null hypothesis and an alternative hypothesis.

2. Assuming the null hypothesis is true, state an expected result of the experiment. The statement of the expected result takes the form of a range of values that you would expect the mean to fall within. This is stated in terms of standard scores in a sampling distribution; for example, between a standard score of -1.96 and $+1.96$.

3. Collect the data and compute the sample mean. Using the known value of the standard error of the mean (we will consider the case where it is unknown in a later section), translate the sample mean into a standard score to determine how many standard errors it is from μ *assuming the null hypothesis is true*. This is accomplished by the following formula:

$$z = \frac{\overline{X} - \mu}{\sigma_{\overline{X}}}$$ [7.1]

where z is the standard score, $\overline{X}$ is the sample mean (105), μ is the population mean assuming the null hypothesis is true (100), and $\sigma_{\overline{X}}$ is the standard error of the mean (2.5).

4. Compare the observed result (2.0) with the expected result (between -1.96 and $+1.96$). If the observed result does not correspond with the expected result, reject the null hypothesis. Otherwise, do not reject the null hypothesis.

An investigator was interested in whether the mean score on an aptitude test for the type of students who attended a rural elementary school was different from the typical score of students in general. A typical score was represented by a score of 50, and the standard deviation of the population, σ, was 10.0. A random sample of twenty-five students was obtained and the test administered. The mean score for the sample was 56.0. Test the viability of the hypothesis that $\mu = 50$.

Answer. This problem has one extra step relative to the problem developed in the text. We include this step because it will sometimes be encountered in actual applications. The step concerns the case where we have not been given information about the standard error of the mean, $\sigma_{\bar{X}}$, but rather we have been given the population standard deviation, σ. The calculation of the standard error of the mean is straightforward and uses the formula from Chapter 6:

$$\sigma_{\bar{X}} = \frac{\sigma}{\sqrt{N}} = \frac{10}{\sqrt{25}} = 2.0$$

We begin the problem by explicitly stating the null and alternative hypotheses:

$$H_0: \mu = 50.0 \qquad H_1: \mu \neq 50.0$$

If we define the probability of an unlikely event as being .05 or less, an unexpected result would be a z score that does not occur between -1.96 and $+1.96$. This is determined by use of the z table in Appendix C. The observed z score, using Equation 7.1, is

$$z = \frac{\bar{X} - \mu}{\sigma_{\bar{X}}} = \frac{56.0 - 50.0}{2.0} = 3.0$$

Since 3.0 is outside the expected range, we reject the null hypothesis and conclude that the hypothesis that μ equals 50 is untenable.

7.3 / Defining an Expected and Unexpected Result

A major step in the preceding hypothesis testing procedure was the specification of what constitutes an expected or unexpected result, under the assumption that the null hypothesis is true. As noted, this is typically stated in terms of a range of standard scores. The definition of an unexpected result is made with reference to a probability value known as an **alpha level.** For example, when the alpha level is .05, then a result is defined as being "unexpected" if the probability of obtaining that result is .05 or less, assuming the null hypothesis is true. In the previous example, 95% of the sample means were expected to fall between -1.96 and 1.96 standard errors of μ, assuming $\mu = 100$. The probability of observing a sample mean outside of this range is .05. Using an alpha level of .05. then, an unexpected result is any mean occurring less than -1.96 standard errors or greater than 1.96 standard errors from the value of μ as stated in the null hypothesis. In social science research, it is traditional practice to adopt a .05 alpha level. We will discuss this issue in more detail shortly.

7.4 / Failing to Reject versus Accepting the Null Hypothesis

In the Victor University example, the observed sample mean was inconsistent with a sampling error interpretation of the data and the null hypothesis was rejected. If the observed sample mean was, however, close to 100, say 100.5, would we have accepted the null hypothesis? The answer is no.

When an experimenter obtains a result that is consistent with the null hypothesis (when it falls between the range of -1.96 and $+1.96$ instead of outside of it) technically, he or she does not accept the null hypothesis as being true. Rather, he or she fails to reject the null hypothesis. There is a subtle distinction here, which is very important. In principle, *we can never accept the null hypothesis as being true via our statistical methods; we can only reject it as being untenable.* Consider the null and alternative hypotheses from the Victor University example:

$$H_0: \mu = 100$$

$$H_1: \mu \neq 100$$

Note that the null hypothesis is stated such that the true population mean must equal one and only one value (100), whereas the alternative hypothesis is stated such that the true population mean could potentially equal an infinite number of values (for example, 101, 101.5—anything but 100). Because of sampling error, we can never unambiguously conclude that the true population mean is equal to one specific value based upon sample data. When we observe a highly discrepant sample mean and reject the null hypothesis, we are saying that it is unlikely the true population mean equals 100 and we "accept" the alternative hypothesis. In contrast, if the observed sample mean was not extremely discrepant from the value of 100, we can only say that the sample mean was too close for us to safely conclude that the true population mean does *not* equal 100. We can't say that μ is equal to 100, but we also can't confidently say that it isn't. Thus, we fail to reject the null hypothesis.

7.5 / Type I and Type II Errors

Once an investigator has drawn a conclusion with respect to the null hypothesis, that conclusion can be either correct or in error. Two types of error are possible: (1) rejection of the null hypothesis when it is true (called a **type I error**), and (2) failure to reject the null hypothesis when it is false (called a **type II error**). These are illustrated in Table 7.1. The probability of making a type I error is reflected by the alpha level, in most cases, .05. The probability of making a type II error is traditionally called **beta** and is represented by the Greek letter β.

TABLE 7.1 TWO TYPES OF ERRORS IN HYPOTHESIS TESTING

	True State of Affairs	
	H_0 Is True	H_0 Is False
Reject H_0	Type I Error (Probability = α)	Correct Decision (Probability = $1 - \beta$)
Do not reject H_0	Correct Decision (Probability = $1 - \alpha$)	Type II Error (Probability = β)

Decision on the Basis of the Statistical Analysis

The nature of decision errors can be explained with reference to Figure 7.1. Consider a population of scores in which the null hypothesis that $\mu = 100$ is true. Distribution A represents a sampling distribution based on $N = 50$. The point marked "100" represents the population mean. A type I error would occur if we selected a sample from this population and concluded, based on the sample mean, that μ did not equal 100. If the alpha level was .05, this would occur, on the average, only five times out of 100, since only 5% of the sample means would occur outside the range -1.96 to 1.96 standard errors. Thus, the probability of a type I error equals the alpha level, and is indicated by the shaded areas of Figure 7.1.

Now consider the probability of a type II error. Suppose that the mean of a population of scores is 105 and we are testing the null hypothesis that $\mu = 100$. Distribution B in Figure 7.1 presents the sampling distribution of the population based on $N = 50$. The point marked "105" represents the true population mean. Note that if the observed sample mean occurred anywhere within the cross-hatched area, we would fail to reject the null hypothesis when it is, in fact, false. This cross-hatched area represents the probability of a type II error, or β. $1 - \beta$ defines the probability that an investigator will correctly reject the null hypothesis when it is false, and this probability is called the **power** of the test.

The above concepts can be illustrated intuitively with an electronics example. Suppose you are listening to a set of earphones and trying to decide if you hear a

Figure 7.1 Type I and Type II Errors

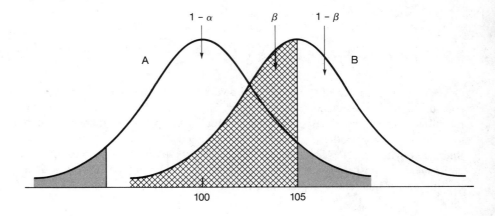

particular signal. There is static on the earphones, which makes this difficult for you. You have been told that you should hear the signal within 30 seconds. One type of error you could make is to say you heard the signal, when, in fact, it did not occur. This is analogous to a type I error. Suppose that making such an error would lead to devastating consequences. You would want to be very sure of yourself. Only if you are very certain you heard the signal would you say you heard it. This is similar to setting a low alpha level (for example, .05) in an experiment.

On the other hand, there is another error you could make; namely, saying you did not hear the signal, when, in fact, it was there. This corresponds to a type II error. The ability not to miss the signal corresponds to the power of a statistical test. If you have a very sensitive ear, you might be certain of your ability not to miss the signal. However, if you do not have a sensitive ear, you will be more likely to make this type of error.

Notice that the value of the alpha level directly affects the power of the statistical test. If you are very conservative about saying you heard the signal (setting a low alpha level), then this increases the likelihood that you will say the signal is not there when, in fact, it is (making a type II error). Factors other than the alpha level that influence the power of a statistical test are discussed in the next section.

7.6 / Selection of Alpha and the Power of Statistical Tests

As previously noted, the alpha level in an experiment reflects the probability of making a type I error. The tradition of adopting a conservative alpha level in social science research evolved from experimental settings where a given kind of error was very important and had to be avoided. An example of such an experimental setting is that of testing a new drug for medical purposes, with the aim of ensuring that the drug is safe for the normal adult population. In this case, deciding that a drug is safe when, in fact, it tends to produce adverse reactions in a large proportion of adults, is an error that is certainly to be avoided. Under these circumstances, the hypothesis that the "medicine is unsafe," or its statistical equivalent, would be cast as the null hypothesis, and a small alpha level selcted so as to avoid making the costly error. With a conservative alpha level, the medical researcher takes little risk of concluding that the drug is safe (H_1) when actually it is not (H_0). Thus, the practice of setting conservative alpha levels evolved from situations where one kind of error was extremely important and had to be avoided if possible. By casting such errors as type I errors in the hypothesis testing framework and then setting a low alpha level, the risk of committing the error could be minimized.

Several researchers (Cohen, 1969; Greenwald, 1975) have argued that social scientists have been preoccupied with type I errors at the expense of type II errors. The alpha level directly affects the power of the statistical test (and hence the probability of making a type II error), with more conservative alpha levels yielding less powerful tests. The argument is that for some social science research, it is hard to justify that a certain error should have the drastic character implied by a low alpha level. It is not necessarily worse, the argument goes, to falsely conclude that there *is* a difference between a mean and a hypothesized value than it is to falsely conclude that there isn't a difference. The issue of setting one's alpha level is one that enjoys much controversy. Interested readers are referred to Kirk (1972) for a detailed discussion of this issue.

In terms of the power of a statistical test, not only is the alpha level important, but so is the sample size: the larger the sample size, the more powerful the statistical test will be, other things being equal. This can be seen with reference to the formula for a *z* score in the inferential test (Equation 7.1) using a sample mean:

$$z = \frac{\overline{X} - \mu}{\sigma_{\overline{X}}} = \frac{\overline{X} - \mu}{\dfrac{\sigma}{\sqrt{N}}}$$

In this equation, as *N* becomes larger, *z* will also become larger, other things being equal. As *z* becomes larger, it is more likely that we will reject H_0, if it is false, thereby increasing the power of the statistical test.

In sum, an experimenter typically has control over the alpha level he or she selects as well as the sample size to be used in the experiment. The power of one's test can be increased by selecting larger sample sizes and larger alpha levels. Selecting a larger sample size must be evaluated relative to the practical concerns of the increased costs in obtaining more individuals in the investigation. In addition, increasing the alpha level must be evaluated relative to the importance of making a type I versus a type II error. There does come a point when increasing the power of a test is of diminishing value, since at that point the test is sufficiently powerful to draw a conclusion with an appropriate degree of confidence. Statisticians have developed procedures for estimating the power of a test in specific experimental applications and these are discussed in Cohen (1969). Cohen's treatment provides procedures for estimating minimal sample sizes that should be used in an investigation in order to achieve a specified level of statistical power. We will return to this topic in future chapters.

As a rough guide, investigators generally attempt to achieve statistical power (the probability of correctly rejecting the null hypothesis when it is false) in the range of .80 to .95, depending upon the nature of the theoretical proposition being investigated. The alpha level is traditionally set at .05.

7.7 / Directional versus Nondirectional Tests

In the preceding examples, the alternative hypothesis has always been *nondirectional*. It has stated that the true population mean is *either* higher *or* lower than the value specified by the null hypothesis. A *directional* alternative hypothesis, in contrast, is one that specifies that a population mean is different from a given value and also indicates the direction of that difference. For example,

$$H_1: \mu > 100$$

represents an alternative hypothesis that the population mean is *greater* than a score of 100.

A **directional test** is one designed to detect differences from a hypothesized population mean score (for example, $\mu = 100$) in one direction only. Suppose, for example, that a counselor is concerned about the reading skills of freshman students in her college due to poor preparation. She decides to administer a well-known reading test to a sample of incoming freshmen; if they tend to score lower than the national average on the test (corresponding to a score of 112), she will consider instituting a remedial program. In this case, the investigator would state a directional alternative hypothesis concerning the population mean of incoming freshmen:

$$H_1: \mu < 112$$

That is, the investigator wants the null hypothesis to be rejected only if the mean reading test score is less than the national average of 112, since this will be a signal to introduce a remedial reading program. The counselor's concern is only with detecting a mean *lower* than the national average.

Figure 7.2a presents a **rejection region** that is associated with only the lower end of a sampling distribution. Any score occurring in the rejection region would lead to the rejection of the null hypothesis. Assuming an alpha level of .05, the rejection region is defined by scores of less than -1.645. Use Appendix C to verify that 5% of the scores in a normal distribution are less than -1.645.

Figure 7.2b presents rejection regions for a **nondirectional test**, or one that is designed to detect differences either above or below the mean. In this case, the .05 alpha level is "split" such that .025 of the scores occur at the upper end of the distribution and .025 of the scores occur at the lower end of the distribution. The z scores that define the rejection regions are now -1.96 and 1.96. Verify this using Appendix C. Note in Figure 7.2 that the directional test is more sensitive than the nondirectional test to a decline of freshman scores below the national average. If the null hypothesis is false due to lower scores than the national average, the observed sample mean is more likely to fall in the broader rejection region of the directional test (scores less than -1.645) than the nondirectional test (scores less than -1.96).

We can summarize appropriate use of a directional as opposed to a non-directional test as follows: When there is exclusive concern that the population mean differs from a value in a specified direction, use a directional test. This concern is stated a priori, before performing any data analysis. (Never compute the necessary statistics and then, based upon the results, decide that a directional hypothesis should or should not be used. This defeats the logic of the statistical test.) If concern is not with a specific direction of difference, but rather with either higher or lower discrepancies, use a nondirectional test.

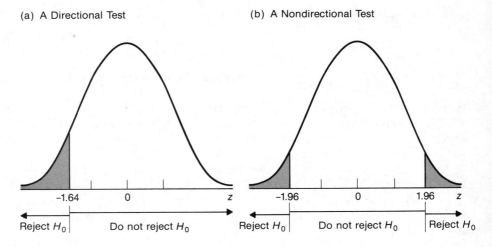

Figure 7.2 Rejection Regions for Directional and Nondirectional Alternative Hypotheses (From Witte, 1980)

(a) A Directional Test

(b) A Nondirectional Test

*Rejection regions are the darkened areas.

A directional test is often referred to as a **one-tailed test,** since it focuses on one "tail" of the distribution (see Figure 7.2a). A nondirectional test is often referred to as a **two-tailed test.**

7.8 / Using Estimated Standard Errors: The *t* Test

A major property of the statistical test developed in the previous sections was that the standard error of the mean, $\sigma_{\bar{X}}$, was known. It is more common that the standard error of the mean is not known, and that it must be estimated from sample data.

Suppose an investigator administered to a random sample of freshmen students an attitude scale designed to measure attitudes toward living in dormitories. Scores on the scale could range from 1 to 7, with increasing numbers representing increasing levels of favorability towards living in dormitories. A score of one indicates a very unfavorable attitude, a score of 4 a neutral attitude, and a score of 7 a very favorable attitude. The investigator is interested in whether the mean attitude score for the population was different from 4, the score representing a neutral feeling. The null and alternative hypotheses are

$$H_0: \mu = 4$$

$$H_1: \mu \neq 4$$

Note that a two-tailed alternative hypothesis is stated since the investigator is interested in whether the attitude is negative (less than 4) or positive (greater than 4).

The data collected from the sample of one hundred freshmen students yielded $\overline{X} = 4.51$ and $\hat{s} = 1.94$. Although the observed mean is consistent with the alternative hypothesis, a test for the possibility that sampling error could have yielded the result is necessary. But, unlike before, we do not know the value of the standard error of the mean, $\sigma_{\overline{X}}$. We can estimate it from the sample data, using the formula in Chapter 6, which is restated here:

$$\hat{s}_{\overline{X}} = \frac{\hat{s}}{\sqrt{N}}$$

$$\hat{s}_{\overline{X}} = \frac{1.94}{\sqrt{100}} = .194$$

Given this estimate, you might reason that we can modify Equation 7.1 from

$$z = \frac{\overline{X} - \mu}{\sigma_{\overline{X}}}$$

to

$$z = \frac{\overline{X} - \mu}{\hat{s}_{\overline{X}}}$$

That is, we simply substitute the estimated standard error for the known standard error and do everything we did previously. Unfortunately, some complications result from this. Since $\hat{s}_{\overline{X}}$ is calculated from sample data, it is subject to sampling error, whereas $\sigma_{\overline{X}}$ is not. Stated differently, the estimate of the standard error $\hat{s}_{\overline{X}}$, will vary from sample to sample. One could select several different samples and get several different values of $\hat{s}_{\overline{X}}$. Thus, it is possible that one could observe different z scores, even for samples with the same mean, since $\hat{s}_{\overline{X}}$ will differ from sample to sample.

Statisticians have developed procedures for dealing with this problem. Although the mathematics are complex, the idea can be stated in general terms: When $\sigma_{\overline{X}}$ is known and you calculate a z score for the sample mean, you are calculating the *exact* number of standard errors that your sample mean is from the mean of the sampling distribution, μ. If you substitute $\hat{s}_{\overline{X}}$ for $\sigma_{\overline{X}}$ and calculate a z score, you are calculating the *estimated* number of standard errors that your sample mean is from μ. The probability that a sample mean will be a certain number of *actual* standard errors from μ is not the same as the probability that a sample mean will be a certain number of *estimated* standard errors from μ. A distribution other than the normal distribution is called for to determine this probability, and it is called the *t* **distribution.**

The *t* distribution is very similar to the normal distribution in standardized form. It is bell-shaped and symmetrical, and has a mean of zero. However, the nature of the distribution is influenced by the number of scores (N) in the sample. Thus, unlike the normal distribution, there is a different t distribution for every sample size. Figure 7.3 presents a comparison of the t and normal distributions for three different sample sizes. When the sample size is large ($N>30$), the normal and t distributions are very similar. However, when the sample size is small, the shapes of the two distributions differ. For example, given an interval occurring at the extreme ends of the distribution, the proportion of scores occurring in that interval will be larger in the t distribution than in the normal distribution (see Figure 7.3). Stated in terms of probability, this means that the probability associated with an interval of scores at the extreme ends of the distribution will be greater in the t distribution than in the normal distribution. The smaller the sample size, the greater this discrepancy will be. This indicates the danger of using a t value as if it were a z value, since extreme values of t are more likely than comparable values of z.

A **t value,** then, is analogous to a standard score except that it represents the number of *estimated* standard errors a sample mean is from μ. The formula for a t value is

$$t = \frac{\overline{X} - \mu}{\hat{s}_{\overline{X}}}$$ [7.2]

which is the same as the z formula but with $\hat{s}_{\overline{X}}$ substituted for $\sigma_{\overline{X}}$ in the denominator. In the example with the attitude scores,

$$t = \frac{4.51 - 4.00}{.194} = 2.63$$

Just as statisticians have studied the normal distribution extensively, a considerable amount of information is known about t distributions. For example, it is possible to specify the probability of obtaining a t value greater than or equal to any number, just as we did in the previous section with the normal distribution. As noted earlier, there is a different t distribution for every sample size. Technically, there is a different t distribution depending upon the degrees of freedom associated with the t, which in this case equals $N - 1$. Because of the complexity of the t distribution,

Figure 7.3 Comparison of t and Normal Distribution for Three Sample Sizes

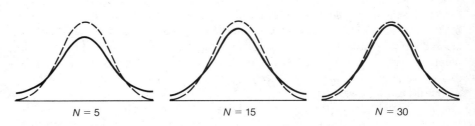

| $N = 5$ | $N = 15$ | $N = 30$ |

Solid line represents t distribution; dotted line represents normal distribution.

Appendix E presents a table of critical t values (that is, those that define rejection regions) corresponding only to selected alpha levels. Instructions for using this table are presented in Appendix E. In the attitude-toward-dormitories example, the critical value of t that defines the rejection region is 1.96 (α = .05, nondirectional test, df = 99). Thus, scores less than -1.96 or greater than 1.96 would lead us to reject the null hypothesis. The observed value of t was 2.63, and we would therefore reject the null hypothesis. Attitudes toward living in dormitories, on the average, were positive.

Numerical Example. Several years ago the maximally permissible speed limit on U.S. highways for motor vehicles was changed from 65 miles per hour to 55 miles per hour. One of the reasons for this change was the fuel savings that would result if people drove an average of 55 mph. To test if people were driving an average of 55 mph, one county performed an investigation in which the speed of twenty-five drivers was monitored in selected highway locations. For this sample of twenty-five drivers, the following data were observed:

$$\overline{X} \text{ speed} = 58.0$$

$$\hat{s} = 5.0$$

The average speed for this sample was 58.0 miles per hour. The population from which this sample was drawn, as conceptualized by the investigator, was drivers on U.S. highways in this particular county. Can we conclude from these data that the average speed of this population is, in fact, greater than the 55 mph standard? We begin by specifying our two hypotheses:

$$H_0: \mu = 55$$

$$H_1: \mu > 55$$

The alternative hypothesis dictates a one-tailed test since we are only interested in whether people drive faster, on the average, than 55 mph. With an alpha level of .05, the critical t value, taken from Appendix E, is 2.064. This is based on $N - 1$ or 24 degrees of freedom. If the calculated t value is greater than or equal to 2.064, then the null hypothesis will be rejected. We now calculate the t value:

$$t = \frac{\overline{X} - \mu}{\hat{s}_{\overline{X}}} \text{ and } \hat{s}_{\overline{X}} = \frac{\hat{s}}{\sqrt{N}}$$

$$\hat{s}_{\overline{X}} = \frac{5}{\sqrt{25}} = \frac{5}{5} = 1.00$$

$$t = \frac{58 - 55}{1.0} = \frac{3.0}{1.0} = 3.00$$

The t value does exceed 2.064. We therefore reject the null hypothesis and tentatively conclude that the type of driver in this county, on the average, drives faster than the 55 mph speed limit.

Although the statistical analyses for this study are consistent with the conclusion that the type of driver in the county drives faster than 55 miles per hour, the conclusion is not definitive. Beyond the statistical analyses, one must consider the research design in order to draw an appropriate conclusion. For example, if the measurements of speed were taken during one day only, perhaps there was something unique about that particular day relative to other days that caused drivers to speed. Driving speeds are affected by weather conditions such as temperature and wind current. Or maybe the day on which the measurements were taken was a holiday, resulting in an unusual number of ''out-of-county'' drivers passing through the county on their way home from a long trip. In the final analysis, interpretation of one's data is a function of the results of statistical analyses *and* the research design used to collect the data. Issues related to this will be considered in detail in the next chapter.

7.9 / Assumptions Underlying the Validity of the t Test

The inferential test that we have developed in this chapter (the t test for a hypothesized population mean value) is based on several assumptions about the nature of the population and measurements taken in a study. Unless these assumptions are met, the validity of the test is questionable and the inferences one draws may not be accurate. The assumptions are

1. The scores on the variable are normally distributed in the population.

2. The observations are independent* and randomly sampled from the relevant population.

Statisticians have shown that minor to moderate violations of the first assumption will not seriously affect the validity of the inferential test (Pearson and Please, 1975). However, marked violations, expecially with small sample sizes ($N < 10$), can create problems and, when these occur, other statisitcal techniques are required. A complete discussion of testing the above assumptions is presented in Winer (1971).

See Method of Presentation on page 139.

7.10 / Confidence Intervals When $\sigma_{\bar{X}}$ Is Unknown

The logic of the t distribution and the t test allows us to consider the calculation of confidence intervals when $\sigma_{\bar{X}}$ is unknown and must be estimated. Recall that confidence intervals allow us to specify a range of values within which the true

---

*This means, in most applications, that the scores on the variable must come from different individuals.

Method of Presentation

The statistical test just outlined is called a *t test* because it makes use of the *t* distribution. There are many different types of *t* tests and this one is used for testing a sample mean (for example, 58) against a hypothesized population mean value (for example, $\mu = 55$). When reporting the results of such tests, the null and alternative hypotheses are seldom formally written out. This is because the steps outlined above are implied whenever the test is reported, and formally writing out each step would be redundant and consume too much journal space. Unless stated otherwise, a nondirectional test is usually assumed. A *t* test like the one currently under consideration is typically reported as follows:

A *t* test was performed comparing the sample mean against a hypothesized population mean of 55. The sample mean of 58.0 was found to be significantly different from the value of 55.0 ($t = 3.00$, df = 24, $p < .05$, directional test), suggesting that the mean speed is greater than 55.

The first sentence states the type of test that was performed and the value of the hypothesized population mean. The second sentence begins by presenting the value of the observed sample mean (58.0). The phrase *significantly different from* is statistical jargon meaning that the null hypothesis was rejected. If the null hypothesis had not been rejected, the phrase might read *was found to be nonsignificant when compared with the value of 55,* or something similar to this. Although the terms *significant* and *nonsignificant* frequently occur in the literature, it is better to use the terms **statistically significant** or **statistically nonsignificant**. This makes it explicit that the use of the word *significant* is with respect to the statistics per se and not beyond this. The author has seen several instances in the popular press where studies are quoted as reporting "significant" findings when all that was intended by the researchers was to signify rejection of a null hypothesis. The real-world implications and significance of a particular rejection of a null hypothesis in a study is a very different matter than the statistical rejection of a null hypothesis. A statement to the effect that a null hypothesis about the fairness of a coin was rejected, although "statistically significant," certainly is not significant in terms of its real-world implications for the social sciences. Thus, it is best to use the terms *statistically significant* or *statistically nonsignificant.*

Toward the end of the second sentence, in parentheses, selected aspects of the statistical analysis are presented. The first is the computed *t* value using Equation 7.2. The second is the degrees of freedom associated with the *t* distribution. Sometimes these are presented within parentheses when reporting the *t* value, such as $t(24) = 3.00$. The third is a statement concerning the alpha level (.05), where the *p* stands for the *probability* of obtaining the observed *t* under the assumption that the null hypothesis is true. The less than sign indicates that the probability of obtaining the *t* was less than .05. Had the *t* test been nonsignificant (that is, the null hypothesis not rejected), the information in the parentheses would appear as follows ($t = 1.03$, df = 24, ns). Both the *t* and the degrees of freedom are reported, as before. However the ns (which stands for nonsignificant) indicates that the probability of obtaining the *t* score was *not* less than the traditional alpha level of .05. If an alpha level other than .05 is selected, then justification for this should be provided in the text.

population mean falls, with a given probability. Recall also from Chapter 6 that the formula for computing a confidence interval when $\sigma_{\bar{X}}$ is known is

$$\bar{X} - (z_{.05})(\sigma_{\bar{X}}) \text{ to } \bar{X} + (z_{.05})(\sigma_{\bar{X}})$$

The structure of the formula is the same when $\sigma_{\bar{X}}$ is unknown, except the t distribution is used in place of the normal distribution and $\hat{s}_{\bar{X}}$ is used in place of $\sigma_{\bar{X}}$. Thus,

$$\bar{X} - (t_{.05})(\hat{s}_{\bar{X}}) \text{ to } \bar{X} + (t_{.05})(\hat{s}_{\bar{X}}) \qquad [7.3]$$

where $t_{.05}$ represents the t value defining the 95% confidence limits, and all other terms are as previously defined. In the example on speeding, the 95% confidence intervals for a nondirectional test would be

$$58.0 - (1.645)(1.00) \text{ to } 58.0 + (1.645)(1.00)$$

$$= 56.355 \text{ to } 59.645$$

This means that there is a .95 probability that the true value of the population mean is within the range of 56.355 to 59.645.

The calculation of confidence intervals bears a relationship to aspects of hypothesis testing. Consider the case of testing the following null hypothesis:

$$H_0: \mu = 100$$

Suppose this hypothesis is tested for $\alpha = .05$, nondirectional test, and that the null hypothesis is rejected. The sample data leading to the rejection of the null hypothesis is $\bar{X} = 110$, $N = 25$, $\hat{s} = 10$. The 95% confidence interval about the mean is

$$\hat{s}_{\bar{X}} = \frac{\hat{s}}{\sqrt{N}} = 2.0$$

$$110 - (2.064)(2.0) \text{ to } 110 + (2.064)(2.0)$$

$$105.87 \text{ to } 114.13$$

Note that the value of μ in the null hypothesis ($\mu = 100$) is *not* contained within this interval. As it turns out, any null hypothesis that specifies a population mean value outside of the above interval would be rejected based upon the sample data. Further, any null hypothesis that specified a population mean within the interval would fail to be rejected. Because of this, confidence intervals provide the researcher with more information than the standard hypothesis testing procedure discussed in the present chapter. Specifically, the interval indicates to the investigator all null hypotheses that would or would not be rejected. Consequently, several social scientists have advocated reporting confidence intervals rather than the results of a formal hypothesis test about a specific value of μ when providing the results of statistical tests. Nevertheless, the large majority of research reports you encounter will use the formal hypothesis testing procedures presented earlier rather than the strategy of interval estimation.

7.11 / Summary

In this chapter, we have considered the basic logic underlying hypothesis testing. In doing so, we have introduced two statistical tests for testing a hypothesized mean value. These tests are used whenever an investigator wants to test the validity of an assertion that a given population has a specific mean score on some variable. The z test is used when the standard error of the mean is known, whereas the t test is used when the standard error of the mean must be estimated.

The logic of hypothesis testing begins with the specification of a null and an alternative hypothesis. The null hypothesis is the hypothesis that is tentatively accepted as being true for purposes of the statistical test. The alternative hypothesis can be either directional or nondirectional, depending on the nature of the question being asked. Next, an alpha level, usually .05, is specified and, based on this and N, a rejection region is specified. The data are collected and analyzed by converting the observed sample mean into a t value. This t value is compared with the "critical t," which defines the rejection region, and a conclusion is drawn. The null hypothesis is never accepted. Rather, we either reject it or fail to reject it. When drawing a conclusion, we also recognize that there is a possibility of error. Two kinds of errors are (1) falsely rejecting the null hypothesis (type I error), and (2) failing to reject a false null hypothesis (type II error). The probability of making a type I error is indicated by alpha and the probability of making a type II error is indicated by beta. The probability an investigator will correctly reject the null hypothesis is called the power of the test and is indicated by $1 - \beta$. The t distribution can be used to express confidence intervals about the mean when $\sigma_{\bar{x}}$ must be estimated.

EXERCISES

1. Define each of the following:

   (a) null hypothesis
   (b) alternative hypothesis
   (c) directional hypothesis
   (d) alpha level

2. For a t test of a hypothesized population mean value, state the critical value(s) of a t that would be used to reject the null hypothesis under each of the following conditions:

   (a) directional test, $N = 20$, positive direction
   (b) nondirectional test, $N = 20$

(c) directional test, $N = 10$, positive direction 1.833

(d) nondirectional test, $N = 10$ 2.262

3. Why is the t distribution used instead of the normal distribution when testing a null hypothesis based on sample statistics?

4. Under what conditions does the t distribution closely approximate the normal distribution?

5. What are the major assumptions underlying the t test of a hypothesized population mean value?

6. Why can't we ever "accept" the null hypothesis in traditional statistical tests?

7. Define each of the following:

(a) type I error

(b) type II error

(c) alpha

(d) beta

(e) power

8. What factors influence the power of a statistical test?

9. Suppose you wanted to test whether a sample was drawn from a population with $\mu = 100$ on an intelligence test. Calculate the t value and evaluate the viability of the null hypothesis for each of the following cases:

(a) $\bar{X} = 101$, $\hat{s} = 10$, $N = 10,000$ 10   rej.

(b) $\bar{X} = 101$, $\hat{s} = 10$, $N = 100$ 1   don't

(c) $\bar{X} = 101$, $\hat{s} = 2$, $N = 100$ 5   rej.

In each of the above cases, the difference of the sample mean ($\bar{X} = 101$) from the hypothesized population mean ($\mu = 100$) was the same. Why is the result in (a) different from the result in (b)? Why is the result in (b) different from the result in (c)?

10. Population trends in the United States have shown a dramatic decrease in fertility rates during the past five years. Presently, the fertility rate is below zero population growth levels. The fertility rate corresponding to zero population growth is 2.11 (that is, if couples average 2.11 children, then the size of the population will remain stable). One factor that has been shown to be related to family size is that of religion, with Catholics tending to have more children than Protestants. A researcher was interested in whether Catholics in the United States were currently having children at a rate

correspondent with zero population growth. Data on the number of children were obtained for a representative sample of Catholics and the following was observed: $\bar{X} = 3.11$, $\hat{s} = 2$, and $N = 144$. Test the viability of the hypothesis that Catholics are currently reproducing at a zero population growth rate. Report your results using the principles developed in the Method of Presentation section.

11. A large number of studies have investigated the effects of marijuana on human physiology and behavior. It is commonly believed by laypersons that marijuana does affect physiological processes in the human. One of the more common beliefs is that smoking marijuana makes one hungry (gives you the "munchies"). Weil, Zinberg, and Nelson (1968) examined this influence empirically. The hypothetical data presented below are representative of the results of this study.

Ten subjects participated in the study. Each was administered a high dose of marijuana by smoking a potent marijuana cigarette. The subjects were all males, 21 to 26 years of age, all of whom smoked tobacco cigarettes regularly but had never tried marijuana. The precise physiological mechanism that causes hunger is not well understood by psychologists. However, one factor that is often associated with hunger is blood sugar. Several theorists have suggested that this mechanism may be the cause of marijuana-induced hunger. The present experiment measured the level of blood sugar for each subject before they smoked marijuana and then again 15 minutes after having smoked marijuana. A "change score" was then computed for each subject by subtracting the amount of sugar in the blood before smoking marijuana from the amount of sugar in the blood after smoking marijuana. If no difference in blood sugar level is found (that is, if the change scores tend to equal 0), then two possibilities exist: (1) some other mechanism is causing the hunger, and/or (2) it may be that individuals don't really become more hungry after smoking marijuana, but rather, they find eating to be more pleasurable and, hence, eat more. The data regarding blood sugar levels (measured in mg) are presented below. If marijuana has no effect on hunger, one would expect the mean change score in the population to be zero. Test the viability of this and report your results using principles in the Method of Presentation section.

Individual	Change Score (Before-After)
1	14
2	− 2
3	6
4	− 2
5	− 2
6	− 2
7	− 6
8	− 2
9	− 18
10	− 6

12. For Exercise 10, compute the 95% and 99% confidence intervals and interpret them.

13. For Exercise 11, compute the 95% and 99% confidence intervals and interpret them.

14. A psychologist administers an intelligence test to a sample of thirty students in a school. Based upon previous studies, the psychologist has good reason to believe that the population variance, σ^2, for this school equals 116. The mean score in the sample was 121. Assuming an alpha level of .05 and a two-tailed test, test the viability that the true population mean equals 110.

15. Compute the 95% and 99% confidence intervals for Exercise 14 and interpret them.

The Analysis of Bivariate Relationships

Experimental Design and Statistical Preliminaries for Analyzing Bivariate Relationships

Section I outlined basic statistical concepts that will be used extensively in Sections II and III of this book. This section is concerned with analyzing relationships between two variables, or **bivariate relationships.** We will examine both inferential and descriptive statistics in this context.

8.1 / Principles of Research Design: Statistical Implications

In order to facilitate an understanding of the use of statistics in interpreting research, it will be instructive to review general principles that guide research design. As noted in Chapter 1, statistics and experimental design are highly interwoven with one another, and a consideration of design principles will lead to a better understanding of the statistics considered in later chapters.

Two Strategies of Research. When studying the relationship between two variables, the investigator is essentially interested in determining how the different values of one variable are associated with the values of another variable. If one were interested in studying the relationship between a person's gender (male versus female) and his or her mathematical ability, the focus would be on whether the different values on the variable of gender are associated with different values on the variable of mathematical ability (for example, whether males have higher mathematical aptitude than females).

In general, social scientists use two strategies for assessing the relationship between variables. First, they may use what is referred to as an **experimental strategy,** where a set of procedures (or manipulations) is performed in order to *create* different values of the independent variable in the research participants. The relationship of these manipulations to the dependent variable is then examined. For example, if the independent variable is test anxiety, the researcher might create two different values on this variable by telling half the research participants that their performance on a test is very important and will reveal many aspects of their personal competencies (thus creating high test anxiety), while telling the other half that the test is unimportant and won't reflect on them personally (thus creating low test anxiety). In this instance, the variable of test anxiety has two values and each research participant can be distinguished with respect to which value describes that participant.

An **observational strategy,** in contrast, does not involve the process of actively creating values on an independent variable, but rather involves using a set of procedures that will allow one to measure differences in values that naturally exist in the research participants. In a questionnaire, a person's gender might be measured based upon responses to the question, "What is your sex, male or female?" In this instance the researcher is not using a set of manipulations to create values on the variable, but rather is measuring the values that naturally exist.

Research strategies are not exclusively experimental or observational. Many investigators use both techniques in assessing different independent variables. A study might investigate the effects of a person's gender *and* test anxiety on test performance. Sex would be indexed by an observational strategy and test anxiety could be indexed by the manipulations noted above. The effects of both of these variables on test performance might then be examined.

A dependent variable is always measured in the "observational" sense noted above. This is because we are trying to determine how the dependent measure responds to the manipulations of the independent variable or covaries with the naturally existing values of the independent variable.

Controlling for Alternative Explanations: Random Assignment to Experimental Groups. A major goal of experimental design is to control for alternative explanations. Consider an experiment in which an investigator is interested in the effects of alcohol on reaction time. Two groups of subjects from a small college are used. One-half respond to a reaction time task while under the influence of alcohol and the other half respond to the same reaction task while not under the influence of alcohol. The task involves pressing a button when a certain slide appears in a series of slides shown sequentially. The investigator computes the mean reaction time in each group. Suppose in the alcohol condition that the reaction time mean was 2.45 seconds whereas in the no alcohol group the reaction time mean was .98 seconds. It appears that alcohol has had an effect. However, there are alternative explanations. First, it may be the case that the alcohol did *not* affect reaction time, and that the difference between means is simply the result of subjects in the alcohol condition having slower reaction times than the other subjects, independent of alcohol. If the study had been conducted without giving alcohol to any of the subjects, perhaps the results would still have yielded means of 2.45 and .98, respectively.

In order to control for this possibility, investigators will typically assign subjects to groups using random assignment procedures. If the condition in which a given subject participates is determined on a completely random basis, then it is no more likely that subjects assigned to the alcohol condition will have slower reaction times than those assigned to the no alcohol condition, independent of alcohol. Thus, **random assignment** helps to control for alternative explanations such as the above.

It is important to note that random assignment does *not guarantee* that the various groups will not differ, a priori, on reaction time. Rather, it is *unlikely* that they will. There is always the chance that, just by a fluke and even with random assignment, the various groups will differ on the dependent variable. In general, the larger the number of subjects randomly assigned to the two groups, the less likely it is that this will happen. Consequently, one should interpret studies with small numbers of subjects with caution.

Reducing Sampling Error. A second approach to controlling for alternative explanations focuses on sampling error. Consider the experiment on alcohol and reaction time. In this study, the investigator conceptualizes the two groups of subjects as random samples from two populations: (1) a population of individuals who are under the influence of alcohol and (2) a population of similar individuals who are not under the influence of alcohol. The two populations are assumed to be similar to each other in all respects except one, the presence or absence of alcohol. This is a reasonable assumption given the random assignment of subjects to experimental conditions. If the alcohol has no effect on reaction time, then the two populations are, for all intents and purposes, similar in *all* respects and one would expect the mean reaction time scores for the two populations to be the same. Stated in formal terms, we would expect

$$\mu_A = \mu_{NA}$$

where μ_A represents the population mean reaction time for individuals under the influence of alcohol, and μ_{NA} represents the population mean reaction time for individuals not under the influence of alcohol. If the alcohol *does* have an effect on reaction time, then we would expect the population means to be different, or

$$\mu_A \neq \mu_{NA}$$

The subjects in the experiment are conceptualized as random samples from these two populations. The mean reaction time in the alcohol condition was 2.45 and the mean reaction time in the no alcohol condition was .98. This would appear to be consistent with the notion that $\mu_A \neq \mu_{NA}$. However, we know from Chapter 6 that a sample mean does not always equal the population mean, due to sampling error. Perhaps the two population means are equal and the results of the experiment are nothing more than the result of sampling error. Ideally, we would like to minimize sampling error so as to rule out this interpretation as an alternative explanation. In Chapter 9, we will consider *statistical* methods an investigator can use to assess the impact of sampling error. In the present chapter, we will consider procedures an investigator can use to minimize its impact.

Recall from Chapter 6 that two factors that influence the accuracy of a sample mean as an estimate of μ are the size of the sample (N) and the variance of scores in the population (σ^2). One way an investigator can reduce sampling error is to increase the sample size for the various groups. In general, the larger the N in each group, the less the sampling error. Obviously, there are practical limitations to the number of subjects that can participate in an experiment. Research is expensive and the time and effort involved in collecting data can be quite consuming. As such, researchers will sometimes have to settle for relatively small sample sizes.

A second procedure for reducing sampling error is to define the population so that the variance of scores in the groups (σ^2) will be relatively small. Consider the alcohol-reaction time example. There is some evidence to indicate that males, in general, have slightly faster reaction times than females on tasks similar to that used in the experiment. The population of individuals in the alcohol condition consists of both males and females as does the population of individuals in the no alcohol condition. The presence of both males and females within the population yields more variability in reaction time scores than if males were considered separately or if females were considered separately. This can be illustrated with the following hypothetical set of reaction time scores:

Females	Males	Females and Males Combined	
1.3	1.1	1.3	1.1
1.4	1.2	1.4	1.2
1.5	1.3	1.5	1.3
$\mu = 1.4$	$\mu = 1.2$	$\mu = 1.3$	
$\sigma^2 = .007$	$\sigma^2 = .007$	$\sigma^2 = .017$	

Note that the male reaction time is slightly faster, on the average, than the female reaction time. Note also that the variance of the combined groups ($\sigma^2 = .017$) is greater than either group considered separately ($\sigma_M^2 = .007$ and $\sigma_F^2 = .007$).

One procedure for reducing sampling error would be to restrict the experiment to female subjects. This would have the effect of yielding a smaller population variance within the experimental and control groups relative to a study using both males and females. Unfortunately, in reducing σ^2, we have had to restrict the generalizability of the results of the study to females only. Social scientists frequently find themselves in such trade-off situations. The researcher must weigh the theoretical and applied benefits of reducing sampling error at the cost of reducing the generalizability of one's results.

Control of Variables. A large amount of social science research is designed to determine the causal relationship between variables; that is, does variable X cause variable Y? In order to draw unambiguous inferences it is necessary to control other variables in the research setting. In the previous sections we have been implicitly considering two basic types of variables a researcher must control. In this section, we will make the nature of these variables explicit.

One type of variable that is sought to be controlled by a researcher is a *confounding variable*. Suppose a researcher wanted to test for the existence of sex discrimination and did so by examining the relationship between the sex of a job applicant and a person's decision to hire that individual. An experiment might be designed in which the resumés of 50 applicants are gathered from personnel files, 25 males and 25 females. A group of personnel directors is then asked to read each resumé and rate the likelihood that they would hire each applicant on a 20 point rating scale. Suppose that the appropriate statistical analysis indicated there was, in fact, a relationship between the applicant's sex and the likelihood of hiring ratings, with males being more likely to be hired than females. Would this be evidence for sex discrimination? Not necessarily. The strongest evidence for sex discrimination would occur when a male is chosen over a female even when the two are equally qualified for the job. In the above experiment, it may have been the case that the males were, in fact, more qualified for the position than the females. Rather than the judgments being a function of the applicant's sex, they may instead have simply reflected the qualifications of the applicant, which just happened to be related to the applicant's sex. If this were the case, the test for sex discrimination would be ambiguous since the result could be attributed to either sex discrimination *or* differences in the quality of applicants. In this experiment, the qualifications of the applicant represents a **confounding variable.** *A confounding variable is one that is related to the independent variable (the presumed cause) and that influences the dependent variable (the presumed effect), rendering a causal inference ambiguous.* In the earlier example on alcohol and reaction time, the failure to randomly assign subjects to conditions could result in a priori differences in reaction times due to individual differences. Random assignment is one method of controlling confounding variables defined by individual differences in reaction times.

A second type of variable that must be controlled in research aimed at drawing causal inferences is a **disturbance variable.** *A disturbance variable is one that is unrelated to the independent variable (and hence, is not confounded with it) but that influences the dependent variable.* As a result, a disturbance variable increases sampling error by influencing variability (σ^2) within groups. In the alcohol-reaction time study, the sex of the subject was an example of a disturbance variable. Disturbance variables serve to obscure or mask a relationship that exists between the independent and dependent variables. To use an electronics analogy, they create "noise" in a system where we are trying to detect a "signal" and the more "noise" there is, the harder it is to detect the "signal."

There are several procedures that social scientists use to control for confounding and disturbance variables. Three of the most common strategies are (1) holding a variable constant, (2) matching, and (3) random assignment to experimental groups. Consider the following example. In the sociological literature, a relationship has been established between a woman's religion and how many children she wants to have in her completed family (that is, her ideal family size). Generally speaking, Catholics want more children than Protestants, who in turn want more children than Jews. Sociologists have interpreted this in terms of the effects of the religious doctrine that these women are exposed to. Another interpretation is possible, however. Catholics tend to come from larger families than Protestants, who tend to come from larger families than Jews. It may not be religion that influences family size desires, but rather the fact that those people who are raised in large families prefer large families and those people who are raised in small families prefer small families. Thus, in today's society, it may not be religion that is influencing family size desires but rather the size of the family in which one is raised. Religion and the size of the family raised in are confounded with each other. How might family size background be controlled?

We saw earlier for the alcohol-reaction time example how **holding a variable constant** could be used to control for a disturbance variable (sex of the subject). This procedure can also be used to control for a confounding variable, such as family size background. For example, a study might be undertaken with Catholics, Protestants, and Jews who all come from families with two children. In this case, family size background and religion are *not* related since family size background has been held constant. Differences in ideal family size between the three religious groups can not be attributed to differences in family size background since everyone in the study comes from the same-sized family. The variables are no longer confounded.

As noted earlier, the major disadvantage of holding a variable constant is that it may restrict the generalizability of the results. If the study is conducted only on individuals who come from families with two children, would the results generalize to individuals who come from families with five children? Perhaps religion influences family size desires when one comes from a relatively small family (two children), but not when one comes from a relatively large family (five children). When we hold a variable constant, we have no way of knowing the extent to which the results will generalize across the different levels of the variable that is held constant. One way to circumvent this would be to design an experiment whereby one studied the effects of religion at each of several levels of background family size (for example, one child, two children, three children, and so on), separately. In fact, this procedure is part of an experimental strategy called *factorial designs,* which are discussed in Chapter 17.

A second strategy that is used to control for confounding variables is a technique called **matching.** In this approach, an individual in one group is "matched" with an individual in each of the other groups so that all individuals have the same value on the confounding variable. This strategy is different from holding a variable constant, since within a group the confounding variable can vary considerably. However, for each individual in one group, there is a comparable individual in each of the other groups who has the same value on the confounding variable. As an example, fifteen subjects might be selected for purposes of study, five in each group, who have the following family size backgrounds:

Catholics	Jews	Protestants
3	3	3
4	4	4
2	2	2
3	3	3
1	1	1
$\overline{X} = 2.6$	$\overline{X} = 2.6$	$\overline{X} = 2.6$

Note that the average family size background is identical in all three groups. As such, any differences in the mean *ideal* family size among the three groups cannot be attributed to differences in family size background. Again, religion and family size background are now unconfounded. Further, the problem of restricted generalizability of results that was encountered with holding a variable constant is not applicable since a range of family size backgrounds is used.

Unfortunately, there are some disadvantages to this approach. In addition to being difficult to accomplish on a practical level, matching has the effect of turning a confounding variable into a disturbance variable! By letting the confounding variable vary within each group, we are unwittingly creating within-group variability that will reflect an increase in sampling error. Because of this and other problems with matching strategies (see Thorndike, 1944), this procedure should be used sparingly and with considerable care.

A final strategy for dealing with potential confounding variables is random assignment to experimental groups. This has already been discussed but two additional points should be made here. First, random assignment does not control for disturbance variables. It is only a procedure for controlling *some* confounding variables. Second, random assignment is feasible only when an investigator is using a manipulative experimental strategy to "create" values on an independent variable. If a person's sex is the independent variable, we can't randomly assign subjects to the conditions "male" and "female." By definition, males are males and females are females. Such independent variables will always be confounded with all other variables that are naturally related to them.

An Electronics Analogy. Many of the concepts discussed thus far can be illustrated intuitively using the electronics analogy from Chapter 7. Suppose you are listening to a set of earphones and trying to decide if you hear a particular signal.

There is a good deal of static on the earphones. In experimental design, the static corresponds to disturbance variables and your goal is to eliminate them to the extent that you can. Suppose you do some repairs and eliminate a large portion of the static. This is analogous to controlling for disturbance variables. There is still another set of problems, however. In addition to the signal you must detect, there are two other signals just like it. If you hear them, you will think your signal has occurred when, in fact, it hasn't. These other signals represent confounding variables. Another mechanical device you hook up eliminates one of these signals altogether, and turns the other one into static (a disturbance variable). The additional static is relatively minor, so you decide you have done everything possible and get ready to listen for the signal.

You are first, however, confronted with yet another problem, as discussed in Chapter 7. There are two possible errors you can make: (1) claim you heard the signal when, in fact, it did not occur, or (2) claim you did not hear the signal when, in fact, it did occur. These correspond to type I and type II errors. Suppose you determine that falsely saying the signal was present would be very detrimental, and that falsely saying it did not occur would be of little import. In this case, you would want to ensure you minimize the first type of error. This would correspond to setting a low alpha level (.05) in your experiment. With this in mind, you proceed to listen.

Problems with Inferring Causation. The statistics that will be developed in this text are designed to indicate to a researcher whether there is a relationship between two variables in the context of the research conducted. It must be emphasized that these statistics say nothing about whether two variables are *causally* related. It is entirely possible for two variables to be related to one another, but for no causal relationship to exist between them. A good example of this is shoe size and verbal ability. It turns out that a random sample of people living in the United States would reveal a moderate relationship between shoe size and verbal ability: The bigger one's foot, the greater the verbal ability. Does a causal relationship exist between these variables? Certainly not. It turns out that this relationship is due to a confounding variable, age. A large proportion of a random sample of people in the United States would be children. When children are very young, they have small feet and also have little verbal ability. As they grow older, their feet become larger and their verbal skills increase. If we were to hold age constant and remove its influence, there would be no relationship between shoe size and verbal ability.

Most statistics and research design books emphasize this point only when considering a specific type of statistical analysis called correlation analysis. However, the issue holds for *all* statistics that assess the relationship between variables. *The ability to make a causal inference between two variables is a function of the research design, not the statistical technique used to analyze the data that are yielded by that research design.* The fact that two variables are related and do not necessarily imply causality is important to keep in mind when interpreting statistics in the context of social science research.

Box 8.1 Confounding Variables

The identification of confounding and disturbance variables is critical for evaluating any research design. Huck and Sandler (1979) have presented an interesting and very readable collection of one hundred experiments that have received attention in the popular press or through advertising channels. They elaborate, as well, confounding and disturbance variables that could affect the interpretation of each experiment. An example from their book illustrates their approach and underscores the importance of controlling such variables.

Problem. The following story appeared about an advertisement in a weekly news magazine as well as in the local newspapers—you may have seen it yourself. It seems that the Pepsi-Cola Company decided that Coke's three-to-one lead in Dallas was no longer acceptable, so they commissioned a taste-preference study. The participants were chosen from Coke drinkers in the Dallas area and asked to express a preference for a glass of Coke or a glass of Pepsi. The glasses were not labeled "Coke" and "Pepsi" because of the obvious bias that might be associated with a cola's brand name. Rather, in an attempt to administer the two treatments (the two beverages) in a blind fashion, the Coke glass was simply marked with a "Q" and the Pepsi glass with an "M." Results indicated that more than half chose Pepsi over Coke. Besides a possible difference in taste, can you think of any other possible explanation for the observed preference of Pepsi over Coke?

Solution. After seeing the results of the Pepsi experiment, the Coca-Cola Company conducted the same study, except that Coke was put in both glasses. Participants preferred the letter "M" over the letter "Q," thus creating the plausible rival hypothesis that letter preference rather than taste preference could easily have accounted for the original results. Since no statistical tests were given, another plausible rival hypothesis is that of instability; that is, we don't know whether "more than half" means 51% or 99% or how much confidence we should place in the finding. Flipping a coin 100 times is almost sure to result in either heads or tails occurring more than half the time.

Strangest of all was the fact that the same design error of using one letter exclusively for each brand was repeated in a second study conducted by Pepsi. In a feeble attempt to demonstrate that their initial results were not biased by the use of an "M" or a "Q," Pepsi duplicated their first study, this time using an "L" for Pepsi and an "S" for Coke! Clearly, these three studies indicate that there is sometimes more in advertisements than meets the eye (or the taste buds).

Between- versus Within-Subjects Variables. When deciding on the type of statistical analysis one should use to analyze the relationship between two variables, it is necessary to introduce a distinction concerning how a given variable is measured. This distinction also has implications for how an investigator might design an experiment, and hence, is considered here. For purposes of exposition, we will develop the distinction in the context of a qualitative independent variable, and then extend the analysis to other situations in later chapters. Consider an experiment where the investigator is interested in the relationship between two variables, type of drug and learning. Specifically, the investigator has two types of drugs, A and B, and she wants to know if the two drugs differentially affect performance on a learning task. Fifty subjects are randomly assigned to one of two conditions. In the first condition, subjects are injected with drug A and then read a list of 15 words. They are asked to recall as many of the words as possible. A learning score is derived by counting the number of words correctly recalled (hence, scores can range from 0 to 15). In the second condition, subjects are injected with drug B and then read the same list of 15 words. The recall task is also administered and then comparisons between the two groups made to determine the relative effects of the drugs on learning.

In this experiment, the investigator is studying the relationship between two variables: (1) type of drug and (2) learning as measured on a recall task. Type of drug is a qualitative variable and the learning measure represents a quantitative variable. The qualitative variable is such that subjects who received drug A did *not* receive drug B and those who received drug B did *not* receive drug A. The two groups included different individuals; that is, they were **independent groups.**

Now consider a similar experiment that is conducted in a slightly different fashion. A group of fifty subjects is injected with drug A and then given the learning-recall task. One month later, the same fifty subjects return to the experiment and are injected with drug B and given the learning-recall task. The performance of these subjects under the influence of drug A is then compared with their later performance while under the influence of drug B. Note that in this experiment the fifty subjects who received drug A also received drug B. The values on the qualitative variable did not consist of independent groups but rather consisted of **correlated groups.**

Both of these experimental strategies are reasonable. However, different statistical procedures are used to test the relationship between the variables, depending upon whether the qualitative independent variable consists of correlated groups or independent groups. The reason for this will be discussed in later chapters. Another term that has been used to describe an independent group variable is the term **between-subjects variable.** This is because the values of the variable are "split up" between subjects instead of occurring completely within a subject. Another term for the correlated groups variable is a **within-subjects variable.**

In the above example, the experimenter had an option as to how to design her experiment. She could either use a **between-subjects design** or a **within-subjects design.** However, for many variables, there is no choice. Consider the variable of sex, which has the values male and female. This variable is, by definition, a between-subjects variable. Individuals who are in the group called "male" by definition cannot also be in the group called "female."

The relative advantages and disadvantages of designing an experiment using a between-subjects versus within-subjects design can be illustrated in the context of the example relating type of drug (A or B) to learning. The major advantage of within-subjects design concerns the control of variables that may lead to type I or type II errors. In the ideal experiment, the conditions would be such that the individuals in the two groups (drug A versus drug B) are identical in all respects except one—the type of drug they are given. If differences between the two groups occur in learning, then there would be one and only one logical explanation: The difference in drugs caused the differences in learning. In the between-subjects design, individuals are randomly assigned to one of the two conditions. The random assignment constitutes an attempt to "equalize" the two groups on all variables except that of the drug. By randomly assigning individuals, it is unlikely that one condition will, for example, contain more intelligent individuals (on the average) than the other condition. But the key word here is "unlikely." Although it is unlikely that between-group differences in intelligence will occur, it could happen, due to chance factors. When this does occur, the differential intelligence in the two conditions could make one drug appear superior when it is not. Or, it might offset the effects of the superior drug, making it appear as if there is no difference between them when there is.

In contrast, this cannot happen for the within-subjects design. The same individuals who receive drug A also receive drug B. By using the same individuals, intelligence must be the same in both conditions (unless it changed in the one month separating the administration of drugs). This is true not only of intelligence but all individual variables that might otherwise render interpretation ambiguous. Thus, the within-subjects design can offer considerably more experimental control than the between-subjects design.

This last statement must be qualified by a number of additional considerations. One problem with repeated measures designs (another term used for within-subjects designs) is the fact that the treatment in the first condition may have **carry-over effects** and influence performance in the second condition. For example, the effects of drug A may not have worn off completely when drug B is administered. This could make interpretation of the experiment ambiguous. Carry-over effects are not necessarily restricted to the independent variable of interest. For example, performance on the dependent measure during the second condition in our hypothetical experiment might be better than in the first condition. Instead of reflecting any difference in drugs, this may simply reflect the fact that the subjects are taking the recall test for the second time and are more familiar with it. This increased familiarity, not the drugs, could produce learning differences.

When an investigator is confident that no carry-over effects will occur, a within-subjects experimental design is usually superior to a between-subjects design. When carry-over effects are possible, a between-subjects design is more appropriate. Additional advantages of the within-subjects design will be considered in Chapter 10.

There is another type of research design, called a **matched-subjects design,** where different individuals are used in two or more conditions, but they are treated as if a within-subjects design is in force. In this case, a given subject in one condition is "paired-up" or "matched" with a subject in another condition who has characteristics similar to the first subject. The data are then treated as if these different individuals represent the same subject. For reasons discussed earlier as well as others (Thorndike, 1949), there are some serious problems with this strategy and its use is recommended only under restricted circumstances.

Study Exercise 8.1

For each experiment, indicate whether the qualitative variable is a between-subjects variable or a within-subjects variable.

EXPERIMENT I

A researcher was interested in the effects of religion on family size decisions. She interviewed a total of 150 people, 50 of whom were Catholic, 50 of whom were Protestant, and 50 of whom were Jewish. Each person was asked the number of children they wanted to have in their completed family. The mean number of desired children was computed for each of the three groups.

Answer. This experiment is concerned with the relationship between religion (a qualitative variable) and the number of desired children (a quantitative variable). Religion is a between-subjects variable with three values (Catholic, Protestant, and Jewish). It is a between-subjects variable because the 50 individuals who are Catholic are not the same group of individuals who are Protestant who, in turn, are not the same group of individuals who are Jewish.

EXPERIMENT II

An investigator was interested in the effects of music on problem-solving performance. Two hypotheses were possible: (1) music helps to relax an individual and should therefore facilitate problem-solving performance, or (2) music serves to distract the individual and, hence, it should interfere with problem-solving performance. One hundred individuals each tried to solve ten problems with soft background music playing. Three weeks later, these same individuals returned and tried to solve ten similar problems. This time, however, there was no background music. The number of problems correctly solved by each individual under each of the two conditions separately was computed and compared.

Answer. In this experiment, the independent variable is a qualitative variable with two values (background music vs. no background music). The dependent variable is the number of problems solved and constitutes a quantitative variable. The qualitative variable represents a within-subjects variable. This is because the same one hundred individuals participated in both the background music and no-music conditions.

8.2 / Choosing the Appropriate Statistical Test to Analyze a Relationship: A Preview

Parametric Versus Nonparametric Statistics. All of the inferential statistical techniques described in this section that use quantitative dependent variables are usually applied when these variables are measured on approximately an interval level. These techniques rely heavily on the analysis of means, variances, and standard deviations and are called **parametric statistics.** Each requires assumptions about the distribution of scores within the population, assumptions that will be made explicit as the techniques are introduced. In contrast, there is also a class of statistics called **nonparametric statistics** that do not require assumptions about distributional properties of scores and that are appropriate for ordinal (and hence, also interval and ratio) data.

There is currently some controversy among social scientists concerning the use of parametric and nonparametric statistics. Those arguing for nonparametric analyses hold that measures used in the social sciences often depart radically from interval level characteristics. Further, the application of parametric analyses, they argue, is inappropriate when distributional assumptions are not met. Those arguing for parametric analyses note that parametric-based analyses are more refined and powerful than current nonparametric methods. They argue that many of these techniques are *robust* to violations of the distributional assumptions required of them. The term **robust** means that even though the assumptions of the technique are violated, the frequency of type I and type II errors is relatively unaffected when compared with conditions under which the assumptions are met. Because of this, some statisticians, such as Bohrnstedt and Carter (1971), have concluded "that when one has a variable which is measured at the ordinal level, parametric statistics not only can be, but should be, applied."

A complete discussion of this issue is beyond the scope of this book and interested readers are referred to Stevens (1951), Bohrnstedt and Carter (1971), Lord (1953), Boneau (1960), and an excellent collection of readings by Kirk (1972).

Robustness of Statistical Tests. The concept of robustness is important in inferential statistics. As noted above, robustness refers to the extent that conclusions drawn on the basis of a statistical test (for example, rejection of the null hypothesis) are unaffected by violations of the assumptions underlying the test. Mathematicians who develop inferential tests use assumptions for different reasons. Sometimes an assumption will be used to simplify mathematical derivations so as to make the statistics more *manageable.* Other times, an assumption will be used because it characterizes what is likely to be the case in the real world; that is, the assumption is *credible.* As an example, a statistical test that assumes intelligence test scores occur with equal frequency, would have little applicability to most real-world problems. The assumption of a normal distribution, with a large proportion of central scores and few deviating scores, is much more credible.

Assessing the robustness of a statistical test can be extremely difficult. Such assessments often make assumptions that may be more credible, but that are also less manageable. As you might imagine, the results of studies investigating the robustness of inferential tests are quite complex. Consequently, in future chapters where the robustness of a specific test is being considered, it will be impossible to discuss robustness with any degree of precision. Rather, general characterizations of the robustness of the test will be provided as well as technical references that have investigated the robustness issue.

The general strategy used by statisticians to test robustness can be illustrated with reference to the *t* test discussed in Chapter 7. The following steps would be followed:

1. Given a true null hypothesis and an alpha level of .05, the critical value in the *t* distribution is determined so that only 5% of the *t* scores obtained exceed this critical value *when the assumptions of the test have been satisfied.* The alpha level in this case is called the **nominal significance level** and equals .05.

2. By empirical or mathematical means, the *actual* proportion of *t* values that exceed this critical value when the null hypothesis is true and one of the assumptions is *not* satisfied is determined. This proportion represents the **"actual" significance level,** when an assumption is violated.

3. The actual significance level and the nominal significance level are then compared. If the two are very close, then the test is robust to the violation of the assumption in question. If there is a large discrepancy between the two, then the test is not robust.

Selection of a Statistical Test. It is impossible to state any precise rules for determining the type of statistical test to apply in a given situation. It is always possible to find an exception to the rule whereby a different form of analysis might be more appropriate. Nevertheless, we can specify the most typical research designs you will encounter and the most typical type of analysis used in the context of these designs. In Chapters 9–15, we will consider the most common statistical tests for analyzing bivariate relationships and, in each chapter, specify the factors one considers in deciding to apply the test. As we will see, the essence of the proposed framework rests on distinguishing between qualitative and quantitative variables and within-subjects and between-subjects designs. The required steps are: (1) identify the independent and dependent variables; (2) classify each as being either a qualitative or a quantitative variable; (3) classify the independent variable as either a between-subjects or within-subjects variable; and (4) note the number of values that occur for each variable. At the beginning of each of the following chapters, we will specify the conditions that typically apply to the statistical test being considered. Chapter 16 will formally discuss procedures for choosing a statistical test to analyze one's data.

The statistics will focus on three questions concerning a bivariate relationship:

1. Given sample data, can we infer that a relationship exists between two variables in a population?

2. If so, what is the *strength* of the relationship?

3. If so, what is the *nature* of the relationship?

8.3 / Summary

The results of statistical analyses must be interpreted in the context of the research design used to generate data. Social scientists generally use two types of research design, an experimental strategy and an observational strategy. A major goal of research is to control for alternative explanations. This involves the control of confounding variables and disturbance variables. Techniques for controlling confounding and disturbance variables include random assignment to groups, holding a variable constant, and matching. In addition, between-subjects or within-subjects experimental designs can be used for purposes of control.

In Chapters 9 through 15, statistics for analyzing bivariate relationships will be examined. An important feature of these statistics is the assumptions they must make about the nature of data. To the extent that conclusions drawn from a statistical test are unaffected by the violation of these assumptions, the test is said to be robust. This issue will be examined for each test considered in these chapters.

EXERCISES

1. Characterize what is meant by the phrase "robust to violations of assumptions."

2. What is the difference between an experimental research strategy and an observational research strategy?

3. There is a moderately strong relationship between a person's height and the length of the person's hair. In general, the shorter one's hair, the taller one tends to be. Obviously, there is not a causal relationship between length of hair and how tall someone grows. What do you think is the variable that produces this "spurious relationship"?

4. Characterize a disturbance variable. How is it different from a confounding variable?

For each experiment described below, indicate the independent variable and the dependent variable. State whether each is qualitative or quantitative. Indicate if the independent variable is between-subjects or within-subjects in nature.

5. *Experiment I*. Morrow and Davidson (1976) studied the effects of race on family size decisions. These investigators interviewed a total of 300 people, 100 of whom were black, 100 of whom were Mexican-Americans, and 100 of whom were white. Each person was asked the number of children they wanted to have in their completed family. The average number of desired children was compared for the three groups.

6. *Experiment II*. Sears (1969) has reviewed studies on the relationship between an individual's sex and his or her political party preference. In one investigation, a group of 150 individuals was interviewed, 75 males and 75 females. Respondents were asked whether they considered themselves to be Democrats, Republicans, or Independents. The frequency with which males identified with the three classifications was compared with the corresponding frequency for females.

7. *Experiment III*. Steiner (1972) has reviewed studies on the effects of the presence of others on problem-solving performance. In one study, a group of 100 individuals was used in the experiment. At the first session, each subject was seated in a room alone and given a math problem to solve. The amount of time it took the subject to solve the problem was measured. Two weeks later, the subject returned and solved another problem, but this time there was an observer present watching her. The amount of time it took to solve the problem was again measured. For each of these two conditions, observer present vs. observer absent, the mean problem-solving time was computed across the 100 subjects. These means were then compared.

8. *Experiment IV*. Harvath (1943) was interested in the effects of noise on problem-solving performance. A group of 150 individuals participated in the study. Subjects were randomly assigned to two conditions. 75 subjects tried to solve 30 problems while a steady "buzz" was present in the background. The other 75 subjects tried to solve the same 30 problems but with no background noise. The number of correct problems was computed for each subject. The average number of correct solutions was compared for people in the noise condition with those in the no-noise condition.

9. What are the advantages of a within-subjects experimental design as compared to a between-subjects experimental design? What are the disadvantages?

10. Design an experiment to test the relationship between two variables (you choose the two) where one of the variables is a within-subjects variable.

Independent Groups *t* Test

9.1 / Use of the Independent Groups *t* Test

The statistical technique developed in this chapter is called the **independent groups *t* test.** It is typically used to analyze the relationship between two variables when

1. the dependent variable is quantitative in nature and is measured on a scale that approximates interval characteristics;

2. the independent variable is *between-subjects* in nature (note: it can be either qualitative or quantitative)*; and,

3. the independent variable has two and only two values.

Let us consider an example of an experiment that meets these conditions. When a friend describes a stranger to you, you may form an impression as to whether you would like that individual. One question of interest to social scientists has been whether the order in which information is provided about a stranger influences the kind of impression you form about him. Consider the following experiment: A group of twenty subjects is read a verbal description of a stranger and then asked to rate on a 7 point scale the extent to which they think they would like or dislike that person. The scale is such that the higher the number, the more the stranger would be liked. The stranger is described by six adjectives: intelligent, sincere, honest, conceited, rude, and nervous. The first three characteristics are positive in nature while the last three characteristics are negative. Ten of the subjects are randomly selected and given the description in the order listed above; that is, first the positive traits and then the negative traits. The other ten subjects are given the same description, but in reverse order; that is, first the negative traits and then the positive traits. Thus, we have two conditions reflecting a different order of presentation:

Group 1 (pro–con): intelligent, sincere, honest, conceited, rude, nervous

Group 2 (con–pro): nervous, rude, conceited, honest, sincere, intelligent

*Matched-subjects designs are not applicable. See Chapter 10 for the appropriate statistical method for matched-subjects experimental designs.

If the order of information has no effect on the type of impression formed, we would expect that, on the average, the likeability ratings would not differ between the two groups. However, if the order of information does matter, then the average likeability ratings should differ between the groups.

In this experiment, the order in which trait information is presented is the independent variable, and the ratings of perceived liking represent the dependent variable. The independent variable has two values (pro–con versus con–pro) and is between-subjects in nature. It is a qualitative variable whereas the dependent variable is quantitative in nature. Given these conditions, the independent groups t test is the statistical technique that will typically be used to analyze the relationship between the variables. The data for the experiment are presented in Table 9.1.

9.2 / Inference of a Relationship Using the Independent Groups t Test

Null and Alternative Hypotheses. The first question to be addressed is whether a relationship exists between the independent variable and the dependent variable. We begin by stating this question in terms of null and alternative hypotheses. This is accomplished with reference to population means. Because we are interested in generalizing the results of our study beyond just those subjects who participated in it, we conceptualize the subjects as representing a random sample from an infinitely large population of similar individuals. In the experiment, there are two populations of interest (as conceptualized by the investigator): (1) individuals who have had a list of trait descriptions read in an order from pro to con, and (2) individuals who have had a list of trait descriptions read in an order from con to pro. Given that individuals were randomly assigned to groups, and if we assume that the order of information does *not* matter, then we would expect the population means for the two groups to be equal:

$$H_0: \mu_1 = \mu_2 \qquad \text{[9.1]}$$

The null hypothesis posits no relationship between the independent variable (order of information) and the dependent variable (likeableness). It states that it doesn't matter what the value of the independent variable is, because as this value changes (from pro–con to con–pro), the mean score on the dependent variable does not. The alternative hypothesis would be that there *is* a relationship between the two variables since the value of the independent variable *does* influence the average score on the dependent variable:

$$H_1: \mu_1 \neq \mu_2 \qquad \text{[9.2]}$$

We have now restated the question in the context of two competing hypotheses, the null hypothesis (there is no relationship between the variables) and an alternative

TABLE 9.1 DATA FOR ORDER-OF-INFORMATION EXPERIMENT

Pro–Con Condition	Con–Pro Condition
7	1
4	5
3	2
5	5
5	3
6	3
5	3
5	2
4	4
6	2

$$\Sigma Y_1 = 50 \qquad \Sigma Y_2 = 30$$
$$\overline{Y}_1 = 5.0 \qquad \overline{Y}_2 = 3.0$$
$$SS_1 = 12.0 \qquad SS_2 = 16.0$$
$$\hat{s}_1^2 = 1.33 \qquad \hat{s}_2^2 = 1.78$$
$$\hat{s}_1 = 1.15 \qquad \hat{s}_2 = 1.33$$

hypothesis (there is a relationship between the variables). The next step is to choose between these two hypotheses. Examine the sample means in Table 9.1. The mean likeability score for the pro–con group was 5.0 while the mean likeability score for the con–pro group was 3.0. The two means are not equal and they would appear to be consistent with the alternative hypothesis. However, we know from Chapter 6 that a sample mean may not reflect the true value of its population mean due to sampling error. Thus, the observed difference between the two sample means may not reflect the influence of the order of information on likeability, but rather may reflect sampling error. Our task is to determine whether this is a reasonable interpretation of the observed sample difference between means.

Sampling Distribution of the Difference Between Two Independent Means. In order to test the sampling error interpretation, we will use logic directly analogous to that developed in Chapter 7 on hypothesis testing. To accomplish this, we must develop the concept of a **sampling distribution of the difference between two independent means.** Consider two large populations whose mean scores on a variable are equal, $\mu_1 = \mu_2$. Now suppose we select a random sample of size 10 from each population and compute the mean score in each sample as well as the difference between the means. We might find a result such as that illustrated in the first row of Table 9.2. Suppose we repeat this process again, yielding a second set of means and a second difference between two means, as shown in Table 9.2. In principle, we could do this for all possible random samples of size 10, yielding a distribution of mean differences. Table 9.2 presents some example differences that might be observed. This distribution of the differences between two means represents a sampling distribution of the difference between two independent means. It is directly analogous to a sampling distribution of the mean. However, now the concern is with a distribution of scores representing the difference between two means. As we did in Chapter 6, we can compute the mean and standard deviation of this sampling distribution.* If we were to do so, we would find that many of the same properties hold for a sampling distribution of the difference between means as that for a sampling distribution of the mean. One of these is that

> *The mean of a sampling distribution of the difference between means is always equal to the difference between the population means.*

Consider the case where two population means are equal and hence their difference is zero ($\mu_1 - \mu_2 = 0$). If we were to generate a sampling distribution of the difference between means for these populations, the mean of the differences would equal zero. The underlying principle is much the same as developed in Chapter 6. When we select samples and compute the difference between means, some of our samples will overestimate the true mean difference while others will underestimate it. When we average all of these, the underestimations will cancel the over-estimations, with the result being the true population difference between means.

The standard deviation of the sampling distribution of the difference between two independent means is called the **standard error of the difference between two independent means.** Like the standard error of the mean, it reflects how much sampling error we can expect, on the average. Statisticians have developed a formula that allows us to compute the standard error of the difference between two means when the population standard deviations are known. The formula is

$$\sigma_{\bar{Y}_1 - \bar{Y}_2} = \sqrt{\frac{\sigma_1^2}{n_1} + \frac{\sigma_2^2}{n_2}} \qquad [9.3]$$

where $\sigma_{\bar{Y}_1 - \bar{Y}_2}$ is the standard error of the difference between two independent means, σ_1 is the population standard deviation for group 1, σ_2 is the population standard deviation for group 2, n_1 is the sample size for group 1, and n_2 is the sample size

---

*In practice, we would never actually do this. However, as developed in Chapter 7, these concepts are central to hypothesis testing.

TABLE 9.2 SELECTED SAMPLE MEANS AND MEAN DIFFERENCES FROM A SAMPLING
DISTRIBUTION OF THE DIFFERENCE BETWEEN TWO MEANS

$\overline{Y}_1$	$\overline{Y}_2$	$\overline{Y}_1 - \overline{Y}_2$
5.0	3.0	2.0
4.3	3.6	.7
5.6	6.4	− .8
4.8	5.4	− .6
4.7	4.7	0
5.2	5.9	− .7
6.0	5.7	.3
4.3	3.9	.4
5.2	6.5	− 1.3
6.1	7.5	− 1.4
5.0	3.9	1.1
5.2	5.9	− .7
4.7	4.1	.6
4.9	4.4	.5
5.0	5.9	− .9
6.2	6.4	− .2
4.3	2.5	1.8
5.1	6.4	− 1.3
5.7	6.6	− .9
5.1	4.1	1.0
4.9	5.4	− .5
4.9	5.4	− .5

for group 2.* Note the similarities of this formula to the one for the standard error
of the mean, which is repeated here:

$$\sigma_{\overline{Y}} = \frac{\sigma}{\sqrt{N}}$$

*When more than one group is involved, the lower case n refers to the sample size of a particular group,
whereas capital N refers to the total number of subjects in the study, that is, $N = n_1 + n_2$. Note that in
this design, n_1 does not have to equal n_2.

As with the standard error of the mean, the standard error of the difference between two independent means is influenced by two factors: (1) the size of the samples (n_1 and n_2) and (2) the amount of variability in the populations, (σ_1^2 and σ_2^2). The same principles discussed in Chapter 6 apply here.

If we know the values of σ_1^2, σ_2^2, n_1, and n_2, we can compute the value of the standard error in Equation 9.3. In practice, we typically know the value of n_1 and n_2, but we do not know the values of σ_1^2 and σ_2^2. It is therefore necessary to estimate them from sample data. For the purposes of the test we will ultimately apply, it is necessary to assume that the two population variances are equal:

$$\sigma_1^2 = \sigma_2^2 = \sigma^2$$

This assumption, known as the assumption of **homogeneity of variance,** redefines the estimation problem. Instead of estimating σ_1^2 and σ_2^2 separately, our goal is to estimate σ^2, the variance of both populations. This can best be accomplished by combining the estimates from the two samples. By pooling the estimates from the two independent samples, we increase the degrees of freedom and thereby obtain a better estimate of σ^2. The simplest method of pooling the two variance estimates is to compute their unweighted mean. This is, in fact, what is done unless n_1 and n_2 are not equal to one another. If one of the groups has a larger sample size than the other, it makes sense to give the variance estimate from the group with the larger n more "weight" in determining the pooled variance estimate. This is because the larger the sample size, the better the estimate. This is accomplished by the following equation:

$$\hat{s}_p^2 = \frac{(n_1 - 1)\, \hat{s}_1^2 + (n_2 - 1)\, \hat{s}_2^2}{(n_1 + n_2 - 2)} \qquad \textbf{[9.4]}$$

where $\hat{s}_p^2$ represents the pooled estimate of σ^2, and all other terms are as previously defined. Examine the right-hand side of Equation 9.4. We have multiplied each sample variance by its degrees of freedom and then divided this quantity by the total number of degrees of freedom, or

$$\hat{s}_p^2 = \frac{\mathrm{df}_1\, \hat{s}_1^2 + \mathrm{df}_2\, \hat{s}_2^2}{\mathrm{df}} \qquad \textbf{[9.5]}$$

where df_1 represents the degrees of freedom for the variance in sample 1, df_2 represents the degrees of freedom for the variance in sample 2, and df represents the total degrees of freedom, or $\mathrm{df}_1 + \mathrm{df}_2$. This has been done so that the contributions of $\hat{s}_1^2$ and $\hat{s}_2^2$ to $\hat{s}_p^2$ will be proportional to their degrees of freedom. For example, if $\hat{s}_1^2$ has twice as many degrees of freedom as $\hat{s}_2^2$, it will contribute twice as much to $\hat{s}_p^2$. Equations 9.4 and 9.5 characterize $\hat{s}_p^2$ as the mean of $\hat{s}_1^2$ and $\hat{s}_2^2$, after these have been adjusted for their respective degrees of freedom.

Assuming that σ_1^2 equals σ_2^2, we can now substitute $\hat{s}_p^2$ into Equation 9.3 to get the estimated standard error of the difference between two independent means:

$$\hat{s}_{\bar{Y}_1 - \bar{Y}_2} = \sqrt{\frac{\hat{s}_p^2}{n_1} + \frac{\hat{s}_p^2}{n_2}} \qquad [9.6]$$

where $\hat{s}_{\bar{Y}_1 - \bar{Y}_2}$ represents the estimated standard error of the difference between two independent means, and all other terms are as previously defined. In the example on the effects of order of information on likeability ratings,

$$\hat{s}_p^2 = \frac{(10 - 1)(1.33) + (10 - 1)(1.78)}{(10 + 10 - 2)}$$

$$= 1.56$$

and

$$\hat{s}_{\bar{Y}_1 - \bar{Y}_2} = \sqrt{\frac{1.56}{10} + \frac{1.56}{10}} = .559$$

Thus, on the average, mean differences in the sampling distribution tended to be .559 units from the true population difference between the two means, as *estimated* by Equation 9.6.

From a computational perspective, we can compute the standard error of the difference between two independent means without going through the actual pooling steps noted above. An equation that combines Equations 9.5 and 9.6 to yield the standard error is

$$\hat{s}_{\bar{Y}_1 - \bar{Y}_2} = \sqrt{\left(\frac{SS_1 + SS_2}{df}\right)\left(\frac{1}{n_1} + \frac{1}{n_2}\right)} \qquad [9.7]$$

where SS_1 is the sum of squares on the dependent measure for the first group and SS_2 is the sum of squares on the dependent measure for the second group. We will adopt this more efficient computational formula in the remainder of this book.

The *t* test. We now have the background to formally test whether the observed difference between sample means in likeability ratings (5.0 versus 3.0) can be attributed to sampling error or whether it reflects a true relationship between order of information and likeableness. To do so, we will use the same steps outlined in Chapter 7 on hypothesis testing:

1. State a null and an alternative hypothesis.

2. State an expected result based upon the assumption that the null hypothesis is true. To do this we must:

(a) Characterize the mean of the sampling distribution in terms of $\mu_1 - \mu_2$, assuming the null hypothesis is true.

(b) State an alpha level that will be used to define a rejection region, and specify that region using a t distribution with $n_1 + n_2 - 2$ degrees of freedom.

3. Convert the sample difference between means to a t value.

4. Note if this t value falls within the rejection region and reject or fail to reject the null hypothesis accordingly.

Let us proceed through each step.

Step 1 (Specify a null and an alternative hypothesis): The null hypothesis states that there is no relationship between order of information and likeability. The alternative hypothesis states that there is a relationship between the two variables. Expressed in terms of mean scores, these are

$$H_0: \mu_1 = \mu_2$$

$$H_1: \mu_1 \neq \mu_2$$

Step 2a (Characterize the mean of the sampling distribution): If we assume that the null hypothesis is true, then the difference between the two population means is zero, $\mu_1 - \mu_2 = 0$. It follows that the mean of the relevant sampling distribution will equal zero. The estimated standard deviation of the sampling distribution is $\hat{s}_{\bar{Y}_1 - \bar{Y}_2}$. The value of this was computed in the previous section and equals .559.

Step 2b (State an alpha level and define a rejection region): For an alpha level of .05 and a nondirectional test, the rejection region is defined using Appendix E for the t distribution. The degrees of freedom for the t distribution are $n_1 + n_2 - 2 = 10 + 10 - 2 = 18$. The critical values of t are ± 2.10. If the observed t is less than or equal to -2.10 or if it is greater than or equal to $+2.10$, then the null hypothesis will be rejected.

Step 3 (Convert the sample difference between means to a t value): The formula for converting the observed difference between means into a t value is

$$t = \frac{(\bar{Y}_1 - \bar{Y}_2) - (\mu_1 - \mu_2)}{\hat{s}_{\bar{Y}_1 - \bar{Y}_2}} \tag{9.8}$$

where $\bar{Y}_1 - \bar{Y}_2$ is the observed difference between sample means, $\mu_1 - \mu_2$ is the hypothesized difference between the population means, and $\hat{s}_{\bar{Y}_1 - \bar{Y}_2}$ is the estimated standard error of the difference between independent means. In the experiment,

$$\bar{Y}_1 - \bar{Y}_2 = 5.0 - 3.0 = 2.00 \qquad\qquad \mu_1 - \mu_2 = 0$$

$$\hat{s}_{\bar{Y}_1 - \bar{Y}_2} = .559$$

Hence,

$$t = \frac{2.0 - 0}{.559} = 3.58$$

Step 4 (Note if the t value is within the rejection region and draw a conclusion):
The *t* value was 3.58. A *t* value of 3.58 is greater than the critical value of 2.10 that
defines the upper limit of the rejection region. This suggests that the observed mean
difference is too large to be attributed to sampling error and the null hypothesis is
therefore rejected. The order of information is related to the likeability ratings, and
we have answered the first question.

Study Exercise 9.1

For the following experiment, use the appropriate
statistics to test for the existence of a relationship
between the independent and dependent variables: In
order to test for sex discrimination by women against
other women, a researcher had a group of twenty-six
women read an article and rate the article for the
quality of the writing style. The rating was made on
a scale from 1 to 10, with higher numbers indicating
higher quality. Thirteen randomly determined individuals were told that the author was "John
McKay," while the other thirteen were told that the
author was "Joan McKay." The article was identical
for all individuals. The following summary data were
observed.

Male Author	Female Author
$n_1 = 13$	$n_2 = 13$
$\bar{Y}_1 = 7.2$	$\bar{Y}_2 = 6.1$
$SS_1 = 15.0$	$SS_2 = 18.0$

Answer. The null and alternative hypotheses
are

$$H_0\colon \mu_1 = \mu_2 \qquad\qquad H_1\colon \mu_1 \neq \mu_2$$

For an alpha level of .05, and a two-tailed test, and
with a *t* distribution with $13 + 13 - 2 = 24$ degrees

of freedom, the critical value of *t* (from Appendix E)
is 2.06. Thus, if the derived *t* is greater than $+2.06$
or less than -2.06, we will reject the null hypothesis.

The standard error of the difference between two
means is estimated using Equation 9.7:

$$\hat{s}_{\bar{Y}_1 - \bar{Y}_2} = \sqrt{\left(\frac{SS_1 + SS_2}{df}\right)\left(\frac{1}{n_1} + \frac{1}{n_2}\right)}$$

$$\hat{s}_{\bar{Y}_1 - \bar{Y}_2} = \sqrt{\left(\frac{15 + 18}{24}\right)\left(\frac{1}{13} + \frac{1}{13}\right)} = .460$$

The observed and hypothesized mean differences are

$$\bar{Y}_1 - \bar{Y}_2 = 7.2 - 6.1 = 1.1$$

$$\mu_1 - \mu_2 = 0$$

Using Equation 9.8, we compute *t*:

$$t = \frac{1.1 - 0}{.460} = 2.39$$

Since 2.39 is greater than 2.06, we reject the null
hypothesis and conclude that there is a relationship
between the sex of the author and the perceived quality of the essay.

Assumptions of the *t* Test. The above test is based upon the assumption that Equation 9.7 is a reasonable estimate of the standard error of the difference between two independent means and that Formula 9.8 yields scores distributed as *t*. Technically, this is only true under certain conditions. These are

1. The distribution of scores within each population follows a normal distribution.

2. The variances of scores within each population are equal across populations; that is, $\sigma_1^2 = \sigma_2^2$.

3. The samples are randomly and independently selected from their respective populations (which are also independent).

Although these assumptions formally underlie the *t* test, statisticians have shown that the test is relatively robust to violations of some of them. When the sample sizes in the two groups are equal, violations of either of the first two assumptions do not have serious consequences for type I and type II errors. This is especially true when the sample sizes are greater than ten. Boneau (1960) estimated the effects of violating the assumptions by using a technique called a *Monte Carlo procedure*. This involved generating on a computer a large set of scores that represented a population. For the present problem, two such populations were generated, with the restrictions that $\mu_1 = \mu_2$ (that is, the null hypothesis was true). A random sample of size N was selected from each population and a *t* value was computed, using the procedures described in this chapter. The null hypothesis was rejected or not rejected based upon these data. This was repeated 1,000 times and the proportion of times the null hypothesis was rejected was calculated. If $\sigma_1 = \sigma_2$ with normal distributions, and using an alpha level of .05, we would expect this to happen only 5% of the time (remember $\mu_1 = \mu_2$) or for 50 of the 1,000 tests. But what happens if σ_1 does not equal σ_2? Or if the scores are not normally distributed? Boneau created different populations in which the assumptions were violated and examined the effects on the proportion of incorrect rejections of H_0, or type I errors. Table 9.3 presents the estimated probabilities of type I errors based upon Boneau's data for selected violations of the above assumptions. Generally speaking, the robustness of the test is borne out by the data.

TABLE 9.3 ESTIMATED PROBABILITIES OF TYPE I ERRORS WITH A TWO-TAILED t TEST FOR VARIOUS NONNORMAL POPULATIONS AND VALUES OF THE POPULATION VARIANCES[a]

Population 1			Population 2			Proportion of Rejections
Shape	σ_1^2	n_1	Shape	σ_2^2	n_2	
Normal	1	5	Normal	1	5	.053*
Normal	1	15	Normal	1	15	.040*
Normal	1	5	Normal	1	15	.040*
Normal	1	5	Normal	4	5	.064
Normal	1	15	Normal	4	15	.049
Normal	1	5	Normal	4	15	.010
Normal	4	5	Normal	1	15	.160
Exponential†	1	5	Exponential	1	5	.031
Exponential	1	15	Exponential	1	15	.040
Rectangular‡	1	5	Rectangular	1	5	.051
Rectangular	1	15	Rectangular	1	15	.050
Rectangular	1	5	Rectangular	4	5	.071
Normal	1	5	Rectangular	1	5	.056
Normal	1	15	Rectangular	1	15	.056
Exponential	1	5	Normal	1	5	.071
Exponential	1	15	Normal	1	15	.051
Exponential	1	25	Normal	1	25	.046
Exponential	1	5	Rectangular	1	5	.064
Exponential	1	15	Rectangular	1	15	.056
Exponential	1	5	Exponential	4	5	.083

*These cases, for which the t test assumptions are met, were included as checks on the empirical sampling procedure.

†The exponential distribution looks something like the upper third of a normal distribution.

‡The graph of a rectangular distribution is a straight, horizontal line, all values of the variable being equally likely.

[a]From Glass and Stanley (1970)

Boneau did not examine the effects of violating the third assumption concerning independence of scores. This was examined by Lissitz and Chardos (1975) and the *t* test was found to be highly sensitive to violations of the assumption. Thus, the assumption of independence is critical to the independent groups *t* test.

9.3 / Analyzing the Strength of the Relationship

If we reject the null hypothesis, and conclude that a relationship exists between the independent and dependent variables, it becomes meaningful to ask how strong the relationship is (that is, is it a weak relationship, a moderate relationship, or a strong relationship). There are many different statistics one can use to address this question, and statisticians are in disagreement as to which one is best. We will develop the general logic of these approaches using an index called *eta squared* and then discuss the advantages and disadvantages of different measures. Also, we will initially derive eta squared using procedures that are computationally inefficient, but that best illustrate the concept. Computational formulas will then be presented. We will use the experiment on the relationship between order of information and likeability as the example to develop the logic of the approach.

Sum of Squares Total and the Grand Mean.   We begin by introducing some new notation. The second and third columns of Table 9.4 present the condition in which each of the twenty individuals participated and their corresponding score on the dependent variable in the order of information experiment. The logic of our analysis begins by examining the data with respect to the dependent variable. Looking at Table 9.4, we see that there is variability in the likeability ratings—some individuals said they would like the hypothetical stranger better than other individuals. Our goal is to analyze this variability and determine what proportion of it is associated with the independent variable. To accomplish this, we need to derive a numerical index of the amount of variability there is in the likeability ratings. We could use any of the measures discussed in Chapter 3 (sum of squares, variance, standard deviation), but because of certain statistical properties, we will use the sum of squares as the index of variability. Thus, the first step is to compute the sum of squares for the dependent variable across all individuals in the experiment. This sum of squares is called the **sum of squares total** (abbreviated SS_{TOTAL}), since it involves the total number of individuals who participated in the study.

The sum of squares total is computed using the standard formula for a sum of squares. This involves computing the overall mean score, subtracting the mean from each individual's raw score, squaring these deviation scores, and summing them. In

Individual	Conditions	Likeability Rating(Y)	Yn (nullified scores)
1	Pro–Con	7	6
2	Pro–Con	4	3
3	Pro–Con	3	2
4	Pro–Con	5	4
5	Pro–Con	5	4
6	Pro–Con	6	5
7	Pro–Con	5	4
8	Pro–Con	5	4
9	Pro–Con	4	3
10	Pro–Con	6	5
11	Con–Pro	1	2
12	Con–Pro	5	6
13	Con–Pro	2	3
14	Con–Pro	5	6
15	Con–Pro	3	4
16	Con–Pro	3	4
17	Con–Pro	3	4
18	Con–Pro	2	3
19	Con–Pro	4	5
20	Con–Pro	2	3

this case, the mean is called the **grand mean** (abbreviated G), since it also is based upon all individuals in the experiment. Columns 2–4 in Table 9.5 present the calculations for the SS_{TOTAL} and the grand mean in the experiment. We find that

$$G = 4.0$$

$$SS_{TOTAL} = 48.0$$

TABLE 9.5 COMPUTATION OF SS_{TOTAL} AND SS_{ERROR}

		SS_{TOTAL}			SS_{ERROR}	
X	Y	$(Y - \bar{Y})$	$(Y - \bar{Y})^2$	Yn	$(Yn - \bar{Y}n)$	$(Yn - \bar{Y}n)^2$
Pro–Con	7	$7 - 4 = 3$	9	6	$6 - 4 = 2$	4
Pro–Con	4	$4 - 4 = 0$	0	3	$3 - 4 = -1$	1
Pro–Con	3	$3 - 4 = -1$	1	2	$2 - 4 = -2$	4
Pro–Con	5	$5 - 4 = 1$	1	4	$4 - 4 = 0$	0
Pro–Con	5	$5 - 4 = 1$	1	4	$4 - 4 = 0$	0
Pro–Con	6	$6 - 4 = 2$	4	5	$5 - 4 = 1$	1
Pro–Con	5	$5 - 4 = 1$	1	4	$4 - 4 = 0$	0
Pro–Con	5	$5 - 4 = 1$	1	4	$4 - 4 = 0$	0
Pro–Con	4	$4 - 4 = 0$	0	3	$3 - 4 = -1$	1
Pro–Con	6	$6 - 4 = 2$	4	5	$5 - 4 = 1$	1
Con–Pro	1	$1 - 4 = -3$	9	2	$2 - 4 = -2$	4
Con–Pro	5	$5 - 4 = 1$	1	6	$6 - 4 = 2$	4
Con–Pro	2	$2 - 4 = -2$	4	3	$3 - 4 = -1$	1
Con–Pro	5	$5 - 4 = 1$	1	6	$6 - 4 = 2$	4
Con–Pro	3	$3 - 4 = -1$	1	4	$4 - 4 = 0$	0
Con–Pro	3	$3 - 4 = -1$	1	4	$4 - 4 = 0$	0
Con–Pro	3	$3 - 4 = -1$	1	4	$4 - 4 = 0$	0
Con–Pro	2	$2 - 4 = -2$	4	3	$3 - 4 = -1$	1
Con–Pro	4	$4 - 4 = 0$	0	5	$5 - 4 = 1$	1
Con–Pro	2	$2 - 4 = -2$	4	3	$3 - 4 = -1$	1
	$\Sigma Y = 80$		SS = 48	$\Sigma Yn = 80$		SS = 28
	$\bar{Y} = G = 4.0$			$\bar{Y}n = 4.0$		

Treatment Effects and Variance Extraction. The next step in the analysis is to estimate what effect the different values of the independent variable had on the dependent variable. This is accomplished by comparing the mean score in each condition with the grand mean. In the pro–con condition, the mean likeability rating was 5.0 (see Table 9.1). The grand mean, across all individuals in the experiment, was 4.0. Thus, the effect of having traits presented in a pro to con order was to raise the likeability ratings, on the average, one unit above the grand mean ($5 - 4 = 1$). In the con–pro condition, the mean likeability score was 3.0. The effect of having traits presented in the con to pro order was to lower the likeability ratings, on the average, one unit below the grand mean $(3 - 4 = -1)$. These effects are formally called **treatment effects** (abbreviated T), and are defined as the group mean minus the grand mean:

$$T_1 = \overline{Y}_1 - G = 5 - 4 = 1.0$$

$$T_2 = \overline{Y}_2 - G = 3 - 4 = -1.0$$

We can use the estimates of the treatment effects to derive the strength of the relationship between the independent and dependent variables. This will require, as a first step, removing or "nullifying" the influence of the independent variable on the dependent variable. For individuals in the pro–con condition, the effect of the order of information was to raise scores, on the average, one unit *above* the grand mean. If we wanted to nullify this effect, we could *subtract* one unit from each of these individuals' original likeability scores. Consider the first subject in Table 9.4, whose likeability rating was 7. To nullify the effects of being in the pro–con condition, we subtract 1 from his score, yielding a nullified score (Yn) of 6. This process has been repeated for each subject in the pro–con condition in column 4 of Table 9.4.

On the other hand, the effect of order of information for people in the con–pro condition was to lower scores, on the average, one unit *below* the grand mean. If we want to nullify this effect, we simply *add* 1 unit to each of these individuals' original likeability scores. The first subject listed in Table 9.4 in the con–pro condition had a likeability rating of 1. To nullify the effects of being in the con–pro condition, we add 1 to his score, yielding a score of 2. This process also has been repeated for each subject in the con–pro condition in column 4 of Table 9.4.

Column 4 represents scores on the dependent variable with the effects of the independent variable nullified. We have, in essence, extracted or taken out variability that was associated with the effects of the order of information. When you examine these scores, note that there is less variability than in the original scores (column 3). This is because we have removed a certain amount of variability that was associated with the independent variable. Note, however, that there is still variability in the Yn scores, due to factors other than the order of information. In statistical terminology, the remaining variability is called **unexplained variance** or **error variance** and reflects the influence of disturbance variables, as discussed in Chapter 8.

We can derive a numerical index of how much error variance remains after we have removed the influence of the independent variable by computing a sum of squares for the nullified scores. This has been done in columns 5–7 of Table 9.5. Specifically, the mean of the Yn scores was computed and this mean was subtracted from each of the Yn scores. The deviation scores were then squared and summed. In this case we find that

$$SS_{ERROR} = 28.0$$

The sum of squares total represents an index of the total variability in the dependent variable. It equals 48.0. The **sum of squares error** represents an index of the unexplained variability in the dependent variable (that is, variability that remains after the effects of the independent variable are removed). It equals 28.0. The *proportion* of unexplained variance in the dependent variable can be estimated by dividing the sum of squares error by the sum of squares total:

$$\frac{SS_{ERROR}}{SS_{TOTAL}} = \frac{28.0}{48.0} = .58$$

Thus, 58% of the variability in the dependent variable could *not* be explained by the independent variable. If the proportion of variance in the dependent variable that is unexplained is .58, then the proportion of **explained variance** (that is, that which is associated with the independent variable) is 1 minus this, or $1 - .58 = .42$. This statistic, the proportion of explained variance, is called **eta squared** and is represented by the formula

$$Eta^2 = 1 - \frac{SS_{ERROR}}{SS_{TOTAL}} \qquad \textbf{[9.9]}$$

where Eta^2 stands for "Eta squared" and all other terms are as previously defined.* Eta squared indexes the strength of the relationship between the independent and dependent variables since it represents the proportion of variability in the dependent variable that is associated with the independent variable. Eta squared can range from 0 to 1.0. As it approaches 1.0, the relationship between the variables is stronger, and as it approaches 0, the relationship between the variables is weaker.

Standards differ considerably among researchers on the substantive interpretation of eta squared. Much depends upon the specific area of application. In the author's own research area on the relationship between attitudes and behavior, an eta squared less than .20 reflects a "weak" relationship, an eta squared between .20 and .50 reflects a "moderate" relationship, and an eta squared above .50 reflects a "strong" relationship between variables. In other contexts, different interpretations might apply. The behavior of organisms is complex and rarely is there only a small number of variables that determine that behavior. To the extent that a behavior is determined by a multitude of variables, the explanation of as little as 5 to 10 % of

*Some texts use the symbol η^2 to represent eta squared as calculated in Equation 9.9. This symbol will be used in the present text to refer to eta squared in a population, whereas a sample value of eta squared will be represented by Eta^2. It should be noted that eta squared corresponds to the square of a statistic called the *point biserial correlation*. In some applications, researchers prefer the use of a statistic called a biserial correlation rather than the point biserial correlation. This occurs when the independent variable is continuously measurable, but for some reason is reduced to two categories (perhaps because this is the only way the data could be gathered). If it can be assumed that in the population both variables are normally distributed, then the biserial correlation rather than the point biserial correlation will be computed. In general, the biserial correlation will be larger than the point biserial correlation when the respective computational procedures are applied to the same data. For a discussion of the biserial correlation, see Guilford (1965).

the variance in the dependent measure is, in some respects, a considerable amount. Typically, research in the social sciences observes relatively small values of eta squared.

Alternative Measures of the Strength of the Relationship. Eta squared is a statistic that describes the strength of the relationship between two variables *in a set of sample data*. Estimating the strength of the relationship in the *population* requires a slightly different approach. A number of different estimation procedures have been proposed, the application of which is controversial among statisticians. Eta squared is a biased estimator in that it tends to slightly overestimate the strength of the relationship in the population across random samples. Hays (1963, pp. 381–384, 406–407) has argued for a relatively conservative estimate called *omega squared*. The method of estimation one should use is not agreed upon by statisticians. In this book, we will use the measure of eta squared because of its relationship to statistics discussed in later chapters and because of certain desirable statistical properties it possesses (Kennedy, 1970; Kesselman, 1975). However, as will be elaborated shortly, we advocate the use of eta squared in a "nonestimation" context. For a discussion of the statistical properties of various indices of strength of association, see Glass and Hakstian (1969), Carrol and Nordholm (1975), Fisher (1950), Haggard (1958), and Kesselman (1975).

Any index of the strength of the relationship derived from sample data will be subject to sampling error and must be interpreted accordingly. An observed sample eta squared of .50 could result from a population with a value of eta squared that is quite different. The confidence one has in the accuracy of a sample eta squared as an estimate of the corresponding population parameter is a function, in part, of the sample size (the larger the N, the better the estimate). In fact, for Ns less than 30, the amount of sampling error can be rather sizeable (Carrol and Nordholm, 1975). For this reason, indices of the strength of association must be interpreted with caution. In light of the above, we advocate the use of eta squared only as a heuristic index that will aid the researcher in appreciating relationships contained within his or her data. The measure should *not* be used as an estimate of the strength of the relationship in the population since, in our opinion, point estimation procedures are not very useful in this respect. Instead, an interval estimation approach seems more reasonable.

As was done in Chapters 6 and 7, it is possible to calculate confidence intervals about eta squared. These provide the investigator with a range of values of eta squared that, with a specified degree of confidence, contain the population value of eta squared. The mathematics for calculating such confidence intervals are complex and are discussed by Fleishman (1980).

Most of the research examples in this book calculate values of eta squared based upon small sample sizes. The examples use small sample sizes for expositional purposes and are not representative of sample sizes characterizing research in the social sciences. As such, the values of eta squared will be inflated.

Computational Formula. Eta squared can be derived using the computational procedures given in Table 9.5 in conjunction with Equation 9.6. However, there is a simple computational formula for eta squared that can be used instead:

$$\text{Eta}^2 = \frac{t^2}{t^2 + df} \qquad [9.10]$$

where t is the observed t value and df is the degrees of freedom associated with the t. For the experiment on the order of information,

$$\text{Eta}^2 = \frac{3.58^2}{3.58^2 + 18}$$

$$= .42$$

This value agrees with our earlier calculations.

9.4 / Analyzing the Nature of the Relationship

We have now considered the first two questions: (1) Is there a relationship between the independent and dependent variables? and (2) what is the strength of the relationship? The final question concerns the nature of the relationship. In the experiment on the relationship between order of information and likeability for a stranger, the nature of the relationship is determined by examination of the mean scores in the two conditions. For the pro–con condition, the mean likeability score was 5.0, whereas for the con–pro condition, the mean likeability score was 3.0. The nature of the relationship between order of information and likeability is such that when individuals are presented information in an order going from positive to negative, they will tend to form more favorable impressions of the stranger (at least in terms of likeability) than if the order of information goes from negative to positive. This conclusion results from the fact that the mean score in the pro–con condition (5.0) was higher than the mean score in the con–pro condition (3.0).

See Method of Presentation on page 179.

9.5 / Methodological Considerations

Several methodological features should be noted about the experiment on the order of information and likeability rating. These will put the results of the statistical test in proper context. First, the investigator attempted to control for a number of confounding variables by randomly assigning subjects to the two experimental conditions. Without this random assignment, differences between the two groups could be attributed to individual difference factors rather than to the experimental treatment. For example, psychologists have shown that some people have a tendency

Method of Presentation

When reporting the results of an analysis of the relationship between two variables, all three questions outlined at the beginning of this section should be addressed (whether there is a relationship between the variables, the strength of the relationship, and the nature of the relationship). Unfortunately, this is not always done. The results of t tests are usually presented with respect to the first and third questions, with the second question being ignored. A typical report might appear as follows:

A t test was performed comparing the means of the two groups. The t was statistically significant ($t = 2.95$, df $= 26$, $p < .05$), indicating that the mean score for males (28.2) was significantly higher than the mean score for females (22.6).

The first sentence indicates the type of test performed. Actually, an *independent groups t* test was performed, but usually investigators will not state it was "independent groups." Rather, it is understood that this is what was performed. The statement that "the t was statistically significant" means that the null hypothesis was rejected. If it had not been rejected, the t would have been referred to as "nonsignificant." The first value reported in the parentheses is the t value, the second value is the degrees of freedom associated with the t value, and the last value reflects the p value. The remainder of the sentence states the nature of the relationship: "the mean score for males was significantly higher than the mean score for females." If the null hypothesis had not been rejected, then the statement might appear as follows:

A t test was performed comparing the means of the two groups. The t was not statistically significant ($t = 1.03$, df $= 26$, ns), indicating that the mean score for males (28.2) did not differ significantly from the mean score for females (27.9), at the .05 level.

Sometimes investigators will not report the degrees of freedom at all. Of course, the actual values of the mean scores will always be reported, either in a table or in the text itself. In the above examples, they are reported in the parentheses.

Sometimes an investigator reports a p value other than .05, such as $p < .01$, or $p < .001$, within the parentheses containing the statistics. In most cases, this does not mean that an alpha level other than .05 was used as the basis for rejecting the null hypothesis. Rather, the p value reported reflects the probability of obtaining a t value as discrepant as the one computed in the experiment (and reported in the parentheses). The p value and the alpha level are distinct from each other.

When reporting the results of the independent groups t test, all three questions outlined in this section should be addressed. The common practice of excluding information about the strength of the relationship is unfortunate, since some investigators will interpret their data as if a strong relationship exists when, in fact, the relationship is so weak that it is almost trivial. A report of the results of the experiment on order of information and likeability might appear as follows:

A t test was performed comparing the mean likeability rating for the pro–con condition with that for the con–pro condition. The difference between the means was statistically significant ($t = 3.58$, df $= 18$, $p < .05$). The strength of the effect of order of information on likeability, as indexed by eta squared, was .42. The mean likeability rating in the pro–con condition (5.0) was higher than the mean likeability rating in the con–pro condition (3.0).

to view everyone in a positive sense (the optimist) whereas others tend to view everyone in a negative sense (the pessimist). Without random assignment, it is possible that the pro–con condition could have contained more optimists and the con–pro condition more pessimists. With random assignment, this is unlikely.

Another issue concerns the extent to which disturbance variables were controlled in the experiment. Although we did not provide enough of the experimental details, the effects of disturbance variables are to create sampling error, and the amount of sampling error is reflected in the standard error of the difference between two independent means. In this experiment, the overall likeability rating was made on a scale ranging from 1 to 7. Theoretically, we could have observed mean differences between the two groups anywhere from -6 to $+6$. The estimated standard error in the experiment was .558, suggesting that on the average, the difference between any two sample means based upon random samples of size 10 deviated only .558 rating scale units from the true population difference between means. Given a potential range of -6 to $+6$ and the respective sample standard deviations, this suggests a relatively small amount of sampling error.

A second perspective on sampling error is gained by examining eta squared. In this experiment 42% of the variance in the dependent variable was associated with the independent variable and 58% $(100 - 42)$ was due to disturbance variables. The picture that emerges is one of disturbance variables (and the consequent sampling error) not overwhelming any "signal" produced by the independent variable.

The experiment has several limitations from the standpoint of generalizability (or in formal terms, the population that the samples are assumed to be randomly selected from). First, it used college students from one university. Would the results generalize to other types of individuals as well? Second, the experiment used only one pro–con list and one con–pro list. Maybe it was something about the particular adjectives used that produced the experimental outcome. Would similar results occur if different adjectives were used? These and other questions of generalizability can be addressed through additional experimentation. The important point is that you should always consider these issues when trying to draw conclusions from experiments. Statistics and experimental design go hand-in-hand in interpreting the results of an experiment.

9.6 / Numerical Example

In Western society, individuals have long sought methods that will help them to relax. One such method is that of meditation. Maharishi Mahesh Yogi and his followers have introduced a simple relaxation technique called transcendental meditation (TM). An individual who practices TM meditates twice a day for fifteen to twenty minutes. During meditation the individual repeats, subvocally, a mantra, which is a meaningless, two-syllable sound. The mantra is different for each individual and is given to the meditator by a trained teacher of transcendental meditation. Proponents of TM emphasize the importance of the mantra in the meditation process. Recently, a psychiatrist named Benson suggested that the uniqueness of the mantra is not essential to achieve optimal levels of relaxation during meditation and that the same results could be achived by using the word "One" as a mantra. Let us consider a hypothetical experiment that could test this assertion.

The experiment used twenty individuals. Ten were randomly assigned to the TM condition and ten were randomly assigned to the One condition. Each subject practiced meditation using the appropriate technique for two weeks prior to the major test session. During the test session, individuals participated in their regular fifteen minute meditation period. However, during this time a number of physiological indicators of relaxation were measured. One of these was heart rate. In general, the more relaxed an individual, the slower the heart rate will be. The heart rate measures for the twenty individuals appear in Table 9.6 under the columns labeled Y.

The null hypothesis is that individuals who practice TM will not differ in their mean heart rate from individuals who use TM based on a "One" mantra. Stated formally,

$$H_0: \mu_T = \mu_O$$

$$H_1: \mu_T \neq \mu_O$$

In other words, the null hypothesis states that there is no relationship between the type of meditation techniques and relaxation level (as indicated by heart rate). The alternative hypothesis states that there is a relationship between these two variables.

In order to compute the statistics required in the analysis, we need to know the following intermediate values: n_1, n_2, $\overline{Y}_1$, $\overline{Y}_2$, SS_1, SS_2, df, and $\hat{s}_{\overline{Y}_1 - \overline{Y}_2}$. Using formulas developed in Chapter 3 and this chapter, and referring to Table 9.6, we find that

$$n_1 = n_2 = 10$$

$$\overline{Y}_1 = \frac{\Sigma Y_1}{n_1} = \frac{580}{10} = 58.0 \qquad \text{(see Equation 3.1)}$$

$$\overline{Y}_2 = \frac{\Sigma Y_2}{n_2} = \frac{620}{10} = 62.0 \qquad \text{(see Equation 3.1)}$$

$$SS_1 = 184.0 \qquad \text{(see Equation 3.3)}$$

$$SS_2 = 140.0 \qquad \text{(see Equation 3.3)}$$

$$df = n_1 + n_2 - 2 = 10 + 10 - 2 = 18$$

$$\hat{s}_{\overline{Y}_1 - \overline{Y}_2} = \sqrt{\left(\frac{SS_1 + SS_2}{df}\right)\left(\frac{1}{n_1} + \frac{1}{n_2}\right)} = \sqrt{\left(\frac{184.0 + 140.0}{18}\right)\left(\frac{1}{10} + \frac{1}{10}\right)} = 1.897$$

Our first question concerns whether there is a relationship between the type of meditation technique and heart rate. For a t distribution with $n_1 + n_2 - 2 = 18$ degrees of freedom, and an alpha level of .05, nondirectional test, the critical t values are ± 2.10. For the data, the observed t, using Equation 9.8, is

$$t = \frac{(\overline{Y}_1 - \overline{Y}_2) - (\mu_1 - \mu_2)}{\hat{s}_{\overline{Y}_1 - \overline{Y}_2}}$$

where $\overline{Y}_1 - \overline{Y}_2 = 58 - 62 = -4.0$ and $\mu_1 - \mu_2 = 0$, assuming the null hypothesis is true. Thus,

$$t = \frac{-4 - 0}{1.897} = -2.11$$

Since -2.11 is less than -2.10, we reject the null hypothesis and conclude that there is a relationship between the type of meditation used and heart rate.

The strength of the relationship in the sample is indexed by eta squared. Using the computational formula in Equation 9.10, we find that

$$\text{Eta}^2 = \frac{t^2}{t^2 + df}$$

$$= \frac{-2.11^2}{-2.11^2 + 18}$$

$$= .20$$

The proportion of variability in heart rate that is associated with differences in the meditation technique is .20. This represents a weak to moderate relationship.

The nature of the relationship is reflected in the mean scores for the two conditions. Individuals using the TM meditation technique tended to exhibit lower heart rates than individuals using the One meditation technique. The results of the experiment might appear in a research report as follows:

A t test was performed comparing the mean scores in the two meditation conditions. The t was statistically significant ($t = 2.11$, df $= 18$, $p < .05$). The mean heart rate for the TM group (58.0) was lower than the mean heart rate for the "One" group. The strength of the effect, as indexed by eta squared, was .20.

You should think about the above results in terms of the experimental design questions noted previously. Are there any potential confounding variables that haven't been controlled? What has been the role of disturbance variables in the experiment? What kinds of procedures could be used to reduce sampling error? What are the limitations of the experiment in terms of generalizability?

TABLE 9.6 DATA AND SUMMARY STATISTICS FOR MEDITATION EXPERIMENT

TM Condition			One Condition		
Y	$(Y - \bar{Y})$	$(Y - \bar{Y})^2$	Y	$(Y - \bar{Y})$	$(Y - \bar{Y})^2$
58	0	0	62	0	0
62	4	16	57	-5	25
54	-4	16	67	5	25
52	-6	36	66	4	16
64	6	36	58	-4	16
58	0	0	60	-2	4
52	-6	36	64	2	4
64	6	36	62	0	0
56	-2	4	57	-5	25
60	2	4	67	5	25
$\Sigma Y_1 = 580$		$SS_1 = 184$	$\Sigma Y_2 = 620$		$SS_2 = 140$

9.7 / Planning an Investigation Using the Independent Groups *t* Test

When a study has been designed to use two independent groups, the investigator is faced with a number of important decisions. One set of decisions concerns the delineation of confounding and disturbance variables and how to control for these in the context of the study. Another decision concerns the number of subjects that should be sampled for each group. One consideration that influences the latter decision is practicality. One cannot sample more subjects than resources permit. Practical matters aside, there are also scientific considerations that must be taken into account. One major issue concerns the desired power of the statistical test that will be used to analyze the data. Recall from Chapter 7 that the power of a statistical test refers to the probability of rejecting the null hypothesis when it is false. When we say the power of a test is .7, this means that the chances are 7 in 10 we will correctly reject the null hypothesis when it is false. Stated another way, the chances are 3 in 10 we will *not* reject the null hypothesis when it is false.

The power of a statistical test is influenced by three major factors: (1) the value of eta squared *in the population* indexing the strength of the relationship between the two variables, (2) the sample sizes, and (3) the alpha level. Consider first the value of eta squared. If the strength of the relationship in the population is weak, then it is less likely we will detect its presence and hence, it is less likely that we will correctly reject the null hypothesis. The power of the test will be relatively small, everything else being equal. If the strength of the relationship is strong, however, then it is more likely that we will detect the relationship between the independent and dependent variables and, hence, it is more likely that we will correctly reject the null hypothesis. The power of the test will be relatively large. Thus, as eta squared increases, the power of the statistical test increases.

Next, consider sample size. As discussed in Chapters 6 and 7, when sample sizes are relatively large, it is more likely that we will detect a true difference between means (that is, the existence of a relationship) as opposed to when sample sizes are small. Again, the power of the statistical test is affected, with larger sample sizes leading to more powerful tests.

Finally, consider the case of the alpha level. When the alpha level is small (.05), the likelihood that we will reject the null hypothesis is also small since we must be very certain a relationship exists before we are willing to say it does. Consequently, we will be *more* likely to overlook a relationship that does exist, especially if it is a weak one and we have small sample sizes. As the alpha level becomes smaller, the power of a statistical test will decrease.

If a type I error is serious, then the investigator will want to minimize this by setting a small alpha level, such as .05. This means that the major factor the experimenter can use to minimize a type II error is the sample size used in the study. If a type II error is also serious, then one will want to achieve high levels of power, perhaps as high as .95. Larger sample sizes can accomplish this.

Statisticians have developed procedures for estimating sample sizes that are necessary to obtain a desired level of power, given the strength of the relationship in the population and the alpha level. Appendix F.1 presents a set of tables that will be useful in this respect.* Portions of the tables for an alpha level of .05, non-directional test, are reproduced in Table 9.7 for purposes of exposition. The first column of the table presents different levels of power. The values on the top of the table are population values of eta squared. Table entries are required sample sizes *per group* to achieve the corresponding level of power. For example, if the researcher suspects (either on the basis of past research, intuition, or theory) that the strength of the relationship he or she is trying to detect in the population is relatively weak, corresponding to an eta squared of roughly .03, then the necessary sample size to achieve a power of .95, $\alpha = .05$, is 211 per group. If the researcher suspects that the strength of the relationship is much stronger, corresponding to an eta squared of roughly .20, then the necessary sample size to achieve a power of .95 is 27 per group. Examination of Table 9.7 will give you a rough feel for the relationship between power, sample size, and the strength of the relationship one is trying to detect in the population.

It is not always practical to obtain high levels of power by increasing sample sizes, especially when the population relationship is judged to be weak. In the previous example where eta squared equaled .03, the required 211 subjects per group could be costly, especially if the study involved the use of rare organisms such

*These tables are based upon power estimates given by Cohen (1977).

as nonhuman primates. The seriousness of lack of power again depends upon the seriousness of a type II error. When testing for sex discrimination, what would be the consequences of saying discrimination doesn't exist when, in fact, it does? If the consequences are judged to be serious, one would insist upon reasonable levels of power. Some statisticians (Cohen, 1977) have tentatively recommended that we strive to achieve power levels of .80, as a rough guideline. However, this can be revised upward or downward, depending upon the situation.

TABLE 9.7 SAMPLE SIZES NECESSARY TO ACHIEVE SELECTED LEVELS OF POWER FOR ALPHA = .05 AS A FUNCTION OF POPULATION VALUES OF ETA SQUARED

Nondirectional Test, Alpha = .05

Power	*Population Eta Squared*									
	0.01	*0.03*	*0.05*	*0.07*	*0.10*	*0.15*	*0.20*	*0.25*	*0.30*	*0.35*
0.25	84	28	17	12	8	6	5	3	3	3
0.50	193	63	38	27	18	12	9	7	5	5
0.60	246	80	48	34	23	15	11	8	7	6
0.67	287	93	55	39	27	17	12	10	8	6
0.70	310	101	60	42	29	18	13	10	8	7
0.75	348	113	67	47	32	21	15	11	9	7
0.80	393	128	76	53	36	23	17	13	10	8
0.85	450	146	86	61	41	26	19	14	11	9
0.90	526	171	101	71	48	31	22	17	13	11
0.95	651	211	125	87	60	38	27	21	16	13
0.99	920	298	176	123	84	53	38	29	22	18

9.8 / Summary

The independent groups *t* test is used to analyze the relationship between two variables when (1) the dependent variable is quantitative in nature and measured on a scale that approximates interval characteristics, (2) the independent variable is between-subjects in nature, and (3) the independent variable has two and only two values. When testing for the existence of a relationship between the two variables, the difference between two sample means is converted into a *t* value representing an estimated standard score in a sampling distribution of the difference between two means (assuming the null hypothesis is true). This *t* value is compared with the

Box 9.1 Reinforcement or Justification

Social scientists have attempted to isolate psychological factors that influence attitude change. In one experiment, Festinger and Carlsmith (1959) had individuals come into a psychological laboratory and work on a very boring task for one hour. After completing the task, the subject was told that the experiment was over and that it had been a study of the effects of "expectancy" on performance. The subject was told that half of the subjects had been told ahead of time that the task would be interesting while the other half, in the control condition, were not told anything. The purpose of the experiment was to see if this expectancy influenced performance on the task. (In actuality, this was not the purpose of the experiment.) The subject was then told that the experimenter who usually tells subjects that the task is interesting was sick and asked if he would be willing to do it. One-half of the subjects ($n = 20$) were offered $1 to do this, while the other half of the subjects were offered $20 to do this. All subjects complied and did the task. Afterwards, they were given a questionnaire about the original experiment and asked to rate, among other things, how enjoyable the task was on a -5 (extremely boring) to $+5$ (extremely enjoyable) rating scale.

At least two different principles could operate in this situation to influence the enjoyableness rating. The first is a reinforcement principle which would state that subjects who were paid $20 for telling another subject that the task was interesting and enjoyable would receive greater reinforcement for doing so than subjects who were paid only $1 to do it. This greater reinforcement would generalize to their own perceptions of the task, and hence subjects in the $20 condition should change their attitude and rate the task as being more enjoyable than those in the $1 condition. An alternative explanation would predict just the opposite. According to this viewpoint, all subjects performed a behavior (that is, told someone the task was enjoyable) that was in contradiction to what they truly believed (the task was boring). Because of this contradiction, subjects experienced an internal state of "dissonance," which they are motivated to reduce. Subjects in the $20 condition could easily justify their counterattitudinal behavior: they did it for the money. This was not the case, however, for subjects in the $1 condition, where the money was not very much. These subjects would have to reduce the dissonance in another way, and one possibility would be to rationalize that the task was really not all that boring. If this dissonance mechanism were operating, the reverse predictions would be made: subjects in the $1 condition would change their attitudes toward the task so as to rate it as being more enjoyable than those in the $20 condition.

The results of the experiment were analyzed using an independent groups t test. The independent variable was the amount of money paid for telling someone the task was enjoyable ($1 versus $20) and the dependent variable was the perceived enjoyableness of the task (as measured on an 11 point scale). Festinger and Carlsmith reported that the mean rating in the $1 condition was 1.35 while in the $20 condition it was $-.05$. This difference was statistically significant ($t = 2.22$, df $= 38$, $p < .05$) and was consistent with a dissonance interpretation. Festinger and Carlsmith did not report statistics regarding the strength of the relationship between the amount of money offered and the enjoyableness ratings. However, this can be calculated using Equation 9.10 and we find that eta squared $= .11$. This represents a weak effect.

critical values of t, and the decision to reject or not to reject the null hypothesis is made accordingly. The strength of the relationship is analyzed using the eta squared statistic. This statistic represents the proportion of variability in the dependent variable that is associated with the independent variable. Finally, the nature of the relationship is determined by examination of the mean scores in the two conditions characterized by the independent variable.

EXERCISES

Exercises to Review Concepts

1. Under what conditions is the independent groups t test typically used to analyze a bivariate relationship?

2. Match the terms or symbols in column A with the appropriate terms or symbols in column B.

Column A	Column B
$\hat{s}_{\bar{Y}_1 - \bar{Y}_2}$	variance estimate
$\hat{s}_p^2$	standard score in a t distribution
$\hat{s}_{\bar{Y}}$	null hypothesis
$\hat{s}^2$	standard score in a normal distribution
$\mu_1 = \mu_2$	standard error of the mean
$\mu_1 \neq \mu_2$	estimated standard error of the mean
z	standard error of the difference between independent means
t	alternative hypothesis
$\sigma_{\bar{Y}}$	estimated standard error of the difference between two means
$\sigma_{\bar{Y}_1 - \bar{Y}_2}$	pooled variance estimate

3. What are the major assumptions underlying the t test when used to test whether a relationship exists between two variables?

4. Two random samples were taken from their respective populations and the following summary statistics were observed:

Sample A	Sample B
$N = 49$	$N = 49$
$\bar{Y} = 10$	$\bar{Y} = 13$
SS $= 100$	SS $= 120$

(a) Compute the standard error of the mean for sample A.

(b) Compute the standard error of the mean for sample B.

(c) Compute the standard error of the difference between the two means.

(d) Your answer in (c) should be greater than your answer in (a) or (b): The standard error of the difference between independent means will always be greater than the respective standard errors of the mean. Why do you think this is the case?

5. Determine the critical value of t for each of the following conditions:

(a) $n_1 = 10$, $n_2 = 10$, one-tailed test, positive direction

(b) $n_1 = 20$, $n_2 = 20$, two-tailed test

(c) $n_1 = 10$, $n_2 = 10$, two-tailed test

(d) $n_1 = 20$, $n_2 = 20$, one-tailed test, negative direction

(e) $n_1 = 10$, $n_2 = 20$, two-tailed test

An investigator was interested in the relationship between a person's sex and the tendency to hold discriminatory attitudes toward women. An attitude scale was administered to five men and five women. Scores could range from 1 to 10, with higher scores indicating more discriminatory attitudes. The relevant data are presented below.

Males	Females
8	5
7	4
7	4
7	4
6	3

6. Conduct an independent groups *t* test on the above data to test for a relationship between sex and discriminatory attitudes.

7. Compute the grand mean and the sum of squares total for the above data.

8. Compute the treatment effect for males and the treatment effect for females in the above data.

9. Based on your answers in Exercise 8, use the variance extraction procedures discussed in this chapter to nullify the effects of sex on discriminatory attitudes (that is, generate a set of scores on the dependent variable with the effects of sex removed).

10. Compute the sum of squares error on the data derived in Exercise 9. What is the value of eta squared for these data? What is the proportion of unexplained variance?

11. Which is the more conservative measure of the strength of a relationship, omega squared or eta squared?

12. What is the nature of the relationship between a person's sex and his or her discriminatory attitudes in the above experiment?

An investigator wanted to test the effects of alcohol on reaction time. Six subjects were given alcohol to consume until a certain level of intoxication was achieved (as indexed by physiological measures). Another randomly selected group of subjects was not given any alcohol but instead consumed a placebo. All subjects then participated in a reaction-time task. The reaction times, in seconds, for the two groups were as follows:

Alcohol	Placebo
2.0	1.5
2.5	1.0
2.0	1.0
1.5	1.0
2.0	.5

13. Conduct an independent groups *t* test on the above data to test for a relationship between alcoholic consumption and reaction time.

14. Compute the grand mean and the sum of squares total for the above data.

15. Compute the treatment effect for alcohol and the treatment effect for the placebo group.

16. Based on your answers in Exercise 15, use the variance extraction procedures discussed in this chapter to nullify the effects of alcohol on reaction time (that is, generate a set of scores on the dependent variable with the effects of alcohol removed).

17. Compute the sum of squares error on the data derived in Exercise 16. What is the value of eta squared for these data? What is the proportion of unexplained variance?

18. What is the nature of the relationship between alcoholic consumption and reaction time in the above experiment?

19. If an experimenter suspected that the strength of a relationship, as indexed by eta squared, in a population was .10, what sample size should he use in a study using two independent groups in order to achieve a power of .80, for $\alpha = .05$, nondirectional test?

20. Suppose an investigator reported the results of a study using two independent groups with $n = 23$ per group. If the value of eta squared in the population was .10, what would the power of his statistical test be for $\alpha = .05$, nondirectional test? (Hint: use Appendix F.)

Exercises to Apply Concepts

21. Psychologists have studied extensively the effects of early experience on the development of individuals. It has long been recognized that a positive, challenging, and diverse environment (sometimes called an enriched environment) leads to the acquisition of more positive abilities and personality traits than an environment that is relatively impoverished and isolated. Bennett et al. (1964) have suggested that the type of environment may even act to alter the physical characteristics of the brain, and have reported a series of studies to investigate this possibility. In one study, laboratory rats from the same genetic strain were raised in one of two conditions. One-half of the rats were raised in an enriched environment, which involved being housed with ten to twelve other animals in large cages that were equipped with a variety of "toys." Each day these rats were placed in a square field where they were allowed to explore a pattern of barriers that were changed daily. The other half of the rats were raised in an isolated environment. They were caged singly in a dimly lit room where they could not see or touch another animal (although they could hear and smell them). After eighty days all animals were killed for purposes of analyzing the structure of the brain as a function of the type of environment in which the animal was raised. One factor that was examined was the weight of the cortex (reported in milligrams). The hypothetical data presented below are representative of the results of the study. Analyze the data using the appropriate statistical technique, draw a conclusion, and write out the results of the experiment using procedures discussed in the Method of Presentation section.

Enriched Environment	Isolated Environment
685	660
690	642
675	640
660	626
645	612
630	610
635	592

22. McConnell (1966) has reported a series of experiments that have been the subject of considerable controversy concerning the physiological bases of learning and memory. McConnell has suggested that RNA and DNA protein molecules constitute the physiochemical substrate of learning, and that it may be possible to transfer memory, biochemically, between organisms by transferring the relevant RNA and DNA molecules. McConnell's initial experiments were conducted with planaria, a small, worm-like organism. Using classical conditioning procedures, a group of planaria was taught to contract in size whenever they were exposed to a light. McConnell tried several different methods of transferring the relevant RNA and DNA molecules of these trained planaria to other planaria. The only practical method, however, was cannibalism. The trained planaria were chopped up into small pieces and fed to another group of planaria. The trained cannibals represented the experimental group. A control group of untrained cannibals was fed chopped planaria that had not been taught to contract when exposed to the light. The trained and untrained cannibals were then exposed to the light twenty-five times. McConnell counted the number of times each planarian contracted when exposed to the light. If the relevant RNA and DNA molecules had been transferred and appropriately used, then the trained cannibals should exhibit a greater number of contractions than the untrained cannibals. Such a demonstration would at least establish the *possibility* of transferring memory from one organism to another via biochemical means. The hypothetical data reported below are representative of the results of the experiment. Analyze the data using the appropriate statistical technique, draw a conclusion, and write out the results of the experiment using procedures discussed in the Method of Presentation section.

Experimental Group

4	8	10	13	15	21
6	8	10	14	15	22
7	9	11	14	15	
8	10	12	15	17	

Control Group

1	4	5	6	10	16
1	4	6	7	10	19
3	5	6	7	11	
4	5	6	7	11	

Correlated Groups *t* Test

10.1 / Use of the Correlated Groups *t* Test

The statistical technique developed in this chapter is called the **correlated groups *t* test** or the *paired observation t test*. It is typically used to analyze the relationship between two variables when

1. the dependent variable is quantitative in nature and is measured on approximately an interval level;

2. the independent variable is *within-subjects* in nature (it can be either qualitative or quantitative); and

3. the independent variable has two and only two values.

The major difference between applications of the correlated groups *t* test and the independent groups *t* test is that the former is used when the independent variable is within-subjects in nature and the latter is used when the independent variable is between-subjects in nature.* Let us consider an example of an experiment that meets the conditions for the correlated groups *t* test.

An investigator is studying the relationship between two types of drugs, A and B, and learning. Five subjects are administered drug A and then work on a learning task. One month later, the same five subjects are administered drug B and work on the same type of learning task as before. The number of errors made on each task is tabulated and serves as the dependent measure. The investigator has introduced a one month time interval between the two tasks in order to control for potential carry-over effects. We will suggest alternatives to this approach and discuss potential problems with it after the results of the experiment have been presented. The data for the example are presented in the second and third columns of Table 10.1. The experimental design involves a within-subjects independent variable with two values (drug A versus drug B) and a quantitative dependent variable. Hence the correlated groups *t* test is the statistical test that would typically be applied.

*Matched-subject designs would be analyzed as if they represented within-subjects variables.

TABLE 10.1 HYPOTHETICAL DATA FOR THE EFFECTS OF TWO TYPES OF DRUGS
ON LEARNING*

Subject	Drug A	Drug B	$\overline{Y}_i$	Drug A Adjusted	Drug B Adjusted
1	3	7	5.0	2.5	6.5
2	1	5	3.0	2.5	6.5
3	4	4	4.0	4.5	4.5
4	2	8	5.0	1.5	7.5
5	5	6	5.5	4.0	5.0
$\overline{Y} =$	3.00	6.00	4.5	3.0	6.0

*$\overline{Y}_i$ = Mean Y score for subject "i" across conditions.

10.2 / Estimating and Extracting the Influence of Individual Differences

The computational procedures for conducting the correlated groups t test are relatively straightforward. However, we are going to consider this technique in a slightly
unorthodox fashion in order to better illustrate the logic underlying the statistic. The
method of presentation has at least three advantages: (1) it relies heavily on the same
computational procedures developed in Chapter 9 and parallels closely the logic of
the independent groups t test, (2) it illustrates intuitively the interpretational difference in the index of the strength of the relationship in within-subjects designs as
opposed to between-subjects designs, and (3) the variance extraction approach used
generalizes directly to more advanced statistical techniques, such as repeated measures analysis of variance (see Chapter 12). The major disadvantage of the approach
is that it is computationally inefficient. Later in this chapter we will present the more
traditional and computationally efficient analysis of the correlated groups t test.

Overview. As noted in Chapter 8, one goal of experimental design is to control for
disturbance variables; that is, variables that influence the dependent variable and that
are unrelated to the independent variable. Disturbance variables cause variability in
the dependent variable and make it more difficult to test if a relationship exists
between the independent and dependent variables, because of the "noise" they
create.

One major source of "noise" in the present kind of experiment is the different
backgrounds and abilities of the individuals participating in the experiment. For
example, some individuals are more intelligent than others and this would be
expected to influence performance on the learning task. If we knew the influence of

these factors on the dependent variable, then their influence could be separated from the effects of the drug, and the sensitivity of the experiment could be improved. However, if their influence cannot be estimated, then these differences in background and ability remain as uncontrolled sources of variability and become part of the experimental error. One advantage of the within-subjects design is that it allows us to estimate this source of variability and extract it from the data.

Extraction Procedures. Consider an individual who tries to solve three math problems, each worth 5 points on an exam. The individual obtains a 5 on the first problem, a 1 on the second problem, and a 3 on the third problem. Another individual worked on the same three problems and scored a 5 on the first one, a 5 on the second one, and a 2 on the third. Which individual has more math ability? In the absence of any other information, your best guess in answering this question is to compare the average scores of the two individuals. The first individual's average score was $(5 + 1 + 3)/3 = 3.0$ and the second individual's average score was $(5 + 5 + 2)/3 = 4.0$. Using the average score across questions as an index of ability, you might conclude that the second individual has more math ability than the first individual.

In within-subjects designs, estimates of the influence of individual difference variables are derived in a similar fashion. For each individual, the average score across conditions is computed. This has been done for the experiment in column 4 of Table 10.1. The first subject had a score of 3 in the A condition and 7 in the B condition. The mean of these two scores is $(7 + 3)/2 = 5$. We can see by inspection of column 4 that some people had higher average scores than others. These differences reflect differences in the individuals' backgrounds (for example, intelligence, familiarity with the learning task, and so on).

We will now develop the logic of extracting this source of variability. In order to do so, we must first compute a grand mean. This involves summing *all* of the scores in columns 2 and 3 of Table 10.1 $(3 + 1 + 4 + 2 + 5 + 7 + 5 + 4 + 8 + 6)$ and dividing by the number of scores summed. In this case,

$$G = \frac{45}{10} = 4.5$$

Thus, the average score across all subjects and across both experimental conditions was 4.5.

Consider the first subject. His average score across conditions was 5.0. The grand mean, which serves as a reference point for all subjects, was 4.5. If the average score across conditions is an index of the individual's background, then the effect of this background was to raise the individual's score one-half a unit above the grand mean $(5.0 - 4.5 = .5)$. Using the same logic we used to extract variation in scores in Chapter 9, we can nullify this effect by subtracting .5 units from the scores of this individual. This has been done in columns 5 and 6 of Table 10.1. The score in condition A was 3 and this is adjusted to 2.5 $(3 - .5 = 2.5)$. The score in condition B was 7 and this is adjusted to 6.5 $(7 - .5 = 6.5)$. The new scores have had the effect of the individual's background removed or nullified. This same approach is taken for each subject. For subject 2, the average response across conditions was 3.0. This is 1.5 units below the grand mean $(3.0 - 4.5 = -1.5)$. The effect of this individual's background was to hold performance down 1.5 units from the overall average. To nullify the effects of this background, we simply add 1.5 to the two scores of the subject. And so on.

We now have an adjusted data set in which the effects of individual differences in background have been removed. Compare the adjusted data in Table 10.1 with the original data. Note that there is less variability in the scores in the adjusted data. This is because we have removed a source of variability, namely individual abilities and background. If you were to compute the average score across conditions for each subject in the adjusted data, you would find that every subject would have the *same* average score. Again, this is because we have extracted variability due to individual differences. The test of the effects of the drugs on learning will now be much more sensitive, since the "noise" created by these individual differences has been eliminated.

The scores comprising the adjusted data serve as the basis for our analysis of the relationship between the independent and dependent variables. In fact, the logic and computational procedures from this point on are *identical* to those developed in Chapter 9 on the independent groups t test, with one minor exception. The exception is that the relevant t distribution is distributed with $n - 1$ degrees of freedom (where $n = n_1 = n_2$, since in this type of design n_1 will always equal n_2) instead of $n_1 + n_2 - 2$ degrees of freedom. Since there is no need to develop the logic underlying the t test a second time, we will analyze the adjusted data using the same computational procedures outlined in Chapter 9, for the drug experiment.

In order to compute the statistics required in the analysis, we will need to know the following intermediate values: n, $\overline{Y}_1$, $\overline{Y}_2$, SS_1, SS_2, df, and $\hat{s}_{\overline{Y}_1 - \overline{Y}_2}$ for the adjusted data. Using the formulas referenced in Chapter 9 and referring to Table 10.2, we find that

$$n = n_1 = n_2 = 5$$

$$\overline{Y}_1 = \frac{\Sigma Y_1}{n_1} = \frac{15}{5} = 3.0$$

$$\overline{Y}_2 = \frac{\Sigma Y_2}{n_2} = \frac{30}{5} = 6.0$$

$$SS_1 = 6.00$$

$$SS_2 = 6.00$$

$$df = n - 1 = 5 - 1 = 4$$

$$\hat{s}_{\overline{Y}_1 - \overline{Y}_2} = \sqrt{\left(\frac{SS_1 + SS_2}{df}\right)\left(\frac{1}{n_1} + \frac{1}{n_2}\right)} = \sqrt{\left(\frac{6.00 + 6.00}{4}\right)\left(\frac{1}{5} + \frac{1}{5}\right)} = 1.095$$

Estimating and Extracting the Influence of Individual Differences **193**

TABLE 10.2 ADJUSTED DATA AND SUMMARY STATISTICS FOR DRUG EXPERIMENT

Drug A				Drug B		
Y	$Y - \overline{Y}$	$(Y - \overline{Y})^2$		Y	$Y - \overline{Y}$	$(Y - \overline{Y})^2$
2.5	$-.5$	.25		6.5	.5	.25
2.5	$-.5$	.25		6.5	.5	.25
4.5	1.5	2.25		4.5	-1.5	2.25
1.5	-1.5	2.25		7.5	1.5	2.25
4.0	1.0	1.00		5.0	1.0	1.00
$\Sigma Y_1 = 15.0$		$SS_1 = 6.00$		$\Sigma Y_2 = 30.0$		$SS_2 = 6.00$

10.3 / Inference of a Relationship Using the Correlated Groups *t* Test

The first question concerns the existence of a relationship between the type of drug administered and learning.

Null and Alternative Hypotheses. We begin by formally stating our question in the context of a null and an alternative hypothesis. In this case, they are

$$H_0: \mu_A = \mu_B$$

$$H_1: \mu_A \neq \mu_B$$

The null hypothesis states that there is no relationship between the type of drug and learning. If the effects of the two drugs on learning are the same, then one would expect the average learning score under the influence of drug A to be the same as the average learning score under the influence of drug B. The alternative hypothesis states that there is a relationship between the type of drug and learning, that is, which drug is used does matter and the population means will, accordingly, not be equal.

The Sampling Distribution and the *t* Test. In the correlated groups *t* test, the relevant sampling distribution of the difference between two means is analyzed in terms of a *t* distribution with $n - 1$ degrees of freedom. The standard error derived from this sampling distribution was estimated above as being 1.095. If we assume that the null hypothesis is true, then the difference between the population means is $\mu_1 - \mu_2 = 0$, and this represents the mean of the sampling distribution. For a *t* distribution with $5 - 1 = 4$ degrees of freedom, and an alpha level of .05, nondirectional test, the critical t values are ± 2.78 (see Appendix E). For the data, the observed t is

$$t = \frac{(\overline{Y}_1 - \overline{Y}_2) - (\mu_1 - \mu_2)}{\hat{s}_{\overline{Y}_1 - \overline{Y}_2}}$$

where $\overline{Y}_1 - \overline{Y}_2 = 3 - 6 = -3$ and $\mu_1 - \mu_2 = 0$. Thus,

$$t = \frac{-3 - 0}{1.095} = -2.74$$

The value of -2.74 is *not* less than the critical value of -2.78, so in this case, we fail to reject the null hypothesis. We cannot confidently conclude that a relationship exists between the independent and dependent variables.

Assumptions of the *t* Test. The assumptions underlying the validity of the t test are similar (but not identical) to those for the independent groups t test as discussed in Chapter 9. These are:

1. The distribution of scores within each population follows a normal distribution; and

2. The sample is randomly selected from the population of interest.

As with the independent groups t test, the t test for correlated groups is relatively robust with respect to violations of the first assumption (Subrahmaniam, Subrahmaniam, and Messeri, 1975).

10.4 / Analyzing the Strength of the Relationship

The formula for computing eta squared for the correlated groups t test is the same as that for the independent groups t test:

$$\text{Eta}^2 = \frac{t^2}{t^2 + df} \qquad \text{[10.1]}$$

It should be kept in mind that this eta squared represents the proportion of variability in the dependent variable that is associated with the independent variable *after the variability caused by individual differences has been removed*. It will be instructive to compute eta squared for the data in our experiment. We find that

$$\text{Eta}^2 = \frac{-2.74^2}{-2.74^2 + 4} = .65$$

An eta squared of .65 represents a relatively strong relationship, yet the *t* test indicated that we could not confidently conclude that a relationship even exists. Although this may appear contradictory, a consideration of the sample size will help to place this in proper perspective. The study was conducted using an extremely small sample size, namely five individuals. This sample is so small that the chances of a large amount of sampling error are extremely high, and an extremely strong relationship must be observed in order for us to be confident that one exists. Given the very small sample size and the amount of variability of scores in the population, a relationship as strong as .65 in the sample is still not strong enough for us to be confident that sampling error is not producing a false relationship in the experiment. The moral is to use large sample sizes when it is practical and feasible. As noted in Chapter 7, larger sample sizes will increase the power of the statistical test.

10.5 / Analyzing the Nature of the Relationship

The analysis of the nature of the relationship is identical to that for the independent groups *t* test and involves examination of the mean scores in the two conditions. Because we failed to reject the null hypothesis, this question is not meaningful for the present data. Had we rejected the null hypothesis, then the nature of the relationship would be that drug B tends to produce more errors in learning ($\overline{Y} = 6.0$) than drug A ($\overline{Y} = 3.0$).

Method of Presentation

The method of presentation for the correlated groups *t* test is identical to that of the independent groups *t* test discussed in Chapter 9. The only exception is that the *t* test will be identified as being either a "correlated groups *t* test" or a "paired observations *t* test." These are different labels given to the same test, as discussed in this chapter.

10.6 / Methodological Considerations

From a methodological viewpoint, a number of points should be made regarding the interpretation of the data from the experiment. These will place the results of the statistical analysis in a proper context. First, consider the potential role of confounding variables. In this study, the investigator attempted to control for carry-over effects such as familiarity with the task, the persistence of the effects of drug A into the administration of drug B, and so on, by having a long time period between the administration of drug A and the administration of drug B. The introduction of the long time interval may, in fact, reduce such carry-over effects. But an additional problem arises. Perhaps during the time interval between the administration of drug A and drug B, the subjects experienced something that improved their performance on the learning task. If the subjects are introductory psychology students, there is a possibility they might learn certain memory aids in the context of their lectures. A procedure that circumvents this problem, and that could have been used here, is **counterbalancing.** In this procedure, one-half of the subjects are given drug A first and then, at some time later when carry-over effects should be minimal, drug B. The other half of the subjects are given drug B first and then drug A. If this is done, then the intervening events and familiarity effects would no longer be related to the administration of drug A and drug B. Any carry-over effects would influence drug A (for the subjects who were given drug A second) and drug B (for subjects who were given drug B second) to an equal extent.* In essence, these confounding variables are turned into disturbance variables, which now create "noise" in the experiment. It is also possible to remove the disturbance influence of these variables using certain advanced statistical techniques. These are, however, beyond the scope of the present text, and are discussed in Winer (1971).

Another problem with the experiment is the lack of a control condition in which subjects perform the learning task while not under the influence of any drugs. Although the *relative* effects of drug A to drug B can be determined from the experiment, the effects of the drugs relative to normal performance cannot. The addition of a control condition would be quite informative. Chapter 12 discusses statistical techniques that could be used to analyze the study if a control condition were included.

In terms of the role of disturbance variables in the experiment, the value of the standard error (1.095) and the value of eta squared (.65) suggest that the role of disturbance variables is not of unreasonable magnitude. However, this must be qualified by the small sample size used in the experiment. As noted earlier, the sample size was so small that we simply cannot have much confidence in the relatively strong effect that seems to be manifested in the study. With small sample sizes, very small magnitudes of sampling error are necessary before we conclude there is a relationship between the independent and dependent variables. This experiment was problematic because the sample size was too small and the power of the statistical test was therefore inadequate.

10.7 / Numerical Example

Developmental psychologists have attempted to specify the age period when infants tend to show signs of fearing strangers. At very early ages (one–two months), young

---

*This is not true when the nature of carry-over effects is different for drug A and drug B.

infants generally will show positive reactions when approached or held by any adult. At some point, however, the infant begins to discriminate adults and will exhibit fear responses (crying) when a stranger approaches as opposed to a familiar person, such as the mother or father. In one experiment, a researcher wanted to compare negative responses to strangers using infants at the age of three months and again at six months. A group of eight infants was involved in a ten minute interaction with a stranger in which the stranger attempted to engage the infant in playful behavior. The interaction was standardized as much as possible and the number of minutes the infant spent crying was used as the measure of negative response to the stranger. The scores on the dependent measure are presented in the second and third columns of Table 10.3.

In this study, the independent variable is the age of the infant (three months versus six months) and the dependent variable is the negative response to the stranger as indicated by the number of minutes spent crying. The independent variable is quantitative in nature and has two values. It is a within-subjects variable since the same infants were involved in both the three month and six month conditions. The dependent variable is quantitative in nature. In this case, the correlated groups t test is an appropriate statistic to apply.

The first step of the analysis involves generating the adjusted data by extracting variability due to individual differences. This involves averaging the scores of each infant across the two age groups, which has been done in column 4 of Table 10.3. The grand mean (G), which is the sum of all sixteen scores divided by 16, is 2.0. The first infant had an average score of 1.5. This is .5 units below the grand mean ($2 - 1.5 = -.5$). The effect of this infant's background is nullified by adding .5

TABLE 10.3 DATA FOR EXPERIMENT ON RESPONSES TO STRANGERS

| | | | | Adjusted Scores | |
| | | | | | |
Infant	3 months	6 months	$\overline{Y}_i$	3 months	6 months
1	0	3	1.5	.5	3.5
2	1	2	1.5	1.5	2.5
3	2	4	3.0	1.0	3.0
4	2	4	3.0	1.0	3.0
5	2	4	3.0	1.0	3.0
6	1	2	1.5	1.5	2.5
7	0	3	1.5	.5	3.5
8	0	2	1.0	1.0	3.0
			$G = 2.0$		

to each of his scores. Thus, 0 becomes .5 and 3 becomes 3.5. This process is performed for each infant, and the adjusted scores are given in columns 5 and 6 of Table 10.3.

The relevant intermediate statistics are generated using the procedures developed in Chapter 9. Referring to Table 10.4, we find that

$$n = n_1 = n_2 = 8$$

$$\overline{Y}_1 = \frac{8}{8} = 1.0$$

$$\overline{Y}_2 = \frac{24}{8} = 3.0$$

$$SS_1 = 1.00$$

$$SS_2 = 1.00$$

$$df = n - 1 = 8 - 1 = 7$$

$$\hat{s}_{\overline{Y}_1 - \overline{Y}_2} = \sqrt{\left(\frac{SS_1 + SS_2}{df}\right)\left(\frac{1}{n_1} + \frac{1}{n_2}\right)} = \sqrt{\left(\frac{1.00 + 1.00}{7}\right)\left(\frac{1}{8} + \frac{1}{8}\right)} = .267$$

The existence of a relationship is tested by converting the observed sample mean difference into a t value and comparing this with a critical t value, which defines the rejection region. For an alpha level of .05, nondirectional test, and a t distribution with 7 degrees of freedom, the critical values of t are ± 2.36. The derived t is

$$t = \frac{(\overline{Y}_1 - \overline{Y}_2) - (\mu_1 - \mu_2)}{\hat{s}_{\overline{Y}_1 - \overline{Y}_2}}$$

where $\overline{Y}_1 - \overline{Y}_2 = 1 - 3 = -2$ and $\mu_1 - \mu_2 = 0$. Thus,

$$t = \frac{-2 - 0}{.267} = -7.49$$

Since -7.49 is less than -2.36, we reject the null hypothesis and conclude that a relationship exists between age and negative response to a stranger.

The strength of the relationship, as indexed by eta squared, is

$$Eta^2 = \frac{t^2}{t^2 + df}$$

$$= \frac{-7.49^2}{-7.49^2 + 7} = .89$$

TABLE 10.4 ADJUSTED DATA AND SUMMARY STATISTICS FOR RESPONSE TO STRANGERS EXPERIMENT

3 months			6 months		
Y	$(Y - \bar{Y})$	$(Y - \bar{Y}^2)$	Y	$(Y - \bar{Y})$	$(Y - \bar{Y}^2)$
.5	$-.5$	.25	3.5	$-.5$	.25
1.5	.5	.25	2.5	.5	.25
1.0	0	0	3.0	0	0
1.0	0	0	3.0	0	0
1.0	0	0	3.0	0	0
1.5	.5	.25	2.5	.5	.25
.5	$-.5$	.25	3.5	$-.5$	.25
1.0	0	0	3.0	0	0
$\Sigma Y_1 = 8.0$		$SS_1 = 1.00$	$\Sigma Y_2 = 24.0$		$SS_2 = 1.00$

The proportion of variability in the dependent variable that is associated with the independent variable (after the effects of individual differences have been removed) for this sample is .89. This represents a strong relationship.

The nature of the relationship is such that infants who are three months old are less likely to exhibit a negative response to a stranger than infants who are six months old. (Mean crying time was 1.0 versus 3.0, respectively.)

The results of the analysis might appear in a research report as follows:

A correlated groups t test was performed comparing the mean crying time in minutes when the infants were 3 months of age as opposed to when they were 6 months of age. The t was statistically significant ($t = 7.49$, df $= 7$, p $< .05$). The mean crying time for infants at the age of 3 months (1.0) was less than the mean crying time for the infants at the age of 6 months (3.0). The strength of the effect, as indexed by eta squared, was .89.

You should think about the above results in terms of the experimental design questions discussed in Chapter 8. Are there any potential confounding variables that haven't been controlled? What has been the role of disturbance variables in the experiment? What kinds of procedures could be used to reduce sampling error? What are the potential limitations of the experiment in terms of generalizability?

10.8 / Planning Investigations Using the Correlated Groups *t* Test

As with the independent groups *t* test, the design of investigations that will use the correlated groups *t* test requires careful consideration of potential confounding and disturbance variables and the procedures one will use to control for them. Sample size selection is also a major issue relative to the power of the statistical test. Appendix F.1 can help select a sample size so as to obtain a desired level of power. Table 10.5 reproduces a portion of this Appendix for purposes of exposition. As in Chapter 9, different levels of power are listed in the first column, and different levels of the population value of eta squared serve as column headings. It should be noted that these values of eta squared are conceptualized as the proportion of variance in the dependent variable associated with the independent variable *after the effects of individual background are removed*. As an example, if the desired level of power is .80 and if the investigator suspects that the strength of the relationship in the population corresponds roughly to an eta squared of .07, then the number of subjects he or she should use in a study for $\alpha = .05$, nondirectional test, is 53.* The level of power an investigator requires will, of course, depend on the seriousness of a type II error.

TABLE 10.5 SAMPLE SIZES NECESSARY TO ACHIEVE SELECTED LEVELS OF POWER FOR ALPHA = .05 AS A FUNCTION OF POPULATION VALUES OF ETA SQUARED

					Population Eta Squared					
Power	*0.01*	*0.03*	*0.05*	*0.07*	*0.10*	*0.15*	*0.20*	*0.25*	*0.30*	*0.35*
0.25	84	28	17	12	8	6	5	3	3	3
0.50	193	63	38	27	18	12	9	7	5	5
0.60	246	80	48	34	23	15	11	8	7	6
0.67	287	93	55	39	27	17	12	10	8	6
0.70	310	101	60	42	29	18	13	10	8	7
0.75	348	113	67	47	32	21	15	11	9	7
0.80	393	128	76	53	36	23	17	13	10	8
0.85	450	146	86	61	41	26	19	14	11	9
0.90	526	171	101	71	48	31	22	17	13	11
0.95	651	211	125	87	60	38	27	21	16	13
0.99	920	298	176	123	84	53	38	29	22	18

*The suggested *n* in Appendix F.1, on the average, represent a slight underestimation of the required sample sizes. The degree of underestimation is sufficiently small that it is of little consequence, unless *n* is < 10.

Psychologists have studied interpersonal relations and friendship patterns in several contexts. One approach to this area has been to document the daily patterns of social behavior that people exhibit through the use of diaries. Nezlek (1978) asked subjects to keep daily diaries concerning the social contacts they had with other people over four two-week periods. One of the variables studied was the number of people that individuals reported they had met. Specifically, Nezlek was interested in comparing the number of same sex contacts to the number of opposite sex contacts that people made. Sixty-three individuals who kept diaries indicated that they contacted on the average 1.54 same sex individuals per day and 1.01 opposite sex individuals per day. A correlated groups t test indicated that this difference was statistically significant ($t = 2.10$, df $= 62$, $p < .05$) such that people tended to make more same sex than opposite sex contacts. The strength of the relationship, as indexed by eta squared, was .07. This represents a weak relationship and indicates, as one would expect, that many factors other than a person's sex influence social contacts.

10.9 / The Analysis of Difference Scores

So far, we have presented one method for illustrating the logic underlying the correlated groups t test. A second, more traditional, approach will be considered in this section. The advantage of the present method of presentation is its simplicity and computational efficiency. You may find it profitable to compare the two orientations in developing the logic of the correlated groups t test.

We will develop the present approach using the same experiment outlined in the beginning of the chapter: An investigator is studying the relationship between two types of drugs, A and B, and learning. Five subjects are administered drug A and then work on a learning task. One month later, the same five subjects are administered drug B and work on the same type of learning task as before. The number of errors made on each task is tabulated and this serves as the dependent measure. The data for the experiment are presented in Table 10.1. The experimental design involves a within-subjects independent variable with two values (drug A versus drug B) and a quantitative dependent variable.

The test for the existence of a relationship begins with a specification of the null and alternative hypotheses:

$$H_0: \mu_A = \mu_B$$

$$H_1: \mu_A \neq \mu_B$$

The null hypothesis states that there is no relationship between the type of drug and learning whereas the alternative hypothesis states that there is a relationship between the variables.

If the null hypothesis is true, and the type of drug does not make a difference, then we would expect performance while under the influence of drug A to be the same as performance while under the influence of drug B. Another way of stating this is that the difference between a subject's score in condition A and his score in condition B should be zero. Column 4 in Table 10.6 reports a difference score (D) for each individual in the experiment. The mean difference score across individuals was -3.0. Note that this is equal to the difference between the means for the two conditions $(3 - 6 = -3)$. The question of interest is whether the mean difference of -3 is sufficiently different from 0 to reject the null hypothesis.

To answer this question, it is necessary to specify the nature of the sampling distribution for the mean of a set of difference scores based on $N = 5$. The sampling distribution should have the same properties as those described in Chapter 7 for testing a sample mean against a hypothesized population mean value. Assuming the null hypothesis is true, the hypothesized population mean value would be zero ($\mu_{\bar{D}} = 0$). The sampling distribution would be analyzed in terms of a t distribution with $N - 1$ degrees of freedom where N represents the number of subjects in the experiment. The estimated standard error of the mean is $\hat{s}_D/\sqrt{N}$. The computation of $\hat{s}_D$ is presented in Table 10.6 and yields a value of 2.45. The estimated standard error is $2.45/\sqrt{5} = 1.094$. Note that the latter value is within rounding error of the

TABLE 10.6 RAW DATA AND SUMMARY STATISTICS FOR DRUG EXPERIMENT

Individual	Drug A	Drug B	Difference (D)	D^2
1	3	7	-4	16
2	1	5	-4	16
3	4	4	0	0
4	2	8	-6	36
5	5	6	-1	1

$$\Sigma Y = 15 \qquad \Sigma Y = 18 \qquad \Sigma D = -15 \qquad \Sigma D^2 = 69$$
$$\bar{Y}_1 = 3.0 \qquad \bar{Y}_2 = 6.0 \qquad \bar{D} = -3.0$$
$$SS_D = 69 - \frac{(-15)^2}{5} = 24$$
$$\hat{s}_D^2 = \frac{SS}{N-1} = \frac{24}{4} = 6.0$$
$$\hat{s}_D = \sqrt{\hat{s}_D^2} = \sqrt{6} = 2.45$$

value computed earlier in the chapter. The derived t score is thus

$$t = \frac{\overline{D} - \mu_{\overline{D}}}{\hat{s}_{\overline{D}}}$$

$$= \frac{-3 - 0}{1.094}$$

$$= -2.74$$

For an alpha level of .05, nondirectional test, the critical values of t (see Appendix E) are ± 2.78. Since -2.74 is not less than -2.78, we fail to reject the null hypothesis. We can not confidently conclude that a relationship exists between the type of drug administered and learning.

The derivation and interpretation of the strength of the relationship using eta squared and the nature of the relationship is the same as in the case of the independent groups t test. However, as explained previously, eta squared does not reflect the proportion of total variability in the dependent variable that is associated with the independent variable. Rather, it refers to the proportion of explained variance *after the effects of individual differences have been removed.*

10.10 / Summary

The correlated groups t test is typically applied when (1) the dependent variable is quantitative in nature and is measured on approximately an interval scale, (2) the independent variable is *within-subjects* in nature, and (3) the independent variable has two and only two values. The logic of the correlated groups t test is very similar to that of the independent groups t test. The major difference is that the correlated groups t test extracts the influence of individual differences on the dependent variable, thereby yielding a more sensitive analysis of the relationship between the independent and dependent variables.

EXERCISES

Exercises to Review Concepts

1. Under what conditions is the correlated groups t test typically used to analyze a bivariate relationship?

2. What are the major assumptions underlying the correlated groups t test in terms of inferring that a relationship exists between two variables?

3. Determine the critical value of t for each of the following conditions in which a correlated groups t test is to be computed:

(a) $N = 10$, one-tailed test, positive direction

(b) $N = 10$, two-tailed test

(c) $N = 20$, one-tailed test, negative direction

(d) $N = 20$, two-tailed test

4. For the following data in which subjects participated in each of two conditions, A and B, extract the effects of individual differences due to background (that is, generate a set of adjusted scores).

Subject	Condition A	Condition B
1	10	12
2	11	13
3	12	14
4	13	17
5	14	14

5. Compute the mean scores for condition A and condition B in the original data. Compute the corresponding mean scores in the adjusted data. The respective mean scores should be the same (for example, the mean for A in the original data set should equal the mean for A in the adjusted data set). Why do you think this is the case? (Hint: Think about the individual differences that have been extracted as disturbance variables rather than confounding variables, as discussed in Chapter 8.)

Consider the following scores on a learning test that was administered under one of two conditions, A or B.

A	B
10	22
15	17
10	12
5	9
0	0

6. Analyze these data as if the independent variable was between-subjects in nature (that is, answer all three questions about a relationship).

7. Analyze these data as if the independent variable was within-subjects in nature (that is, answer all three questions about a relationship).

8. Compare your results in Exercise 6 with those in Exercise 7. How does this illustrate the advantage of within-subjects experimental designs?

For each of the following experiments, indicate whether you would use an independent groups t test or a correlated groups t test to analyze the relationship between the independent and dependent variables (assume the underlying assumptions of the statistical tests have been satisfied).

9. *Experiment I*. Frieze et al. (1978) report a study that examined the relationship between a person's sex and mathematical abilities as measured by the Graduate Record Exam. The mean score on the quantitative test was compared for males and females who took the exam during the year 1972.

10. *Experiment II*. Smith (1970) studied the effects of a drug, ethosuximide, on verbal skills of children with learning problems. Smith observed that previous research has shown a similarity in EEG patterns of children with learning problems and those who have a mild form of epilepsy, called petit-mal. Smith speculated that ethosuximide, a drug used to treat petit-mal, might therefore serve as a "learning facilitator" for children with learning problems. A group of children participated in the study for six weeks. During the first three weeks, some of the children were injected with ethosuximide and others were given a placebo. During the second three weeks, the reverse was given (that is, children who had been given a placebo were now given ethosuximide and vice versa). Each child was given a verbal skills test after the first three week period and again after six weeks. For all children, the mean score on this test after they had the placebo was compared with the mean score after they had the drug ethosuximide.

11. *Experiment III*. A consumer psychologist was interested in the relative preferences of consumers for two types of headache remedies, Anacin and Tylenol. Five hundred individuals were interviewed and rated each brand on a 10 point rating scale, with progressively higher numbers indicating a more positive attitude toward that brand.

12. *Experiment IV*. Jensen (1973) has reviewed research on the relationship between scores on an intelligence test and an individual's race. In one study, the average scores of blacks and whites were compared.

13. *Experiment V*. Gallup (1976) has reported how individuals' attitudes toward nuclear energy have changed over a period of ten years. The same individuals were interviewed in 1965 and 1975 and during each interview, the respondents indicated their attitude toward nuclear energy on a 5 point scale, with progressively higher numbers indicating a more positive attitude.

Exercises to Apply Concepts

14. The process of decision making has been studied extensively by psychologists. One area of inquiry has been the effects of making a decision on the subsequent evaluation of the decision alternatives. Brehm (1956) has suggested that when an individual is forced to choose between two equally attractive alternatives, the individual will justify his or her decision, after the choice has been made, by downgrading the unchosen alternative and upgrading the chosen one. In one experiment, subjects rated the desirability of two household products. The products were pretested so that the experimenter knew they would be rated with about equal desirability. Subjects then read marketing reports on each product and were asked to choose one as payment for participation in the study. Finally, subjects were told that their first rating might be considered a first impression and that the experimenters wanted to get a second rating since the subject now had more time to think about the products. The ratings were made on an 8 point scale, where 1 indicated low desirability, and 8 indicated high desirability. We will consider only the results for the unchosen alternative. The hypothetical data presented below are representative of the results of this study. Analyze the data, draw a conclusion, and write up your results using the principles given in the Method of Presentation section.

Individual	Rating Before Choice	Rating After Choice
1	8	4
2	7	3
3	7	5
4	4	6
5	8	4
6	6	4
7	6	2
8	5	5
9	5	3
10	4	4

15. As noted in the Exercises in Chapter 7, investigators have studied the effects of marijuana on human physiology. One common belief held by laypersons is that marijuana affects pupil size. Weil et al (1968) studied ten subjects. Each was administered a high dose of marijuana by smoking a potent marijuana cigarette. The subjects were all males, 21 to 26 years of age, all of whom smoked tobacco cigarettes regularly but had never tried marijuana. In this study, pupil size was measured with a millimeter rule under constant illumination with eyes focused on an object at a constant distance. Pupil size was measured before and after smoking marijuana. The data are presented below. Analyze these data, comparing the "before" mean with the "after" mean, draw a conclusion, and write up your results using the principles given in the Method of Presentation section.

Individual	Before Marijuana	After Marijuana
1	5	7
2	7	5
3	6	8
4	7	5
5	6	6
6	5	7
7	3	9
8	3	5
9	5	9
10	3	9

Between-Subjects One-Way Analysis of Variance

11.1 / Use of One-Way Analysis of Variance

Between-subjects one-way analysis of variance is typically used to analyze the relationship between two variables when:

1. the dependent variable is quantitative in nature and is measured on approximately an interval level;

2. the independent variable is between-subjects in nature (it can be either qualitative or quantitative); and

3. the independent variable has three or more groups or values.

One-way between-subjects analysis of variance is used under the same circumstances as the independent groups t test (Chapter 9) except that the independent variable has more than two values. Also, there is one qualification to rule 2. When the independent variable is qualitative in nature, the one-way analysis of variance is almost always used (given that the conditions of rules 1 and 3 are also met). However, when the independent variable is quantitative in nature, there are situations where an alternative analysis is applied, called correlation analysis. The distinction between the approaches, and which one to use, will be discussed in Chapter 13, after the logic of correlation analysis has been developed. Let us consider an example of an investigation that meets the conditions for a one-way analysis of variance.*

An investigator was interested in the relationship between a person's religion and what he or she considered the ideal number of children to have in a family. Twenty-one people were interviewed, seven Catholics, seven Protestants, and seven Jews. Each was asked what he or she considered to be the ideal number of children.

---

*The present chapter discusses the most common type of analysis of variance in the social science literature, called a fixed effects analysis. For a discussion of other possibilities, see Hayes (1963).

Their responses are presented in Table 11.1. In this investigation, religion is the independent variable and has three values. It is a qualitative variable and is between-subjects in nature. The ideal number of children is the dependent variable and it is quantitative in nature.

TABLE 11.1 DATA FOR INVESTIGATION ON RELIGION AND IDEAL NUMBER OF CHILDREN

	Catholics	Protestants	Jews
	4	1	1
	3	2	1
	2	0	2
	3	2	0
	3	4	2
	4	3	0
	2	2	1
$\Sigma Y = $	21	14	7
$\bar{Y} = $	3.0	2.0	1.0

11.2 / Inference of a Relationship Using One-Way Analysis of Variance

Null and Alternative Hypotheses. The first question concerns whether a relationship exists between religion and ideal family size. The null hypothesis states that there is no relationship between religion and the ideal number of children. It is expressed in terms of population mean scores, as we have done in previous chapters:

$$H_0: \mu_C = \mu_P = \mu_J$$

The alternative hypothesis states that there is a relationship between religion and the ideal number of children and that the three population means are not all equal to one another. Unlike previous chapters, however, H_1 cannot be summarized in a single mathematical statement. This is because there are a number of different ways in which the population means can pattern themselves so that they are not all equal to one another. Some examples are

$$\mu_C = \mu_P > \mu_J$$

$$\mu_C > \mu_P > \mu_J$$

$$\mu_C > \mu_P = \mu_J$$

The alternative hypothesis does not distinguish among these different possibilities. It simply states that a relationship exists such that the three population means are *not* all equal to one another. The question of the exact patterning of means is addressed in the context of the nature of the relationship. Thus, the alternative hypothesis is

H_1: The population means are not all equal

The basic problem is to choose between the null and alternative hypotheses. If we look at the mean scores for the samples, we see that they are not equal. The Catholic mean was 3.0, the Protestant mean was 2.0, and the Jewish mean was 1.0. This would seem to support the alternative hypothesis. However, we know that the differences in sample means could reflect sampling error. We therefore want to test the viability of a sampling error interpretation.

The logic for testing this interpretation will be similar to that discussed in Chapter 7. First, we will assume that the null hypothesis is true. We will then state an expected result based on this assumption. Next, we will compute a sample statistic (in this case, the statistic is a *ratio of variances,* which we will discuss shortly). The statistic will be treated as a score in a sampling distribution where the null hypothesis is assumed to be true. If the score is within the rejection region (as defined by our alpha level), then we will reject the null hypothesis and conclude that there is a relationship between the two variables. Otherwise, we will fail to reject the null hypothesis.

Between- and Within-Group Variability. For the data in Table 11.1, we can distinguish two types of variability. The first concerns the differences in the mean scores between the three groups, or in more formal terms, **between-group variability.** If the mean scores in the three groups were all equal, then there would be no between-group variability. The more different the three means are from one another, the more between-group variability there is.

Why do mean differences exist? There are two factors that contribute to between-group variability. The first is sampling error: Even if the population means were, in fact, equal, we might observe between-group variability in the samples because of sampling error. A second factor that contributes to between-group variability is the effect of the independent variable on the dependent variable: If religion influences the ideal number of children, then this will tend to make the sample means different from one another. Thus, between-group variability in mean scores on the dependent variable reflects two things (1) the effects of sampling error, and (2) the effects of the independent variable on the dependent variable.

The second type of variability concerns the variability of scores *within* each of the three groups, or **within-group variability.** Examine the seven scores for Catholics in Table 11.1. Note that there is variability in the scores; some Catholics have a higher ideal number of children than other Catholics. Since religion is held constant, the variability in the scores cannot be attributed to it. Other factors are operating to cause the variability in scores (such as how religious the person is, and other such disturbance variables). The fact that scores vary within a group suggests

Inference of a Relationship Using One-Way Analysis of Variance **209**

that there is variability in scores in the *population* represented by that group. Recall from Chapter 6 that when the variability of scores in a population is large, we expect more sampling error than when the population variability is small. As such, the larger the variability of scores within a group, the larger the amount of variability in the population and the larger the amount of sampling error we can expect. Thus, within-group variability reflects sampling error. Note that within-group variability is *not* influenced by the effects of the independent variable on the dependent variable. Since the variability of scores is considered within each group *separately,* the values of the independent variable are held constant and cannot be a source of variation within groups.

If between-group variability reflects both sampling error *and* the effects of the independent variable, and if within-group variability reflects just sampling error, then the ratio of these two variances (or the **variance ratio**) can be used to test the sampling error interpretation discussed in the previous section:

$$\frac{\text{Between-Group Variability}}{\text{Within-Group Variability}} \qquad [11.1]$$

Consider the case where the null hypothesis is true. In this case, the independent variable has no effect on the dependent variable, and the between-group variability only reflects sampling error. Since within-group variability also reflects sampling error, then the ratio in Formula 11.1 is simply dividing one estimate of sampling error by another estimate of sampling error. The result should be a value which, over the long run, will approach 1.0 (since any number divided by itself is 1.0). In contrast, if the null hypothesis is not true, between-group variability will reflect both the effects of sampling error and the effects of the independent variable. In this case, we would expect the variance ratio, over the long run, to be greater than 1.0.

Computation of Between-Group Variability and Within-Group Variability. It will be necessary to derive numerical estimates of between- and within-group variability. This is accomplished by conceptualizing an individual's score in terms of two components, the mean score of the group in which he or she is a member, and how far his or her score deviates from that mean. The first individual in the Catholic group had a score of 4. The mean for this group was 3 and she deviated $4 - 3 = +1$ units from the mean. Her score can be represented as

$$4 = 3 + 1$$

We can write a general equation that reflects this relationship:

$$Y = \overline{Y}_j + d \qquad [11.2]$$

where Y represents an individual's score, $\overline{Y}_j$ is the mean score of the group in which she is a member (group "j"), and d is the deviation of the score from the group mean $(Y - \overline{Y}_j)$. It is possible to express the score of every individual in the experiment in this manner, which has been done in Table 11.2.

TABLE 11.2 BREAKDOWN OF SCORES ON THE DEPENDENT VARIABLE

Individual	Religion	Y	$=$	$\overline{Y}_j$	$+$	$d = Y - \overline{Y}_j$
1	C	4	=	3	+	1
2	C	3	=	3	+	0
3	C	2	=	3	+	−1
4	C	3	=	3	+	0
5	C	3	=	3	+	0
6	C	4	=	3	+	1
7	C	2	=	3	+	−1
8	P	1	=	2	+	−1
9	P	2	=	2	+	0
10	P	0	=	2	+	−2
11	P	2	=	2	+	0
12	P	4	=	2	+	2
13	P	3	=	2	+	1
14	P	2	=	2	+	0
15	J	1	=	1	+	0
16	J	1	=	1	+	0
17	J	2	=	1	+	1
18	J	0	=	1	+	−1
19	J	2	=	1	+	1
20	J	0	=	1	+	−1
21	J	1	=	1	+	0

The column labeled Y is the score of each individual on the dependent variable. We can compute an index of how much variability there is in the scores by computing a sum of squares for this column of numbers. That is, we compute the mean of this column of numbers, subtract the mean from each score, square these deviation scores, and then sum them. This sum of squares is called the **sum of squares total,** because it reflects the total variability in the dependent variable across all individuals. It is identical to the sum of squares total as computed in Chapter 9. In this case, the sum of squares total equals 32.0.

Examine the scores in the column labeled $\overline{Y}_j$ in Table 11.2. The sources of variability in these scores are the differences in means between the groups. If all three means of the groups had been equal, every score would be the same. The more the three means differ, the more the scores will tend to be different from each other. A sum of squares of this column would therefore reflect between-group variability.

The mean of this column of numbers is subtracted from each score and the deviation scores are squared and then summed. This yields an index called the **sum of squares between.** In this example, the sum of squares between, abbreviated $SS_{BETWEEN}$, equals 14.0.

Examine the scores in the column labeled *d*. Each of these scores represents a deviation from the group mean, or how far an individual's score deviates from the average score *within the group*. If there was little within-group variability, all of these numbers would be close to zero and would tend to equal one another. If there was considerable variability within-group, there would be considerable variability in these scores. A sum of squares of this column of numbers would therefore reflect how much within-group variability there is. This sum of squares is called the **sum of squares within** and equals 18.0. It is abbreviated SS_{WITHIN}. Again, the mean of the column is subtracted from each number and these deviation scores are squared and then summed. Another name given to the sum of squares within is the *sum of squares error* because it reflects the effects of disturbance variables and hence, sampling error.

Note that the sum of squares total equals the sum of squares between plus the sum of squares within:

$$SS_{TOTAL} = SS_{BETWEEN} + SS_{WITHIN} \qquad [11.3]$$

$$32 = 14 + 18$$

This is consistent with the idea of expressing each individual's score in terms of the group mean and the deviation from the mean, as we did in Table 11.2. Thus, just as each individual's score can be expressed in terms of (1) a group mean and (2) a deviation from the group mean, the total *variability* in scores can be expressed in terms of (1) the *variability* between the group means and (2) the *variability* of deviations from the group means.

The Variance Ratio. The variance ratio that is computed to test the null hypothesis is the between-group variability divided by the within-group variability. As noted earlier, the larger the variance ratio, the less likely it is the null hypothesis is true. The particular variance ratio used by statisticians does not use measures of sums of squares, but rather measures of variance, or mean squares (these are simply the sum of squares divided by its corresponding degrees of freedom). The reason variances are used instead of sums of squares will be made explicit shortly. We will first consider the computation of the relevant variances. The formulas are

$$MS_{BETWEEN} = \frac{SS_{BETWEEN}}{df_{BETWEEN}} \qquad [11.4]$$

$$MS_{WITHIN} = \frac{SS_{WITHIN}}{df_{WITHIN}} \qquad [11.5]$$

where $MS_{BETWEEN}$ stands for the mean square for between-group variability (also commonly called the **mean square between**), MS_{WITHIN} stands for the mean square for the within-group variability (also commonly called the **mean square within**), $df_{BETWEEN}$ is the degrees of freedom associated with the sum of squares between, and df_{WITHIN} is the degrees of freedom associated with the sum of squares within.

The degrees of freedom can be computed from the following formulas:

$$df_{BETWEEN} = k - 1 \qquad \qquad \textbf{[11.6]}$$

$$df_{WITHIN} = N - k \qquad \qquad \textbf{[11.7]}$$

where k represents the number of groups or values of the independent variable (in our example, $k = 3$) and N represents the total number of subjects in the experiment (in our example, $N = 21$).* Thus

$$df_{BETWEEN} = 3 - 1 = 2$$

$$df_{WITHIN} = 21 - 3 = 18$$

and using Equations 11.4 and 11.5,

$$MS_{BETWEEN} = \frac{14}{2} = 7.0$$

$$MS_{WITHIN} = \frac{18}{18} = 1.0$$

The variance ratio, called the **_F_ ratio** (in honor of the statistician Sir Ronald Fisher, who developed this approach), is

$$F = \frac{MS_{BETWEEN}}{MS_{WITHIN}} \qquad \qquad \textbf{[11.8]}$$

which in our example is

$$F = \frac{7.0}{1.0} = 7.0$$

To recapitulate, the F ratio is a ratio between two variances. In the context of the one-way analysis of variance, it is the mean square between (an index of between-group variability) divided by the mean square within (an index of within-group variability). The mean square between and mean square within are derived by computing the sum of squares between and the sum of squares within and dividing

---

*The logic underlying these formulas is presented in Appendix 11.1.

each by the appropriate degrees of freedom. If the null hypothesis is true, we would expect the F ratio, over the long run, to approach 1.0. If the null hypothesis is not true, we would expect the F ratio, over the long run, to be greater than 1.0. Let us now consider this in more detail.

Sampling Distribution of a Variance Ratio. Consider three populations A, B, and C, where $\mu_A = \mu_B = \mu_C$. Suppose we select a random sample of thirty individuals from population A, a random sample of thirty individuals from population B, and a random sample of thirty individuals from population C. Using the procedures described in the previous section, we could compute a variance ratio:

$$F = \frac{MS_{BETWEEN}}{MS_{WITHIN}}$$

Although we would expect this value to be near 1.00 (because the null hypothesis is true), it may not equal exactly 1.00 because the values for $MS_{BETWEEN}$ and MS_{WITHIN} are independent estimates of variability in the populations, based on sample data. Suppose the variance ratio was found to be 1.52. Suppose we repeated this procedure by randomly selecting samples from the same populations using the same sample sizes, and this time observed a variance ratio of 1.32. We could repeat this procedure a large number of times (in principle, until we have all possible samples based upon sizes of n_1, n_2, and n_3), and the result would be a large number of variance ratios. These variance ratios can be treated as "scores" in a distribution and constitute a **sampling distribution of the variance ratio.** It is analogous to a sampling distribution of the mean, as discussed in Chapter 6.

Examples of the shapes of different sampling distributions are presented in Figure 11.1. The sampling distribution of the variance ratio takes on different forms depending on the degrees of freedom associated with it. In this case the relevant degrees of freedom are the $df_{BETWEEN}$ and the df_{WITHIN}. In all of the distributions, the average (median) value for F is 1.00. This shouldn't be surprising since they are calculated from the case where the null hypothesis is true. Values greater than one become less frequent, and extreme values are highly unlikely. The question then becomes similar to the one posed in Chapter 7. How large (extreme) must the F ratio that we computed from the sample information be before we reject the null hypothesis of no relationship? As it turns out, the sampling distribution of a variance ratio, under certain conditions, corresponds to a well-known theoretical distribution called the **F distribution.** Just as probability statements can be made with respect to scores in the normal and t distributions, so can probability statements be made with respect to scores in the F distribution. It therefore becomes possible to set an alpha level and define a critical value so that if the null hypothesis is true, the probability of obtaining a sample variance ratio equal to or greater than that critical value is less than α. The reason that measures of mean squares are used to define the variance ratio rather than measures of sums of squares is that the former yield a sampling distribution that closely approximates an F distribution whereas a variance ratio based on sums of squares does not.

(a) Distribution with df = 3 and 12

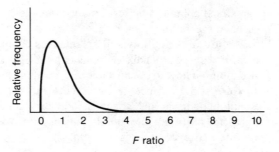

(b) Distribution with df = 2 and 9

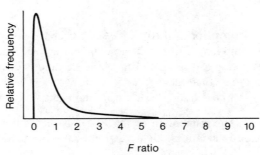

(c) Distribution with df = 3 and 8

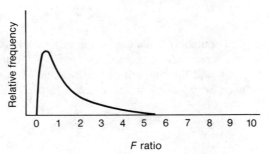

Figure 11.1 Sampling Distribution of Variance Ratios with Different Degrees of Freedom

Before conducting the final test, we will summarize all of our previous calculations in a *summary table:*

Source	SS	df	MS	F
Between	14.0	2	7.0	7.0
Within	18.0	18	1.0	
Total	32.0	20		

The first column of the table indicates the source of the statistics in the other columns. The second column reports the sum of squares between, within, and total. The next column provides the associated degrees of freedom for each sum of

Inference of a Relationship Using One-Way Analysis of Variance   **215**

squares, and column 4 presents the various mean squares (SS/df). The last column presents the F ratio.

The test of the null hypothesis involves comparing the observed F ratio with a critical F value. Appendix G presents a table of critical values and corresponding instructions on how to use the table. For an alpha level of .05, df = 2, 18, the critical value of F is 3.55. Since the observed F ratio (7.0) is greater than 3.55, we reject the null hypothesis and conclude that a relationship exists between a person's religion and his or her ideal number of children. We have answered our first question about the existence of a relationship.

Assumptions of the *F* Test. The above inferential test is based on a number of assumptions. Most of the assumptions are necessary because when they are met, the sampling distribution of the variance ratio is distributed in accord with an F distribution. When the assumptions are violated, however, the sampling distribution of the variance ratio departs from the F distribution.

1. One assumption of the F test is the assumption of **homogeneity of within-group variances.** Given k groups, the assumption can be stated as

$$\sigma_1^2 = \sigma_2^2 = \ . \ . \ . \ \sigma_k^2$$

or the population variances of the groups are assumed to be equal. Several procedures have been suggested for formally testing if this assumption is met in an investigation, the simplest of which is the F_{MAX} test. This involves computing the variance estimate, $\hat{s}^2$, for each group separately:

$$\hat{s}_{\text{C}}^2 = \frac{\text{SS}_{\text{C}}}{n_1 - 1} = \frac{4}{7 - 1} = .67$$

$$\hat{s}_{\text{P}}^2 = \frac{\text{SS}_{\text{P}}}{n_2 - 1} = \frac{10}{7 - 1} = 1.67$$

$$\hat{s}_{\text{J}}^2 = \frac{\text{SS}_{\text{J}}}{n_3 - 1} = \frac{4}{7 - 1} = .67$$

These represent estimates of σ_{C}^2, σ_{P}^2, and σ_{J}^2, which, as noted above, are assumed to be equal. The F_{MAX} statistic is defined as the largest variance estimate divided by the smallest variance estimate, or

$$F_{\text{MAX}} = \frac{1.67}{.67} = 2.49$$

This value is then compared against a critical value of F_{MAX} (see Appendix H) for purposes of testing the null hypothesis that $\sigma_1^2 = \sigma_2^2 = \ . \ . \ . \ = \sigma_k^2$. Using the instructions provided in Appendix H, the critical value for F_{MAX}, $\alpha = .05$, is 8.38. Since 2.49 does not exceed 8.38, we fail to reject the null hypothesis of equal variances. The assumption of equal variances *is* tenable.

Statisticians have recently demonstrated that under certain conditions, moderate violations of the homogeneity of variance assumption do not alter the results of the statistical analysis very much. This is true when the number of subjects in each group is the same and when the number of cases per group is ten or more (see Scheffé, 1959; Boneau, 1960). Thus, a desirable experimental practice, when it makes conceptual sense, is to use equal number of subjects in each group, with n_j equal to at least 10.

A general appreciation of the effects on α of violating the assumption of equal variances can be gained with reference to Table 11.3. The table is read as follows: Given a problem with three groups comprising the independent variable, and $n_1 = 7$, $n_2 = 5$, and $n_3 = 3$, and if the population variances are in the ratio 1:1:3 (for example, $\sigma_1^2 = 10$, $\sigma_2^2 = 10$, $\sigma_3^2 = 30$), then the probability of a type I error is actually .11 rather than the specified α of .05 (see row 6 of Table 11.3).

2. A second major assumption of one-way analysis of variance is that of **normality.** Specifically, it must be assumed that the distribution of scores in the populations of interest is normally distributed. Procedures for addressing this assumption are discussed in Winer (1971). Again, statisticians have demonstrated that moderate violations of the assumption do not affect appreciably type I and type II errors when the number of cases sampled is equal in the various groups and n_j is greater than 10 (Box and Andersen, 1955; Glass, Peckham, and Sanders, 1974). A

TABLE 11.3 EFFECT OF HETEROGENEOUS VARIANCES ON THE PROBABILITY OF TYPE I ERROR IN ONE-WAY BETWEEN-SUBJECTS ANALYSIS OF VARIANCE FOR A NOMINAL SIGNIFICANCE LEVEL, α, OF .05

Number of groups	Ratio of σ^2's	Sample sizes, n_j	Actual probability of type I error
3	1:2:3	5, 5, 5	.056
		3, 9, 3	.056
		7, 5, 3	.092
		3, 5, 7	.040
3	1:1:3	5, 5, 5	.059
		7, 5, 3	.11
		9, 5, 1	.17
		1, 5, 9	.013
5	1:1:1:1:3	5, 5, 5, 5, 5	.074
		9, 5, 5, 5, 1	.14
		1, 5, 5, 5, 9	.025

From Glass and Stanley (1970)

TABLE 11.4 APPROXIMATIONS TO ACTUAL PROBABILITIES OF A TYPE I ERROR IN A ONE-WAY BETWEEN-SUBJECTS ANALYSIS OF VARIANCE FOR VARIOUS NONNORMAL POPULATIONS WHEN $\alpha = .05$

Type of Population	Number of Groups	n per group	Actual Probabilities of a Type I Error
Normal	4	5	.056
Leptokurtic	3	3	.078
Leptokurtic	4	5	.066
Rectangular	3	3	.061
Moderately Skewed	4	5	.052
Extremely Skewed	3	3	.048
Extremely Skewed	4	5	.048
J-Shaped	3	3	.048

general appreciation of the effects on α of violating the assumption of normality can be gained with reference to Table 11.4. More detailed analyses are reported in Glass, Peckham, and Sanders (1972). It can be seen here that the F test is relatively robust with respect to violations of this assumption.

3. A third assumption of one-way analysis of variance is that the subjects comprising each sample are independently and randomly selected, and that the samples are independent. These criteria are critical to the F test.

Computational Formulas. The sum of squares between and the sum of squares within were calculated using the standard sum of squares formula developed in Chapter 3. Because of certain statistical properties, more efficient computational formulas are available. They are considered here. Alternative computational formulas are also presented in Appendix 11.2.

The sum of squares between is most easily computed by defining a treatment effect for each group. Recall from Chapter 9 that a treatment effect is the group mean minus the grand mean

$$T_j = \overline{Y}_j - G$$

where T_j represents the treatment effect for group j, $\overline{Y}_j$ represents the mean score on the dependent variable for group j, and G represents the grand mean or the average score on the dependent variable across all subjects in the experiment. In our example, these are

$$T_C = 3 - 2 = 1.0$$

$$T_P = 2 - 2 = 0$$

$$T_J = 1 - 2 = -1.0$$

The sum of squares between is then computed by the following formula:

$$SS_{BETWEEN} = \sum_{j=1}^{k} T_j^2 n_j \qquad [11.9]$$

where k is the number of groups, n_j is the number of individuals in group j, and all other terms are as previously defined. Equation 11.9 states that each treatment effect should be squared, then multiplied by n_j and the products summed:

$$SS_{BETWEEN} = (1^2)(7) + (0^2)(7) + (-1^2)(7)$$

$$= 7 + 0 + 7$$

$$= 14$$

The sum of squares within can be computed by calculating the sum of squares total and then subtacting from this the sum of squares between, as computed above.* Computation of the sum of squares total can use either the deviational formula or the raw score formula discussed in Chapter 3:

$$SS_{TOTAL} = \Sigma(Y - \bar{Y})^2 = \Sigma Y^2 - \frac{(\Sigma Y)^2}{N}$$

and then,

$$SS_{WITHIN} = SS_{TOTAL} - SS_{BETWEEN} \qquad [11.10]$$

In our example, SS_{TOTAL} was found to be 32.0. Thus,

$$SS_{WITHIN} = 32 - 14$$

$$= 18$$

*For a computational formula that is independent of SS_{TOTAL} and $SS_{BETWEEN}$, see Appendix 11.2.

Study Exercise 11.1

For the following experiment, construct a summary table and test the null hypothesis of equivalent population means.

An investigator wanted to test the relative effects of three types of drugs on learning. Drug A was administered to five individuals who then read a short story. After reading the story, the individuals were given a recall test. Performance on the test was scored from 0 to 10, with larger numbers indicating better performance. Drug B was administered to a second group of five individuals who then performed the same task as individuals in the drug A group. Finally, a group of five individuals performed the task while under the influence of drug C. The data for the experiment are as follows:

Drug A	Drug B	Drug C
6	5	2
5	6	4
7	7	5
8	4	3
4	3	6

Answer. The first step in the analysis is to construct a summary table, which involves computing the appropriate sum of squares, degrees of freedom, mean squares, and F ratio. The sum of squares total is computed in Table 11.5 to be 40.0. The sum of squares between requires computing the mean score for each group and the grand mean. These are

$$\overline{Y}_A = \frac{30}{5} = 6.0$$

$$\overline{Y}_B = \frac{25}{5} = 5.0$$

$$\overline{Y}_C = \frac{20}{5} = 4.0$$

$$G = \frac{75}{15} = 5.0$$

The treatment effects are next computed for each group:

$$T_A = 6 - 5 = 1.0$$

$$T_B = 5 - 5 = 0$$

$$T_C = 4. - 5 = -1.0$$

Using Equation 11.9, we find that

$$SS_{BETWEEN} = (1^2)(5) + (0^2)(5) + (-1^2)(5)$$

$$= 10$$

The sum of squares within is then computed as follows:

$$SS_{WITHIN} = SS_{TOTAL} - SS_{BETWEEN}$$

$$= 40 - 10$$

$$= 30$$

The degrees of freedom for each of the sum of squares is computed using Equations 11.6 and 11.7:

$$df_{BETWEEN} = k - 1 = 3 - 1 = 2$$

$$df_{WITHIN} = N - k = 15 - 3 = 12$$

$$df_{TOTAL} = N - 1 = 15 - 1 = 14$$

The relevant mean squares are computed by dividing the sum of squares by its corresponding degrees of freedom (see Equations 11.4 and 11.5):

$$MS_{BETWEEN} = \frac{SS_{BETWEEN}}{df_{BETWEEN}} = \frac{10}{2} = 5.0$$

$$MS_{WITHIN} = \frac{SS_{WITHIN}}{df_{WITHIN}} = \frac{30}{12} = 2.5$$

Finally, the F ratio is the $MS_{BETWEEN}$ divided by the MS_{WITHIN}:

$$F = \frac{MS_{BETWEEN}}{MS_{WITHIN}} = \frac{5.0}{2.5} = 2.0$$

The summary table appears as follows:

Source	SS	df	MS	F
Between	10	2	5.0	2.0
Within	30	12	2.5	
Total	40	14		

Examining Appendix G for an alpha level of .05, the critical F value, based on 2 and 12 degrees of freedom, is 3.88. Since the observed F value of 2.0 does not exceed this, we fail to reject the null hypothesis and cannot confidently conclude that a relationship exists between the type of drug and learning.

TABLE 11.5 COMPUTATION OF SUM OF SQUARES TOTAL FOR DRUG EXPERIMENT

Y	$(Y - \overline{Y})$	$(Y - \overline{Y})^2$
6	1	1
5	0	0
7	2	4
8	3	9
4	−1	1
5	0	0
6	1	1
7	2	4
4	−1	1
3	−2	4
2	−3	9
4	−1	1
5	0	0
3	−2	4
6	1	1
$\Sigma Y = 75$		$SS_{TOTAL} = 40$
$\overline{Y} = 5.0$		

11.3 / Strength of the Relationship

Earlier, we noted that the sum of squares within is also called the sum of squares error. In fact, it is identical to the sum of squares error used in the index of eta squared, as discussed in Chapter 9. Accordingly, the strength of the relationship between the independent and dependent variables can be computed using the formula for eta squared:

$$\text{Eta}^2 = 1 - \frac{SS_{ERROR}}{SS_{TOTAL}}$$

$$= 1 - \frac{SS_{WITHIN}}{SS_{TOTAL}} = \frac{SS_{BETWEEN}}{SS_{TOTAL}} \qquad [11.11]$$

For the investigation of religion and ideal number of children, the strength of the relationship in the sample data would be

$$\text{Eta}^2 = 1 - \frac{18}{32} = .44$$

In this experiment, differences in religion are associated with 44% of the variation in the ideal number of children.

11.4 / Nature of the Relationship

As noted at the beginning of this chapter, the F test considers the null hypothesis against all possible alternative hypotheses (for example, $\mu_A > \mu_B > \mu_C$, $\mu_A = \mu_B > \mu_C$, $\mu_A > \mu_B = \mu_C$, and so on). If any one of the alternative hypotheses holds, the null hypothesis will be rejected, unless a type I error has occurred. Given this state of affairs, it becomes necessary to conduct additional analyses to determine the exact nature of the relationship between the two variables when three or more groups are involved.*

The procedures for determining the nature of the relationship after the null hypothesis has been rejected are controversial among statisticians. The most common techniques are the Scheffé test, the Newman-Keuls test, Duncan's multiple

---

*It should be noted that the alternative hypothesis for an F test is always nondirectional in nature. In the case of the t test, there were only two directions that discrepancies between null and alternative hypotheses could occur ($\mu_1 > \mu_2$ or $\mu_1 < \mu_2$). With three or more population means, departures can occur in many directions. The result of all departures from the null hypothesis is contained within the upper tail of the F statistic (as defined by the rejection region) and accordingly, the F test is, by definition, nondirectional.

range test, Tukey's honest significant difference (HSD) test, and Fisher's least significant difference (lsd) test. Each of these is discussed by Kirk (1968). The most general technique is the one proposed by Scheffé. However, it also tends to be highly conservative and, for this reason, is sometimes not used by researchers. Probably the most common methods are the Newman-Keuls test and Tukey's HSD test. We will consider the latter because of its ease of presentation.

Tukey's **HSD test** discerns the nature of the relationship by testing a null and alternative hypothesis for all possible pairs of group means. In the study on religion and ideal number of children, there were three groups (Catholic, Protestant, Jewish). The questions of interest in the HSD test are: (1) Is the population mean for Catholics different from the population mean for Protestants? (2) Is the population mean for Catholics different from the population mean for Jews? and (3) Is the population mean for Protestants different from the population mean for Jews? These three questions can be cast in terms of three sets of null and alternative hypotheses:

$$H_0: \mu_C = \mu_P$$
$$H_1: \mu_C \neq \mu_P$$

$$H_0: \mu_C = \mu_J$$
$$H_1: \mu_C \neq \mu_J$$

$$H_0: \mu_P = \mu_J$$
$$H_1: \mu_P \neq \mu_J$$

This is similar to the statement of null and alternative hypotheses for the independent groups t test discussed in Chapter 9. Could we simply test each of the above comparisons by conducting three independent groups t tests? The answer is no. The major problem with doing this is that multiple t tests increase the probability of a type I error beyond the probability specified by the alpha level. Witte (1980) has presented a coin tossing analogy that illustrates the problem:

When a fair coin is tossed only once, the probability of heads equals .50—just as when a single t test is to be conducted at the .05 level of significance, the probability of a type I error equals .05. When a fair coin is tossed three times, however, heads can appear not only on the first toss but on the second or third toss as well, and hence the probability of heads on at least one of the three tosses exceeds .50. By the same token, when a series of three t tests are conducted at the .05 level of significance, a type I error can be committed not only on the first test but on the second or third test as well, and hence the probability of committing a type I error on at least one of the three tests exceeds .05. In fact, the cumulative probability of at least one type I error could be as large as .15 for this series of three t tests.

The Tukey HSD test circumvents this problem and maintains the probability of a type I error at the specified alpha level. The logic underlying the HSD test is presented in Tukey (1953), and is rather complex in nature. We will focus on the mechanics of applying the approach.

In order to facilitate presentation, the summary table for the experiment on religion and ideal number of children is reproduced below:

Source	SS	df	MS	F
Between	14.0	2	7.0	7.0
Within	18.0	18	1.0	
Total	32.0	20		

The HSD test involves the computation of a critical difference, CD, which is defined as follows:

$$CD = q \sqrt{\frac{MS_{WITHIN}}{n}} \qquad [11.12]$$

where MS_{WITHIN} is the mean square within from the summary table, n is the number of subjects within a group, and q is a value obtained from a table, called studentized range values, presented in Appendix I. The value of q is determined with reference to three concepts, the alpha level (in this case, .05), the degrees of freedom within (in this case, 18), and the number of groups, k (in this case, 3). Using the instructions provided in Appendix I, we find that $q = 3.61$. We calculate CD as follows:

$$CD = 3.61 \sqrt{\frac{1.0}{7}} = 1.36$$

Consider the first question from above:

$$H_0: \mu_C = \mu_P$$

$$H_1: \mu_C \neq \mu_P$$

We want to choose between these two competing hypotheses. The rule for doing so in the HSD procedure is the following: If the absolute difference in sample means for the two groups involved in the comparison exceeds the value of CD, then reject the null hypothesis. Otherwise, fail to reject the null hypothesis. In this case, the absolute difference between the sample means is $|\bar{Y}_C - \bar{Y}_P| = |3 - 2| = 1.0$. Since 1.0 does not exceed 1.36, we fail to reject the null hypothesis stated above.

The same logic is then applied to the other two comparisons. We can summarize the results of the entire analysis in a table:

Null Hypothesis Tested	Absolute Difference Between Sample Means	Value of CD	Hypothesis Rejected?
$\mu_C = \mu_P$	$\|3 - 2\| = 1.0$	1.36	No
$\mu_C = \mu_J$	$\|3 - 1\| = 2.0$	1.36	Yes
$\mu_P = \mu_J$	$\|2 - 1\| = 1.0$	1.36	No

The nature of the relationship is now apparent. We can confidently say that Catholics ($\overline{Y} = 3.0$) tend to have a higher ideal number of children than Jews ($\overline{Y} = 1.0$). However, the mean difference in ideal number of children for Catholics relative to Protestants ($\overline{Y} = 2.0$) was not statistically significant, nor was the mean difference between Protestants and Jews. The HSD test provides a clear picture of just how religion is related to the ideal number of children.

Study Exercise 11.2

A researcher tested the relationship between political party preference and attitudes toward legal abortions. Thirty individuals were interviewed, ten Democrats, ten Republicans, and ten Independents. Each completed an attitude scale that measured their attitudes toward legal abortions, yielding an index from 0 to 10, with higher numbers indicating more positive attitudes. Given the following summary table and means, use the HSD test to analyze the nature of the relationship between the two variables.

$$\overline{Y}_D = 7.0 \qquad \overline{Y}_R = 2.0 \qquad \overline{Y}_I = 6.0$$

Source	SS	df	MS	F
Between	420	2	210	14.0
Within	409	27	15	
Total	825	29		

Answer. The critical value of F for testing the null hypothesis, based upon an alpha level of .05 and 2 and 27 degrees of freedom, was 3.35. The first step in applying the HSD test is the calculation of CD:

$$CD = 3.51 \sqrt{\frac{15}{10}} = 4.30$$

Given the value of CD, we can construct a table as before:

Null Hypothesis Tested	Absolute Difference Between Sample Means
$\mu_D = \mu_R$	$\|7 - 2\| = 5.0$
$\mu_D = \mu_I$	$\|7 - 6\| = 1.0$
$\mu_R = \mu_I$	$\|2 - 6\| = 4.0$

Value of CD	Is Null Hypothesis Rejected?
4.30	Yes
4.30	No
4.30	No

The nature of the relationship is such that Democrats, on the average, have more positive attitudes toward legal abortions than Republicans. However, we can not confidently conclude that the attitudes of Independents are different (that is, more or less favorable) from either of these groups.

11.5 / Methodological Considerations

Although the results of the investigation on religion and ideal family size suggest that a relationship exists between the two variables, the data must be interpreted in the context of certain methodological constraints. Consider first the role of confounding variables. Because religion of the individual is not a formal experimental manipulation, subjects in the study could not be randomly assigned to groups. Consequently, all variables that are naturally related to religion will serve as confounding variables. This would include such variables as social class, size of family raised in, and education, to name a few. It is impossible to conclude unambiguously that a causal relationship exists between religion and ideal family size. Religion per se may have no effect on ideal family size and the observed relationship may simply be a function of the causal influence of one or more of the confounding variables.

In terms of disturbance variables, a large number of such variables were uncontrolled in the study. One important disturbance variable is how religious an individual is. Research has shown that ideal family size is influenced by the religiosity of the individual. Subjects within a given religion in this study probably differed in religiosity and this would, in turn, create sampling error. The data used in the text were hypothetical and not representative of data in the social science literature on religion and ideal family size. In practice, the role of disturbance variables is much stronger than represented here.

Eta squared provides an index of the extent to which disturbance variables have influenced the dependent measure. Specifically, one minus eta squared represents the proportion of variance in the dependent variable that can be attributed to disturbance variables. For the present data, this would equal $1 - .44 = .56$. Roughly half of the variance in the dependent measure is due to the influence of disturbance variables. Certainly, this could have been reduced through additional experimental controls.

The results of the study must also be considered in terms of the generalizability of the data. No details were given about who the subjects were. If they were all college students, this would suggest potential limits to the generalizability of the data. Research reports should provide reasonable descriptions of the nature of the sample studied.

Method of Presentation

The results of an analysis of variance are frequently accompanied by a summary table with the following format:

Source	df	MS	F
Between	2	7.0	7.0*
Within	18	1.0	
Total	20		

*$p < .05$

This table is identical in format to the summary table presented in the chapter. However, the various sums of squares have not been reported. This is done to save journal space, since the sum of squares can be derived from the information presented:

$$SS_{BETWEEN} = (MS_{BETWEEN})(df_{BETWEEN})$$

$$SS_{WITHIN} = (MS_{WITHIN})(df_{WITHIN})$$

$$SS_{TOTAL} = SS_{BETWEEN} + SS_{WITHIN}$$

The * by the F refers to the footnote and indicates the p value associated with the F. Usually, when the null hypothesis is not rejected, no * will appear. A report of an analysis of variance might appear as follows:

A one-way analysis of variance was performed comparing the means of the three groups. The F was statistically significant ($F = 7.0$, $df = 2, 18$, $p < .05$). The strength of the effect, as indexed by eta squared, was .44. An HSD test was performed and indicated that the mean score for Catholics (3.0) was statistically significantly different from the mean score for Jews (1.0). The mean score for Protestants (2.0) did not differ significantly from either of these groups.

The first sentence identifies the statistical test that was used and the second sentence states the results of the test for the existence of a relationship. The term "between-subjects" is not used to describe the analysis of variance because it is assumed that a between-subjects analysis was done. As has been the case with other tests, the phrase *statistically significant* means that the null hypothesis was rejected. The numbers within parentheses report the observed F, the degrees of freedom between, the degrees of freedom within, and the p value. Sometimes the df are reported in the context of the F statistic, such as $F(2,18) = 7.0$. The third sentence indicates the strength of the relationship. Unfortunately, this is rarely reported by investigators, but it can always be derived from the summary table. The last two sentences report the results of the HSD test. The intermediate statistics computed when applying the HSD test are usually not given.

It is becoming more common for researchers not to report summary tables in order to conserve journal space. This is problematic since considerable information is lost, especially when the strength of the relationship is not reported. Fortunately, there is a simple computational formula that will allow you to derive eta squared from the information typically provided. This is

$$Eta^2 = \frac{(df_{BETWEEN}) F}{(df_{BETWEEN}) F + df_{WITHIN}} \quad \text{[11.13]}$$

In the example on religion and ideal family size,

$$Eta^2 = \frac{(2)(7)}{(2)(7) + 18} = .44$$

11.6 / Relationship of the *F* Test to the *t* Test

The *F* test considered in this chapter is closely related to the *t* test in Chapter 9. The distinction between the two approaches is more a matter of historical rather than practical significance. For two group designs the *t* distribution bears a mathematical relationship to the *F* distribution such that

$$t^2 = F$$

One could apply all of the procedures developed in this chapter to the problems given in Chapter 9 and exactly the same conclusions would be drawn. The absolute value of the *t* scores computed in those problems would equal $\sqrt{F}$.

11.7 / Numerical Example

An important issue in court cases is the validity of eyewitness testimony. Several social scientists have suggested that the nature of eyewitness accounts can be influenced by many psychological factors. One of these is the way in which a question is phrased to an eyewitness. If subtle phrasing of questions can influence the type of answer an eyewitness gives, then such reports must be interpreted with considerable caution. Consider the following hypothetical experiment.

Fifteen individuals each watched a film of a car accident in which car A ran a stop sign and hit car B. Car A was traveling at a speed of 20 miles per hour. The entire incident was filmed, including arrival of a police officer and citation of the motorist who was in the wrong. After watching the film, each individual was asked to estimate the speed of car A at the moment of impact (subjects did not know the actual speed of the car). The question was asked in three different ways. The first five individuals were asked "How fast was car A going when it hit car B?" The second five individuals were asked "How fast was car A going when it crashed into car B?" The last five individuals were asked "How fast was car A going when it smashed into car B?" Thus, the question was identical for all three groups of individuals except for the verbs hit, crashed, or smashed. The question of interest is whether the speed estimate of the car varies as a function of the type of verb used in asking the question. The speed estimates are provided in Table 11.6.

The null hypothesis is that the type of verb will not influence the mean speed estimates. The alternative hypothesis is that the type of verb will influence the mean speed estimates. Stated formally;

$$H_0: \mu_H = \mu_C = \mu_S$$

H_1: The three population means are not all equal

TABLE 11.6 DATA FOR EXPERIMENT ON EYEWITNESS REPORTS

Hit	Crashed	Smashed
21	25	29
20	24	28
22	26	30
19	23	27
23	27	31
$\Sigma Y_H = 105$	$\Sigma Y_C = 125$	$\Sigma Y_S = 145$
$\bar{Y}_H = 21.0$	$\bar{Y}_C = 25.0$	$\bar{Y}_S = 29.0$

The null hypothesis states that there is no relationship between the type of verb used and speed estimates. The alternative hypothesis states that there is.

The first step in the analysis involves computing mean scores for the three groups for purposes of deriving the sum of squares. The mean score for the "hit" group was 21.0, for the "crashed" group it was 25.0, and for the "smashed" group it was 29.0. The grand mean across all fifteen individuals was 25.0.

The sum of squares total is computed in Table 11.7. The sum of squares total equals 190.0. The sum of squares between is computed using Equation 11.9. This first requires computing the treatment effects, where $T_j = \bar{Y}_j - G$:

$$T_H = 21 - 25 = -4.0$$

$$T_C = 25 - 25 = 0$$

$$T_S = 29 - 25 = 4.0$$

Then,

$$SS_{BETWEEN} = \sum_{j=1}^{k} T_j^2 \, n_j$$

$$= (-4^2)(5) + (0^2)(5) + (4^2)(5)$$

$$= 160$$

The sum of squares within is

$$SS_{WITHIN} = SS_{TOTAL} - SS_{BETWEEN}$$

$$= 190 - 160$$

$$= 30$$

The relevant degrees of freedom are defined by Equations 11.6 and 11.7:

$$df_{BETWEEN} = k - 1 = 3 - 1 = 2$$

$$df_{WITHIN} = N - k = 15 - 3 = 12$$

$$df_{TOTAL} = N - 1 = 15 - 1 = 14$$

TABLE 11.7 COMPUTATION OF SUM OF SQUARES TOTAL FOR EYEWITNESS REPORT EXPERIMENT

Y	$(Y - \overline{Y})$	$(Y - \overline{Y})^2$
21	-4	16
20	-5	25
22	-3	9
19	-6	36
23	-2	4
25	0	0
24	-1	1
26	1	1
23	-2	4
27	2	4
29	4	16
28	3	9
30	5	25
27	2	4
31	6	36
$\Sigma Y = 375$		$SS_{TOTAL} = 190$
$\overline{Y} = 25.0$		

The mean squares are

$$MS_{BETWEEN} = \frac{SS_{BETWEEN}}{df_{BETWEEN}} = \frac{160}{2} = 80$$

$$MS_{WITHIN} = \frac{SS_{WITHIN}}{df_{WITHIN}} = \frac{30}{12} = 2.5$$

The F ratio is

$$F = \frac{MS_{BETWEEN}}{MS_{WITHIN}} = \frac{80}{2.5} = 32.0$$

The summary table appears as follows:

Source	SS	df	MS	F
Between	160	2	80	32.0
Within	30	12	2.5	
Total	190	14		

The first question is whether there is a relationship between the type of verb used and the speed estimate. For an alpha level of .05, and with 2 and 12 degrees of freedom, the critical value of F is 3.88. The calculated value of F was 32.0. This exceeds the critical value, so we reject the null hypothesis and conclude that a relationship does exist between the two variables.

The second question concerns the strength of the relationship. This is indexed by eta squared and is

$$Eta^2 = 1 - \frac{SS_{WITHIN}}{SS_{TOTAL}} = 1 - \frac{30}{190} = .84$$

This represents a strong relationship and indicates that 84% of the variation in speed estimates is associated with the verb used in asking the question.

The final question concerns the nature of the relationship. This requires application of the HSD test. The CD statistic is derived using Equation 11.12:

$$CD = q \sqrt{\frac{MS_{WITHIN}}{n}}$$

$$= 3.77 \sqrt{\frac{2.5}{5}}$$

$$= 2.67$$

We then form a table:

Null Hypothesis Tested	Absolute Difference Between Sample Means	Value of CD	Is Null Hypothesis Rejected?
$\mu_H = \mu_C$	$|21 - 25| = 4$	2.67	Yes
$\mu_H = \mu_S$	$|21 - 29| = 8$	2.67	Yes
$\mu_C = \mu_S$	$|25 - 29| = 4$	2.67	Yes

All of the absolute mean differences exceed the value of 2.67. By inspection of the means we can conclude the following: Speed estimates obtained using the verb "smashed" are higher than speed estimates obtained using the verbs "crashed" or "hit." Further, speed estimates obtained using the verb "crashed" are higher than such estimates using the verb "hit."

The results of the experiment might be reported in a research report as follows:

A one-way analysis of variance was performed comparing the three group means. The F was statistically significant ($F = 32.0$, df $= 2, 12$, $p < .05$). The strength of the relationship, as indexed by eta squared, was .84. An HSD test indicated that the mean score for the "smashed" condition (29.0) was significantly higher, statistically, than the mean score for both the "crashed" condition (25.0) and the "hit" condition (21.0). Further, the mean score for the "crashed" condition was significantly higher than the mean score for the "hit" condition.

You should think about the above data in terms of basic experimental design questions. As stated in previous chapters, what potential confounding variables might be operating? What has been the role of disturbance variables in the experiment? What are the potential limitations of the experiment in terms of generalizability? These are all questions that are critical to drawing an appropriate conclusion from the study.

11.8 / Planning Investigations That Use One-Way Between-Subjects Analysis of Variance

When planning an investigation, Appendix F.2 can help you select the sample sizes that will achieve the desired level of power in a one-way between-subjects analysis of variance. Table 11.8 presents a portion of this Appendix for illustration. The first column presents different levels of power and the column headings are values of eta squared in the population. There is a different table for each experimental design involving different numbers of groups. Table 11.8 represents the case where there are three groups (hence df $= k - 1 = 2$) and an alpha level of .05. As an example,

Box 11.1 Reward versus Punishment

Educational psychologists have studied the effects of positive reinforcement (rewards) and punishment on learning. In the early 1900s, many educators and parents assumed that the threat of punishment or actual punishment itself was the most effective means for motivating children to learn. More recently, the dominant philosophy seems to be rewards as an incentive for learning. Hurlock (1925) reports a study in which fourth and sixth grade students were divided into four groups. Each group took an addition test in class on five successive days. One group, called the control group, was separated from the other students and told to work as usual. Another group was brought to the front of the room each day before the test was given and *praised* for its good work. The third group was also brought to the front of the room, but these students were *reproved* for their poor work. The students in the fourth group were *ignored*. Of interest to Hurlock were the scores students had on the final test of addition and whether the different incentives had different effects on test performance. Since students were randomly assigned to the four conditions, differences in group means would be expected to reflect the effects of the type of incentive on learning.

Hurlock's investigation was conducted before the technique of analysis of variance had been developed (1925). Kerlinger (1973), however, has analyzed the Hurlock data using the one-way analysis of variance technique. The means and summary table are:

	Group			
	Praised	*Reproved*	*Ignored*	*Control*
Mean Test Score	20.22	14.19	12.38	11.35

Source	*SS*	*df*	*MS*	*F*
Between	1260.06	3	420.02	10.08
Within	4249.29	102	41.66	
Total	5509.35	105		

The null hypothesis was rejected (F = 10.08, df = 3,102, p < .05), indicating that the type of incentive was related to test performance. The strength of the relationship, as indexed by eta squared, was .23. An HSD test indicated that the mean score for the praised group was significantly higher than the mean score for each of the other three groups. The difference in mean scores for the other groups, however, were not statistically significant. Thus, the most effective incentive in this experiment was that of reward.

if the desired power level is .80, and if the researcher suspects that the strength of the relationship in the population corresponds to an eta squared value of .15, then the number of subjects that should be sampled in each group is 19. Inspection of Table 11.8 and Appendix F.2 will give you a general appreciation for the relationship between power, sample size, alpha levels, and strength of effects for one-way between-subjects analysis of variance.

Degrees of Freedom = 2
Alpha = 0.05

Power	Population Eta Squared									
	0.01	0.03	0.05	0.07	0.10	0.15	0.20	0.25	0.30	0.35
0.10	22	8	5	4	3	2	2	2	—	—
0.50	165	55	32	23	16	10	8	6	5	4
0.70	255	84	50	35	24	16	11	9	7	6
0.80	319	105	62	44	30	19	14	11	9	7
0.90	417	137	81	57	39	25	18	14	11	9
0.95	511	168	99	69	47	30	22	16	13	11
0.99	708	232	137	96	65	41	29	22	18	14

11.9 / Summary

A one-way analysis of variance is typically used when (1) the dependent variable is quantitative in nature and is measured on approximately an interval level, (2) the independent variable is between-subjects in nature, and (3) the independent variable has three or more values. The essence of analysis of variance is the partitioning of variance into two types of variability, between-group variability and within-group variability. Between-group variability reflects the effects of the independent variable on the dependent variable (if there is an effect) as well as sampling error. Within-group variability reflects just sampling error. The ratio of variance measures (mean squares) based on these sources of variability is called the F ratio. The ratio has a sampling distribution that is distributed as an F distribution and that can be used to test for the existence of a relationship between the independent and dependent variables. The strength of the relationship is analyzed using eta squared, and the nature of the relationship is analyzed using the HSD test. The HSD test ensures that the probability of a type I error is equal to alpha.

APPENDIX 11.1

Degrees of Freedom

Unlike the sum of squares discussed in Chapter 3, the degrees of freedom associated with the sum of squares between is not $N - 1$. This is because the sum of squares between is based upon deviations of the group means from the grand mean (see Equation 11.9). These deviations are the treatment effects for each group. The treatment effects about the grand mean will always sum to zero. Accordingly, if all but one of the treatment effects are known, the unknown treatment effects is not free to vary. Given k groups, there are, thus, $k - 1$ degrees of freedom.

The sum of squares within is based on deviation scores about the group mean. Within each group, there will be $n - 1$ degrees of freedom. This follows from principles developed in Chapter 3. Given k groups, this yields $k(n - 1)$ degrees of freedom, which can be rewritten as $(kn - k)$. Since $N = kn$, the degrees of freedom within is $N - k$.

APPENDIX 11.2

Alternative Computational Formulas

This appendix presents computational formulas for computing the relevant sum of squares in a one-way between-subjects analysis of variance. The formulas are different than those in the text and are typical of computational approaches suggested in other statistics texts. Consider the following data for three groups ($k = 3$) of five ($n_1 = n_2 = n_3 = 5$) individuals each. The total number of subjects is $N = 15$.

A	B	C
7	10	3
6	8	2
4	9	1
3	7	2
5	6	2
$\Sigma = 25$	30	10
R_1	R_2	R_3

The various R represent the sum of scores in each column, respectively. Thus, $R_1 = 25$, $R_2 = 30$, $R_3 = 10$. If we square each score and then sum the columns, we obtain

A	B	C
49	100	9
36	64	4
16	81	1
9	49	4
25	36	4
$\Sigma = 135$	330	22
Q_1	Q_2	Q_3

The various Q represent the sum of the squared scores in each column, respectively. Thus, $Q_1 = 135$, $Q_2 = 330$, $Q_3 = 22$. Let

$$TS = \sum_{j=1}^{k} R_j$$

$$= 25 + 30 + 10$$

$$= 65$$

We need to compute three preliminary terms. They are

$$(I) = \frac{TS^2}{N} = \frac{65^2}{15} = 281.67$$

$$(II) = \sum_{j=1}^{k} Q_j = 135 + 330 + 22 = 487$$

$$(III) = \sum_{j=1}^{k} \left(\frac{R_j^2}{n_j} \right) = \frac{25^2}{5} + \frac{30^2}{5} + \frac{10^2}{5} = 325$$

Then,

$$SS_{BETWEEN} = (III) - (I) = 325 - 281.67 = 43.33$$

$$SS_{WITHIN} = (II) - (III) = 487 - 325 = 162.00$$

$$SS_{TOTAL} = (II) - (I) = 487 - 281.67 = 205.33$$

The various MS and the F ratios are computed from the above using procedures described in the text.

EXERCISES

Exercises to Review Concepts

1. Under what conditions is the one-way analysis of variance typically used to analyze a bivariate relationship?

2. What are the major assumptions underlying the analysis of variance technique in terms of inferring that a relationship exists between two variables?

3. Determine the critical value of F for each of the following conditions in which a one-way analysis of variance is to be computed and where k equals the number of groups or values of the independent variable.

(a) $k = 3, N = 20$

(b) $k = 4, N = 20$

(c) $k = 3, n_1 = 10, n_2 = 10, n_3 = 10$

(d) $k = 5, n_1 = 10, n_2 = 10, n_3 = 10, n_4 = 10$
$n_5 = 10$

4. An experiment was conducted testing the effects of four types of drugs, A, B, C, and D, on learning. Eighty subjects were randomly assigned to one of four conditions, twenty in group A, twenty in group B, twenty in group C, and twenty in group D. All were injected with a drug and then participated in a learning task. Complete the missing entries in the summary table.

Source	SS	df	MS	F
Between	—	—	—	—
Within	152	—	—	
Total	182	—		

5. An investigator wanted to test the effects of marital status on attitudes toward divorce. Thirty individuals were interviewed, ten single, ten married, and ten divorced. An attitude scale was administered to each person measuring their attitude toward divorce. Scores could range from 1 to 12 with higher scores representing more positive attitudes. The means and summary table are as follows:

$$\bar{X}_S = 6.0 \qquad \bar{X}_M = 8.0 \qquad \bar{X}_D = 10.0$$

Source	SS	df	MS	F
Between	80	2	40	8.0
Within	135	27	5	
Total	215	29		

Analyze the relationship between these variables as completely as possible and write out the results of your analysis using the procedures discussed in the Method of Presentation sections.

6. For four groups, A, B, C, and D, list the pairs of means you would test if you were doing an HSD test.

7. Suppose that one of "Nader's Raiders" was interested in comparing the performance of three types of cars, X, Y, and Z. Random samples of five owners were drawn from the list of owners of each model. These owners were asked how many times their cars had undergone major repairs in the past two years. The following data resulted

X	Y	Z
2	5	9
1	4	6
2	3	3
3	4	7
2	4	5

(a) Compute the treatment effects for the groups defining the qualitative variable.

(b) Analyze the relationship between the independent and dependent variables as completely as possible.

8. A study was conducted in which the results of four separate groups (groups A, B, C, and D) were analyzed as a one-way analysis of variance. Complete the remainder of the summary table for this analysis.

Source	SS	df	MS	F
Between	—	—	18	3.60
Within	—	—	—	
Total	—	23		

(a) How many subjects participated in the study?

(b) What proportion of variance in the dependent variable could *not* be explained by the independent variable?

9. What does a "mean square" represent in an analysis of variance?

10. What does the "grand mean" refer to?

11. If the $SS_{TOTAL} = 130$, and $SS_{BETWEEN} = 100$, what must SS_{WITHIN} equal?

12. Consider the following scores in an experiment involving three groups, A, B, and C, with five subjects in each group.

A	B	C
3	5	7
3	5	7
3	5	7
3	5	7
3	5	7

Without formally computing the sum of squares within, what must its value be, based upon inspection of the above scores? Why?

Consider the following scores in which two groups of subjects were tested under one of two conditions, A or B.

A	B
3	9
4	10
5	11
6	12
7	13

13. Analyze these data using the procedures developed in Chapter 9 for the independent groups t test (that is, answer all three questions concerning a relationship).

14. Analyze these data using the procedures developed in the present chapter. (Note: Even though analysis of variance is typically applied in cases where there are three or more groups, for this example, go ahead and use it for analyzing two groups.)

15. Compute the square root of the value of F computed in Exercise 14. Compare this with the value of t computed in Exercise 13. Compute the square root of the critical value of F in Exercise 14. Compare this with the critical value of t in Exercise 13. Compare your values of eta squared in Exercises 13 and 14.

16. Why does within-group variability reflect sampling error only?

For each of the following experiments, indicate the type of statistical technique that would be used to analyze the relationship between the independent and dependent variables (assume that the underlying assumptions of the statistical tests have been satisfied).

17. *Experiment I*. Barron (1965) compared the average intelligence scores for creative individuals in different occupations. Four groups of individuals were administered an intelligence test, creative mathematicians, creative writers, creative psychologists, and creative architects. The mean intelligence scores were compared for the four groups.

    (a) independent groups t test

    (b) correlated groups t test

    (c) one-way analysis of variance

18. *Experiment II*. Steiner (1972) has reviewed studies on the effects of the presence of others on problem-solving performance. In one study, a group of one hundred individuals was used in the experiment. At the first session, each subject was seated in a room alone and given a math problem to solve. The amount of time it took the subject to solve the problem was measured. Two weeks later, the subject returned and solved another problem, but this time there was an observer present watching her. The amount of time it took to solve the problem was again measured. For each of these two conditions, observer present versus observer absent, the mean problem-solving time was computed across the one hundred subjects. These means were then compared.

    (a) independent groups t test

    (b) correlated groups t test

    (c) one-way analysis of variance

19. *Experiment III*. An investigator was interested in the effects of practice on problem-solving performance. A group of 150 individuals participated in the study. Subjects were randomly assigned to two conditions. Seventy-five subjects tried to solve thirty problems after having a ten minute practice session on a similar set of problems. The other seventy-five subjects tried to solve the same thirty problems but with no practice session. The number of correct problems was computed for each subject. The average number of correct solutions was compared for people in the practice condition with those in the no-practice condition.

(a) independent groups t test

(b) correlated groups t test

(c) one-way analysis of variance

20. An investigator planned an experiment using four groups in a one-way between-subjects analysis of variance. For an alpha level of .05 and a suspected population eta squared of .10, how many subjects should be sampled for each group in order to achieve power of .70?

Exercises to Apply Concepts

21. Of recent interest to psychologists has been the study of jury decision making and factors influencing judgments of guilt on the part of jurors. Stephen (1975) has reviewed a number of studies that examine the effects of a defendant's race on guilt verdicts. In one experiment, subjects were presented a transcript of a trial and asked to indicate the probability the defendant was guilty. Judgments were made on a 10 point scale where $0 =$ definitely not guilty through $10 =$ definitely guilty. The higher the number, the higher the probability of guilt. All subjects read the same transcript. However, one-third of the subjects were told that the defendant was white, another third were told the defendant was black, and the other third were told the defendant was Mexican-American. The hypothetical data presented below are representative of the outcome of this experiment. Analyze these data, draw a conclusion, and write out the results of the experiment using procedures discussed in the Method of Presentation section.

White Defendant	Black Defendant	Mexican-American Defendant
6	10	10
7	10	6
2	9	10
3	4	5
5	4	10
0	10	5
1	10	2
0	10	10
6	3	2
0	10	10

22. Ainsworth et. al. (1978) have studied how infants become psychologically attached to their mothers. They have delineated three qualitatively different types of attachment, which Sroufe and Waters (1977) have labeled "security," "avoidance," and "ambivalence." Infants who are securely attached, in general, seek to be near their mother and have contact with her. These infants, when separated from their mother, may or may not exhibit distress and are generally not anxious about being left alone. Infants who are "avoidant" exhibit few tendencies toward maintaining proximity and contact with the mother. Finally, infants who are "ambivalent" display both contact- and proximity-seeking behaviors as well as contact- and interaction-resisting behaviors (hence the term "ambivalent"). When separated from the mother, these infants tend to exhibit anger or become conspicuously passive until the mother returns.

In one experiment, Ainsworth was interested in studying the relationship between the type of attachment exhibited by an infant and maternal behavior. Trained experimenters observed interactions between mothers and infants and rated these interactions between mothers and infants on a number of dimensions. One of these was the extent to which the mother is sensitive or insensitive to the infant's signals and communications. Ratings were made on a scale from 1 to 9 with lower numbers indicating lesser degrees of sensitivity. The hypothesis was that the type of attachment exhibited by an infant may be associated with different levels of sensitivity on the part of the mother. It should be noted that if an association is observed, the causal link between the variables is ambiguous. It may be the case that maternal sensitivity influences the type of attachment relationship an infant forms with its mother. Or it may be that the infant's behavior influences the mother's sensitivity to it. Hypothetical data for this experiment are presented below and are representative of the results of Ainsworth. Analyze these data, draw a conclusion, and write out the results of the study using procedures discussed in the Method of Presentation section.

Security	Avoidance	Ambivalence
5	1	3
9	5	3
9	1	1
5	1	1
9	5	3
5	5	1

12 One-Way Repeated Measures Analysis of Variance

12.1 / Use of One-Way Repeated Measures Analysis of Variance

One-way repeated measures analysis of variance is typically used to analyze the relationship between two variables when

1. the dependent variable is quantitative in nature and is measured on approximately an interval level;

2. the independent variable is within-subjects in nature (it can be either qualitative or quantitative); and

3. the independent variable has three or more values.

The repeated measures analysis of variance is used under the same conditions as the one-way analysis of variance (Chapter 11), except that the independent variable is within-subjects in nature rather than between-subjects in nature. It is an extension of the correlated groups *t* test (Chapter 10), but the independent variable has more than two values. Let us consider an example of an experiment that meets the conditions for a one-way repeated measures analysis of variance.

A consumer psychologist is interested in the effects of brand labels on the perceived taste of wine. He designs an experiment with three conditions. In each condition, an individual tastes a wine, and then rates it for its taste. The rating is made on a 20 point scale, with higher scores indicating that the wine tasted better. In all three conditions the wine is identical. However, in condition A the individual is told it is a French wine, in condition B the individual is told it is an Italian wine, and in condition C the individual is told it is an American wine. The experiment is conducted as a within-subjects design. An individual tastes and rates the wine in condition A, then in condition B, and then in condition C. In order to counteract possible order effects, the type of label given first, second, and third is randomized for each subject. A total of six individuals participated in the experiment. The taste ratings are presented in the second, third, and fourth columns of Table 12.1.

TABLE 12.1 DATA FOR WINE LABEL EXPERIMENT

Indi-vidual	French	Italian	Amer-ican	$\overline{Y}_i$	French	Italian	Amer-ican
					Adjusted Data		
1	15	11	10	12.00	16	12	11
2	16	12	11	13.00	16	12	11
3	17	13	12	14.00	16	12	11
4	14	12	14	13.33	13.67	11.67	13.67
5	16	13	11	13.33	15.67	12.67	10.67
6	18	11	8	12.33	18.67	11.67	8.67
	$\Sigma Y = 96$	72	66	$G = 13.00$	$\Sigma Y = 96$	72	66
	$\overline{Y} = 16$	12	11		$\overline{Y} = 16$	12	11

In this experiment, the independent variable is the type of label, and it has three values (French, American, Italian). It is within-subjects in nature since the same subjects appear in all three conditions. The dependent variable is the perceived quality of the wine. It represents a quantitative variable.

The logic of one-way repeated measures analysis of variance has already been developed in the context of Chapters 10 and 11. In Chapter 10, we developed the idea of extracting variability due to individual differences and noted that an advantage of a within-subjects design is its ability to systematically remove this source of variability from the statistical analysis. A procedure was presented that allowed us to generate adjusted data with the effects of individual differences removed. We then applied the procedures of the independent groups t test to the adjusted data to yield the results of the correlated groups t test. The same can be done here. We use the variance extraction procedures developed in Chaper 10 to generate a set of adjusted data. Then, we apply the procedures of one-way analysis of variance (Chapter 11) to the adjusted data to yield the results for the repeated measures analysis of variance. The only differences occur in how the results are reported and in the estimate of degrees of freedom for the within-group variability. Since it is unnecessary to develop the logic of variance extraction and one-way analysis of variance again, we will turn directly to the mechanics of calculation, using the wine tasting experiment.

12.2 / Extracting the Influence of Individual Differences

The first step involves computation of a grand mean for purposes of extracting individual difference variability. The sum of all eighteen scores in the wine tasting experiment is 234, and the grand mean is 234/18 = 13.0. Next, we compute the mean score across conditions for each subject. This has been done in column 5 of Table 12.1. Note that there is variability in the mean scores across the six subjects. Some subjects, on the average, tended to rate the wines higher than other subjects. This presumably reflects the influence of differences in background of the individuals and we want to remove this influence (which serves as a disturbance variable) in order to obtain a more sensitive statistical test of the relationship between the independent and dependent variables. The mean score across conditions for the first subject was 12.0. The grand mean is 13.0 and, thus, the effect of the individual's background was to lower his scores, on the average, 1 unit from the grand mean (12 − 13 = −1). To nullify this effect, we add 1 unit to his three scores in each condition. This had been done in columns 6, 7, and 8 of Table 12.1. This process is performed for each subject. The second subject had a mean score across conditions of 13.0. This is identical to the grand mean and the background of this individual had no effect relative to the grand mean (13 − 13 = 0). The adjusted scores are therefore the same as the original scores for this subject. For the third subject, the mean score across conditions was 14.0. The effect of this individual's background was to raise his scores, on the average, 1 unit above the grand mean (14 − 13 = 1). To nullify the effect, we subtract 1 from each score obtained by this individual. And so on.

The adjusted data appear in the last three columns of Table 12.1. The data represent each subject's score on the dependent variable with the effects of the individual's background removed. The data are now analyzed using the same procedures that were developed in Chapter 11 for the one-way analysis of variance.

12.3 / Inference of a Relationship Using One-Way Repeated Measures Analysis of Variance

Null and Alternative Hypotheses. The null hypothesis is that the type of label will not influence the perceived taste of the wine. The alternative hypothesis states that there is a relationship between the two variables. Stated formally, the hypotheses are

$$H_0: \mu_F = \mu_I = \mu_A$$

H_1: The three population means are not all equal

Calculation of the Summary Table. The first step requires calculating the three sums of squares (total, between, and within). The sum of squares total is computed using Equation 3.9. The calculations are shown in Table 12.2. We find that

$$SS'_{TOTAL} = 110$$

We have put a ′ by this to indicate we are working with the adjusted data and not the original data.

TABLE 12.2 CALCULATION OF SS'_{TOTAL}

Y	$Y - \overline{Y}$	$(Y - \overline{Y})^2$
16	3.0	9.0
16	3.0	9.0
16	3.0	9.0
13.67	.67	.45
15.67	2.67	7.13
18.67	5.67	32.15
12	−1.0	1.0
12	−1.0	1.0
12	−1.0	1.0
11.67	−1.33	1.77
12.67	−.33	.11
11.67	−1.33	1.77
11	−2.0	4.0
11	−2.0	4.0
11	−2.0	4.0
13.67	.67	.45
10.67	−2.33	5.43
8.67	−4.33	18.75

$\Sigma Y = 234.03$ $SS'_{TOTAL} = 110.00$

$\overline{Y} = 13.00$

The sum of squares between is derived using Equation 11.9. This first requires computing the treatment effects, where $T_j = \overline{Y}_j - G$:

$$T_F' = 16 - 13 = 3$$

$$T_I' = 12 - 13 = -1$$

$$T_A' = 11 - 13 = -2 \qquad .$$

Then,

$$SS'_{BETWEEN} = \sum_{j=1}^{k} T_j^2 n_j$$

$$= (3^2)(6) + (-1^2)(6) + (-2^2)(6)$$

$$= 84$$

The sum of squares within is

$$SS'_{WITHIN} = SS'_{TOTAL} - SS'_{BETWEEN}$$

$$= 110 - 84$$

$$= 26$$

The degrees of freedom are different (at least for the within group and total sum of squares) from those presented in Chapter 11. The relevant formulas are

$$df'_{BETWEEN} = k - 1 \qquad \text{[12.1]}$$

$$df'_{WITHIN} = (n - 1)(k - 1) \qquad \text{[12.2]}$$

$$df'_{TOTAL} = n(k - 1) \qquad \text{[12.3]}$$

where k is the number of groups or values of the independent variable (in this case, $k = 3$) and n is the number of *subjects* in the experiment (in this case, $n = 6$). The logic for the formulas is given in Appendix 12.2 at the end of the chapter. We find that

$$df'_{BETWEEN} = 3 - 1 = 2$$

$$df'_{WITHIN} = (6 - 1)(3 - 1) = 10$$

$$df'_{TOTAL} = 6(3 - 1) = 12$$

The mean squares are

$$MS'_{BETWEEN} = \frac{SS'_{BETWEEN}}{df'_{BETWEEN}} = \frac{84}{2} = 42$$

$$MS'_{WITHIN} = \frac{SS'_{WITHIN}}{df'_{WITHIN}} = \frac{26}{10} = 2.6$$

The F ratio is

$$F = \frac{MS'_{BETWEEN}}{MS'_{WITHIN}} = \frac{42}{2.6} = 16.15$$

Because we are working with adjusted data, different labels are typically used in journal reports for the sources of variability. The labels traditionally used are

$$SS'_{TOTAL} = SS_{WITHIN\text{-}SUBJECTS}$$

$$SS'_{BETWEEN} = SS_{TREATMENTS}$$

$$SS'_{WITHIN} = SS_{ERROR}$$

You should take special care not to confuse **sum of squares within-subjects** with the sum of squares within developed in Chapter 11. The key is the word "subjects" after the word "within." When the term *within-subjects* appears, this refers to the total variability in adjusted data. The term **sum of squares treatments** is similar to between-group variability but with adjusted data, and the term **sum of squares error** is similar to within-group variability, but with the adjusted data. The summary table would appear as follows:

Source	SS	df	MS	F
Treatments	84	2	42.0	16.15
Error	26	10	2.6	
Within-Subjects	110	12		

The test of the null hypothesis involves comparing the observed F ratio (16.15) with a critical F. For an alpha level of .05, and with 2 and 10 degrees of freedom, the critical value of F is 3.15. The observed value of F exceeds the critical value and the null hypothesis of no relationship is rejected.

Assumptions of the F Test. The validity of the above F test rests on three major assumptions:

1. *Homogeneity of within-group variance* (as discussed in Chapter 11). Given k conditions, it is assumed that

$$\sigma_1^2 = \sigma_2^2 = \ldots \sigma_k^2$$

2. *Normality* (as discussed in Chapter 11). The distribution of scores within each of the populations of interest must be normally distributed.

3. *Homogeneity of covariances.* This assumption concerns the extent to which scores in one condition are related to scores in another condition. The covariance is a statistical index of the relationship between two quantitative variables. (The concept of covariance is discussed in Chapter 13.) The assumption essentially states that the covariance (relationship) between population scores in any two conditions must be the same as the covariance (relationship) between population scores for any other two conditions. For example, the covariance (relationship) between subjects' scores in the French and Italian conditions must be the same as the covariance (relationship) between subjects' scores in the French and American condition and between scores in the Italian and American conditions.

Procedures for testing the third assumption are complex and discussed in Myers (1972, p. 177). The F test is relatively robust to violations of the first two assumptions, especially with larger sample sizes ($n > 15$). Violations of the third assumption appear to be more serious and statisticians have developed alternative methods of analysis that are more appropriate when this assumption is violated (see McCall and Appelbaum, 1973; and Collier et. al., 1967).

12.4 / Strength of the Relationship

The strength of the relationship in the sample data is computed using the standard formula for eta squared (see Equation 11.11).

$$\text{Eta}^2 = 1 - \frac{\text{SS}'_{\text{WITHIN}}}{\text{SS}'_{\text{TOTAL}}}$$

$$= 1 - \frac{26}{110}$$

$$= .76$$

This represents a relatively strong relationship between the independent and dependent variables. Remember, however, that this eta squared is the proportion of variability in the dependent variable that is associated with the independent variable *after the effects of individual background have been removed.*

12.5 / Nature of the Relationship

The nature of the relationship is addressed using the HSD test, as discussed in Chapter 11. The first step is to compute the value of the CD statistic, using Equation 11.12.

$$CD = q \sqrt{\frac{MS'_{WITHIN}}{n}}$$

$$= 3.88 \sqrt{\frac{2.6}{6}} = 2.56$$

We then form a table:

Null Hypothesis Tested	Absolute Difference Between Sample Means	Value of CD	Is Null Hypothesis Rejected?
$\mu_F = \mu_I$	$\|16 - 12\| = 4$	2.56	Yes
$\mu_F = \mu_A$	$\|16 - 11\| = 5$	2.56	Yes
$\mu_I = \mu_A$	$\|12 - 11\| = 1$	2.56	No

The nature of the relationship is such that the wine with a French label is perceived as tasting better than the identical wine with either an American or Italian label. The difference between a wine with an American label and an Italian label, however, is not statistically significant.

Method of Presentation

The method of presentation for a one-way repeated measures analysis of variance is identical to that for a one-way analysis of variance, except that the labels used in the summary table are different. The labels in the case of the repeated measures analysis were given earlier in the chapter. Sometimes this type of analysis will be referred to as a "within-subjects analysis of variance" instead of a "repeated measures analysis of variance."

The results of the experiment on labels and the taste of wine might appear in a research report as follows:

A one-way repeated measures analysis of variance was performed relating the type of label used (French versus Italian versus American) and perceived quality of taste. Table 1 presents the summary table for the analysis. The observed F ratio was statistically significant ($F = 16.15$, df $= 2, 10$, $p < .05$). An HSD test revealed that the "French" wine ($\bar{Y} = 16.0$) was rated significantly higher than both the "Italian" wine ($\bar{Y} = 12.0$) and the "American" wine ($\bar{Y} = 11.0$). However, the difference in ratings for the "Italian" and "American" wines was nonsignificant. The strength of the relationship between type of label and taste quality was .76, as indexed by eta squared.

The summary table presented earlier would accompany the report.

12.6 / Planning an Investigation Using One-Way Repeated Measures Analysis of Variance

When planning an investigation, the selection of sample sizes for purposes of achieving desired levels of power in one-way repeated measures analysis of variance can be facilitated through the use of Appendix F.3. Table 12.3 presents a portion of the Appendix for purposes of illustration. The first column presents different levels of power and the column headings are values of eta squared in the population. It should be noted that these values of eta squared are conceptualized as the proportion of variation in the dependent variable associated with the independent variable *after the effects of individual background are removed*. There is a different table for each experimental design involving different numbers of experimental conditions. Table 12.3 represents the case where there are three groups (hence, df $= k - 1 = 2$) and an alpha level of .05. As an example, if the desired power level is .80, and if the researcher suspects that the strength of the relationship in the population corresponds to an eta squared of .15, then the number of subjects that should be used in the study is 28. Inspection of Table 12.3 and Appendix F.3 will give the reader a general appreciation for the relationship between power, sample size, alpha levels, and strength of effects for one-way repeated measures analysis of variance.

TABLE 12.3 SAMPLE SIZES NECESSARY TO ACHIEVE SELECTED LEVELS OF POWER FOR ALPHA $= .05$ AND $K = 3$ AS A FUNCTION OF POPULATION VALUES OF ETA SQUARED

Degrees of Freedom = 2
Alpha Level = 0.05

Power	0.01	0.03	0.05	0.07	0.10	0.15	0.20	0.25	0.30	0.35
0.10	32	11	7	5	4	3	2	2	2	2
0.50	247	81	48	34	23	15	11	8	7	6
0.70	382	125	74	52	36	23	16	13	10	8
0.80	478	157	93	65	44	28	20	15	12	10
0.90	627	206	121	85	58	37	26	20	16	13
0.95	765	251	148	104	70	45	32	24	19	15
0.99	1060	347	204	143	97	62	44	33	26	21

The column headings above Power (0.01 ... 0.35) are values under *Population Eta Squared*.

12.7 / Summary

A one-way repeated measures analysis of variance is typically used when (1) the dependent variable is quantitative in nature and is measured on approximately an interval level, (2) the independent variable is within-subjects in nature, and (3) the independent variable has three or more groups or values. The technique combines the logic of extracting variability due to individual differences (Chapter 10) and that of the one-way analysis of variance (Chapter 11).

APPENDIX 12.1

Computational Formulas

This section presents formulas typically given in textbooks for purposes of computing from raw data the statistics reported in this chapter. We will use as an example the following data in which there are three conditions ($k = 3$), A, B, and C, and five subjects who participated in each condition (that is, $N = 5$).

Individual	A	B	C	Total
1	7	10	3	$20 = P_1$
2	6	8	2	$16 = P_2$
3	4	9	1	$14 = P_3$
4	3	7	2	$12 = P_4$
5	5	6	2	$13 = P_5$
$\Sigma = 25$	30	10	$75 = TS$
	R_1	R_2	R_3	

The various R represent the sum of scores in each condition. Thus, $R_1 = 25$, $R_2 = 30$, $R_3 = 10$. The various P represent the sum of scores across conditions for a given subject. We need to compute four preliminary terms. They are

$$(I) = \frac{TS^2}{kN} = \frac{75^2}{(3)(5)} = 375$$

$$(II) = \text{The sum of all scores squared} = \Sigma Y^2 = 487$$

$$(III) = \frac{\sum_{j=1}^{k} R_j^2}{N} = \frac{25^2 + 30^2 + 10^2}{5} = 465$$

$$(IV) = \frac{\sum_{i=1}^{N} P_i^2}{k} = \frac{20^2 + 16^2 + 14^2 + 12^2 + 13^2}{3}$$

$$= 291.25$$

From these, it is possible to compute the three sum of squares:

$$SS_{\text{WITHIN-SUBJECTS}} = (II) - (IV) = 487 - 291.25$$

$$= 195.75$$

$$SS_{\text{TREATMENTS}} = (III) - (I) = 465 - 375 = 90$$

$$SS_{\text{ERROR}} = (II) - (III) - (IV) + (I)$$

$$= 487 - 465 - 291.25 + 375$$

$$= 105.75$$

The degrees of freedom, mean squares, and F ratio are computed using the formulas presented in this chapter. The summary table would appear as follows:

Source	SS	df	MS	F
Treatments	90.00	2	45.0	3.40
Error	105.75	8	13.22	
Within-Subjects	195.75	10	19.58	

The computation of eta squared and the HSD test would follow procedures outlined in the text.

APPENDIX 12.2

Degrees of Freedom

We will develop the logic of the various degrees of freedom using an experiment with five subjects who participated in each of two conditions, A and B. Although this would dictate a correlated groups t test, the logic is identical and more easily understood using the case where $k = 2$.

The degrees of freedom for the sum of squares within subjects is defined as

$$df_{\text{WITHIN-SUBJECTS}} = N(k - 1)$$

where k is the number of conditions. In our example, there are five subjects in each of two conditions and hence the degrees of freedom are $(5)(2 - 1) = 5$. This means that knowing five of the adjusted scores, the remaining five are not "free to vary" and can be readily derived. This is due to a statistical property of these scores. When the individual difference variability is removed, a subject's score in condition A must deviate the same number of units from the grand mean, G, as that same subject's score in condition B, but in opposite directions. The implications of this are that knowing a subject's score in one condition allows us to specify what the score is in the other. Thus, for each subject there is one degree of freedom; for all subjects, there is a total of $N(k - 1)$ degrees of freedom.

The degrees of freedom for $SS_{\text{TREATMENTS}}$ is defined as

$$df_{\text{TREATMENTS}} = k - 1$$

The logic for this has already been developed in the context of Appendix 11.1. In our example, the degrees of freedom are $2 - 1 = 1$.

The degrees of freedom for SS_{ERROR} is defined as

$$df_{\text{ERROR}} = (N - 1)(k - 1)$$

This formula for the degrees of freedom is different from the df error discussed in Chapter 11. This is because of the difference in the between- and within-subjects designs. It is based on the symmetry of adjusted scores about the grand mean as discussed above. In this context, it was noted that knowing five scores allows us to derive the other five. However, an additional statistical property reduces the degrees of freedom even more in the case of the error term. This is because the error scores within a condition must sum to zero (see Chapters 3 and 11 for the logic underlying this property). Knowing four scores within a group allows us to derive the fifth, and knowing these five scores in one condition allows us to derive the remaining five. In our example, then, there are four degrees of freedom. Applying the above equation, $df_{\text{ERROR}} = (5 - 1)(2 - 1) = (4)(1) = 4$. The logic underlying these degrees of freedom is easily generalized to the case of more than two conditions.

EXERCISES

Exercises to Review Concepts

1. Under what conditions is the one-way repeated measures analysis of variance typically used to analyze a bivariate relationship?

2. What are the major assumptions underlying the one-way repeated measures analysis of variance in terms of inferring that a relationship exists between two variables?

3. A researcher investigated the effects of three types of drugs on learning. Thirty subjects were injected with drug A and then given a learning task. One month later, the same subjects were injected with drug B and given a similiar learning task. Finally, the same subjects were injected with drug C one month later and again given the learning task. Complete the summary table for the experiment.

Source	SS	df	MS	F
Treatments	20	—	—	—
Error	116	—	—	
Within-Subjects	—	—		

4. For a repeated measures analysis of variance with $N = 21$ and five conditions, what would the degrees of freedom treatments, degrees of freedom error, and degrees of freedom within-subjects be?

5. Determine the critical value of F for each of the following situations in which a one-way repeated measures analysis of variance is to be computed and where k = the number of conditions or values of the independent variable.

    (a) $k = 3, N = 20$

    (b) $k = 4, N = 20$

    (c) $k = 3, N = 30$

    (d) $k = 5, N = 60$

6. For the following experiment in which five individuals participated in each of three conditions (A, B, and C), extract the effects of individual differences due to background (that is, generate a matrix of adjusted scores).

Subject	Condition A	Condition B	Condition C
1	5	7	9
2	9	7	5
3	4	5	6
4	6	5	4
5	6	6	6

7. Compute the mean scores for each condition in the original data and the mean score for each condition in the adjusted data. Why are the corresponding means the same in the respective data sets?

8. For a repeated measures analysis of variance, complete the missing entries in the summary table. There are three different values (groups) of the independent variable.

Source	SS	df	MS	F
Treatments	—	—	20	10
Error	—	—	—	
Within-Subjects	—	20		

Consider the following scores on an anxiety test that was administered to a group of subjects under three conditions, A, B, and C.

A	B	C
2	3	4
3	4	8
4	5	6
5	6	7
6	7	5

9. Analyze the data as if the independent variable was between-subjects in nature (that is, answer all three questions about a relationship).

10. Analyze the data as if the independent variable was within-subjects in nature (that is, answer all three questions about a relationship).

11. Compare your results in Exercise 9 with your results in Exercise 10. How does this illustrate the advantage of within-subjects experimental designs?

For each of the following experiments, indicate the type of statistical technique that would be used to analyze the relationship between the independent and dependent variables (assume that the underlying assumptions of the statistical tests have been satisfied).

12. *Experiment I.* Kelman and Hovland (1953) studied the effects of the source of a persuasive communication on attitude change. Three groups of subjects listened to a persuasive message on the treatment of juvenile delinquents. For one group of subjects, the message was attributed to a "positive" source (trustworthy, well-informed), for another group it was attributed to a "negative" source (untrustworthy, poorly informed), and for the third group it was attributed to a "neutral" source. The amount of attitude change was computed for each individual by subtracting indices of attitude after hearing the message from indices of attitude before hearing the message. The mean change scores for the three groups were compared.

    (a) independent groups t test

    (b) correlated groups t test

    (c) one-way analysis of variance

    (d) one-way repeated measures analysis of variance

13. *Experiment II*. A researcher wanted to test whether children differed in their memory capacities at the age of five months versus seven months. A group of forty infants was given a memory task that yielded scores from 1 to 15, with higher scores indicating better memory. The forty infants were given the test at the age of five months and then again at the age of seven months. The mean scores on the test were compared.

    (a) independent groups *t* test

    (b) correlated groups *t* test

    (c) one-way analysis of variance

    (d) one-way repeated measures analysis of variance

14. *Experiment III*. An investigation was undertaken to determine the effects of different types of drugs on driving skills. Thirty subjects participated in a driving simulation task under each of three conditions: (1) while under the influence of small amounts of alcohol, (2) while under the influence of small amounts of marijuana, and (3) while not under the influence of any drug. The simulation task yielded a "driving score," ranging from 1 to 100, with higher scores indicating better driving skills. The mean driving scores were compared for each of the three conditions.

    (a) independent groups *t* test

    (b) correlated groups *t* test

    (c) one-way analysis of variance

    (d) one-way repeated measures analysis of variance

15. *Experiment IV*. Results of surveys are often summarized to the public in a format as follows: "eight out of ten doctors recommend. . ." A researcher was interested in whether the base in which such information was presented influenced its impact on consumer preferences. Two groups of individuals were presented with identical descriptions of an aspirin. However, in one group, subjects were told that "eight out of ten doctors recommended the aspirin," whereas in the other group subjects were told that eighty out of one hundred doctors recommended the aspirin." Although the proportion of doctors recommending the aspirin was identical in both groups, the number of doctors differed. For all subjects a measure of attitude toward the aspirin was obtained on a scale from 1 to 10, with higher scores indicating more favorable attitudes. The mean attitude scores were compared for the two groups.

    (a) independent groups *t* test

    (b) correlated groups *t* test

    (c) one-way analysis of variance

    (d) one-way repeated measures analysis of variance

16. How is the interpretation of eta squared in a repeated measures analysis of variance different from that in a one-way between-subjects analysis of variance?

Exercises to Apply Concepts

17. Population growth trends have a dramatic impact on society, both economically and politically. Social scientists have studied extensively factors that influence population growth. One area of research has been the individual's decision to use birth control. Numerous studies have examined the factors that females take into consideration when choosing a birth control method. Relatively few studies have examined birth control from the standpoint of the male. Probably, male oral contraceptives will be available to the general public in the next ten years. Jaccard (1980) conducted a study to discover those factors that males would be most concerned about when evaluating male oral contraceptives. In this experiment, subjects rated how important each of four factors would be to them in deciding whether to use an oral contraceptive. The ratings were made on a 21 point scale, with progressively higher numbers indicating higher degrees of importance. The data presented below are representative of those obtained by Jaccard. Analyze these data, draw a conclusion, and write up your results using the principles given in the Method of Presentation section.

Individual	Health Risks	Effectiveness	Cost	Convenience
1	20	12	8	8
2	19	15	11	11
3	18	14	10	10
4	17	13	9	9
5	16	16	12	12

18. It has recently been established that many rape victims fail to report their victimization to the police or other public authorities. A recent survey in five large metropolitan areas estimated that only about 50% of most crimes —including rape—are ever reported to the police. Feldman-Summers and Ashworth (1980) conducted an experiment to identify factors that influence a potential rape victim's intention to report her sexual victimization to various agencies and individuals. As part of this study, individuals were asked the likelihood that they would report a rape to the police, her husband/boyfriend/lover, her parents, and a female friend. A separate probability judgment was obtained for each source. The judgment was made on a 7 point scale, with 1 = "very unlikely I would report the crime to the source," through 7 = "very likely I would report the crime to the source." The hypothetical data for this study are presented below and are representative of the experimental outcome. Analyze these data, draw a conclusion, and write up your results using the principles given in the Method of Presentation section.

Individual	Husband/ Boyfriend	Police	Parent	Female Friend
1	7	6	3	4
2	6	5	4	5
3	7	6	1	2
4	6	5	6	7
5	7	6	3	4
6	6	5	4	5

13

Pearson Correlation

13.1 / Use of the Correlation Coefficient

The statistical technique considered in the present chapter, called the **Pearson product-moment correlation coefficient** (or correlation coefficient), is used to analyze the relationship between two quantitative variables. Each variable must be measured on approximately an interval level and the observations on a given variable must be independent. Suppose we were studying the relationship between how traditional an individual is and her ideal family size. For each individual in the study, we can obtain a measure of her traditionalism and her ideal family size. We might find the following:

Individual	Ideal Family Size	Traditionalism
1	3	5
2	2	2
3	2	3
4	5	9
5	3	5

The traditionalism measure is based on a scale that the investigator constructed in which scores can range from 0 to 10. A low score means the person is less traditional and higher scores mean the person is relatively more traditional. Assuming approximately interval level data, the Pearson correlation coefficient would be the statistical technique typically used to analyze the relationship between traditionalism and ideal family size.

The correlation coefficient is a statistic that measures the degree to which two variables are linearly related to one another. Before considering it in detail, we will review the basic characteristics of a linear relationship.

13.2 / The Linear Model

Consider two variables, the number of hours worked (X) and the amount of money paid (Y). A group of four individuals works at a rate of $1 per hour. Their scores on X and Y are as follows:

Individual	X (hours worked)	Y (amount paid)
1	1	1
2	4	4
3	3	3
4	2	2

The relationship between X and Y can be illustrated using a graph called a **scatterplot.** Figure 13.1 presents a scatterplot for X and Y. The abscissa lists the possible X scores and the ordinate lists the possible Y scores. Individual 1 had a score of 1 on X and 1 on Y. We therefore find the point labeled 1 on the X axis and 1 on the Y axis and place a solid dot where the two coincide. This same procedure has been repeated for each individual, plotting the corresponding X and Y scores. We can connect the dots with a straight line, as has been done in Figure 13.1. The relationship between X and Y can be stated as:

$$Y = X$$

The number of dollars paid equals the number of hours worked. There is a linear relationship between Y and X, as indicated by the straight line on the scatterplot.

Suppose the individuals were not paid $1 per hour, but instead were paid $2 an hour. The scores on X and Y would now be

Individual	X (hours worked)	Y (amount paid)
1	1	2
2	4	8
3	3	6
4	2	4

In this case the relationship between X and Y can be stated as

$$Y = 2X$$

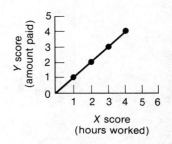

Figure 13.1 Example of a Scatterplot

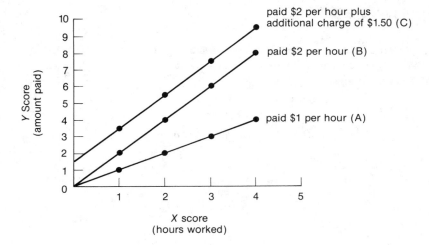

Figure 13.2 Examples of Three Linear Relationships

The number of dollars paid equals two times the number of hours worked. Figure 13.2 presents a scatterplot of these data (line B) as well as the previous data (line A). Notice that we still have a straight line (and hence, a linear relationship) but, in the case of $2 per hour, the line rises much faster than in the case of $1 per hour, that is, the slope of the line is different. One way in which linear relationships can differ is in terms of the *slope* describing the linear relationship. Technically, the **slope** of a line indicates the number of units Y changes as X changes one unit. In the case where people are paid $2 per hour, if an individual works one hour, he is paid $2, and if he works two hours, he is paid $4. When X goes up 1 unit (from 1 to 2 hours), Y goes up 2 units (from 2 to 4 dollars). The slope of this linear relationship is 2.

The slope of a linear relationship can be determined using a simple algebraic formula. This involves first selecting the X and Y scores of any two individuals. We will use individual 1 ($X_1 = 1$, $Y_1 = 2$) and individual 2 ($X_2 = 4$, $Y_2 = 8$). The slope is computed by dividing the difference between the two Y scores by the difference between the two X scores, or in other words, the change in Y scores is divided by the change in X scores:

$$b = \frac{Y_1 - Y_2}{X_1 - X_2}$$ [13.1]

where b is the symbol used to represent a slope. In this case, we find that

$$b = \frac{2 - 8}{1 - 4} = 2$$

A slope can have either a positive value or a negative value. Consider two variables X and Y, with the following scores:

Individual	X	Y
1	2	3
2	1	4
3	4	1
4	3	2

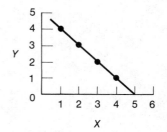

Figure 13.3 Example of a Linear Relationship with a Negative Slope

Using Equation 13.1 for computing a slope, we find it to be

$$b = \frac{3 - 4}{2 - 1} = \frac{-1}{1} = -1.00$$

Figure 13.3 presents a scatterplot of this relationship. The relationship is still linear but now the line changes downward as we move from left to right on the X axis. This downward direction characterizes a negative slope, while an upward direction characterizes a positive slope. A slope of zero would be a horizontal line. In this example,

$$Y = (-1)X$$

and for every unit X increases, Y decreases by 1 unit. A positive slope is referred to as a **direct relationship** between X and Y, and a negative slope is referred to as an **inverse relationship** between X and Y. In the case of a positive slope, when scores on X *increase*, scores on Y *increase*. For a negative slope, when scores on X *increase*, scores on Y *decrease*.

Let us return to the example where an individual is paid $2 per hour worked. Suppose that in addition to this wage, each individual is given a tip of $1.50. Now the relationship between X and Y is

$$Y = 1.50 + 2X$$

Figure 13.2 plots this case for the four individuals (line C). If we compute the slope of this line, it is found to be 2.0, as before. Notice that lines C and A are parallel, but that line C is raised 1.5 units above line A. This point of separation can be measured at the Y axis where $X = 0$. When $X = 0$, the Y value is 1.5 for line C. The point at which the line intersects the Y axis when $X = 0$ is called the **intercept,** and its value is symbolized by the letter *a*. Linear relationships can also differ in terms of the values of their intercepts. The general linear equation (or linear model) is thus

$$Y = a + bX \qquad [13.2]$$

Any linear relationship between two variables can be described in terms of Equation 13.2. There will always be a slope and an intercept that allows you to describe the value of *Y* that will occur when a given value of *X* occurs. For example, suppose you are told that for a group of twenty individuals there is a perfect linear relationship between their grade point average in school (*Y*) and their score on an intelligence test (*X*). Suppose you are also told that the equation describing the relationship is

$$Y = 1.0 + .025X$$

If an individual had a score of 100 on the intelligence test, what would his grade point average be? We would use the above equation to find out:

$$Y = 1.0 + .025 \, (100) = 1.0 + 2.5 = 3.5$$

His grade point average would be a 3.5 (where 4 = A, 3 = B, 2 = C, 1 = D, and 0 = F).

Study Exercise 13.1

Compute the slope of the following two sets of scores.

Set I	
X	Y
2	13
4	7
5	4
6	3

Set II	
X	Y
1	4
3	6
7	10
8	11

Answer. For set I, we take the scores of the first individual ($X_1 = 2$, $Y_1 = 13$) and the second individual ($X_2 = 4$, $Y_2 = 7$). Using Equation 13.1,

$$b = \frac{13 - 7}{2 - 4} = \frac{6}{-2} = -3.0$$

For set II, $Y_1 = 4$, $X_1 = 1$, $Y_2 = 6$, and $X_2 = 3$. Then

$$b = \frac{4 - 6}{1 - 3} = \frac{-2}{-2} = 1.0$$

13.3 / The Correlation Coefficient

It is rare in the social sciences that we observe a perfect linear relationship between two variables. Sometimes the relationship is approximately linear, but seldom is it perfectly linear. Statisticians have developed a statistic, called the **correlation coefficient,** which indicates the extent to which two variables approximate a linear relationship. The correlation coefficient, represented by the letter r, can range from -1.00 to 0 to $+1.00$. A correlation coefficient of 1.00 means the two variables form a perfect linear relationship that is direct in nature (that is, the higher the score an individual obtains on X, the higher the score that individual obtains on Y). A correlation of -1.00 also means the two variables form a perfect linear relationship, but that it is inverse in nature (that is, the higher the score an individual obtains on X, the lower the score that individual obtains on Y). A correlation coefficient of 0.0 means there is no linear relationship between the two variables. Figure 13.4 presents some example scatterplots for correlation coefficients of different magnitudes.

If variables in the social sciences are seldom perfectly linearly related, then an important problem becomes one of defining the line that the relationship approximates. In other words, when describing the relationship between two variables in the context of a linear model, how does one determine the slope and the intercept when the relationship is not perfectly linear? We will now consider procedures for computing a correlation coefficient as well as deriving the slope and intercept of the line that the relationship approximates.

Calculation of the Correlation Coefficient, Slope, and Intercept. Consider Table 13.1, which presents data illustrating (a) a perfect linear relationship, (b) a strong, but not perfect, negative linear relationship, and (c) no linear relationship between two variables, X and Y. For each example, we have converted the raw scores on X and Y to standard scores, z_X and z_Y. Note that when the linear relationship is positive, z scores on X tend to equal the z scores on Y and they are alike in sign (see columns 4 and 5 of Table 13.1). In fact, when the linear relationship is perfect, the standard scores on X and Y will be identical. This means, for example, when an individual scored one standard deviation above the mean on X ($z_X = 1.00$), he also scored one standard deviation above the mean on Y ($z_Y = 1.00$). When the z scores are multiplied by each other, the product is positive. When these products are summed, we obtain a high positive value of $\Sigma z_X z_Y$ (see column 6 of Table 13.1a).

When the linear relationship is negative, the z scores on X will also tend to equal the z scores on Y but they are opposite in sign (see columns 4 and 5 of Table 13.1b). Large positive z scores on X are associated with large negative z scores on Y. In the case of a perfect negative linear relationship, for example, an individual who scores one standard deviation *above* the mean on X ($z_X = 1.00$) will score one standard deviation *below* the mean on Y ($z_Y = -1.00$). When the z scores are multiplied, the

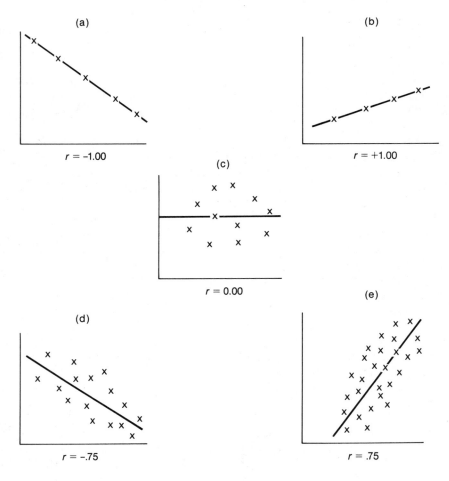

Figure 13.4 Scatterplots for (a) a perfect negative relationship, (b) a perfect positive relationship, (c) no relationship, and (d and e) moderate negative and positive relationships. (From Johnson and Liebert, 1977)

products are negative, and when summed, we obtain a large negative value (see column 6 of Table 13.1b).

Finally, when there is no linear relationship, the z scores on X will bear no necessary relationship to the z scores on Y, either in size or sign. The product of the z scores will be both positive and negative and when summed, these will tend to cancel each other, yielding a value near zero (see column 6 of Table 13.1c).

These observations are consistent with the nature of the correlation coefficient as described earlier. When the correlation is positive, the sum of the product of z scores is positive; when the correlation is negative, the sum of the product of z scores is negative; and when the correlation is zero, the sum of the product of z scores is zero. We can therefore use the sum of the product of the z scores as an index of the

TABLE 13.1 EXAMPLES OF THREE TYPES OF RELATIONSHIPS

(a) Positive Linear Relationship

Individual	X	Y	z_X	z_Y	$z_X z_Y$	
1	8	10	1.42	1.42	2.00	
2	7	9	.71	.71	.50	
3	6	8	.00	.00	.00	$r = \dfrac{5.00}{5} = 1.00$
4	5	7	− .71	− .71	.50	
5	4	6	− 1.42	− 1.42	2.00	
					$\Sigma = 5.00$	

(b) Negative Linear Relationship

Individual	X	Y	z_X	z_Y	$z_X z_Y$	
1	8	7	1.42	− 1.12	− 1.59	
2	7	7	.71	− 1.12	− .80	
3	6	8	.00	.00	.00	$r = \dfrac{-4.78}{5} = -.96$
4	5	9	− .71	1.12	− .80	
5	4	9	− 1.42	1.12	− 1.59	
					$\Sigma = -4.78$	

(c) No Linear Relationship

Individual	X	Y	z_X	z_Y	$z_X z_Y$	
1	3	5	1.12	1.12	1.25	
2	1	3	− 1.12	− 1.12	1.25	
3	2	4	.00	.00	.00	$r = \dfrac{.00}{5} = .00$
4	1	5	− 1.12	1.12	− 1.25	
5	3	3	1.12	− 1.12	− 1.25	
					$\Sigma = 0.00$	

correlation between two variables. There is, however, one complication. When the correlation between two variables is nonzero, the value of the sum of z score products is influenced not only by the size of the correlation, but also by the sample size. In the case of a positive correlation, for example, the larger the number of observations, the greater will be the sum of the products, everything else being equal. Since we want an index of correlation that is independent of N, we can divide

TABLE 13.2 PROOF THAT $\Sigma z^2 = N$

From Chapter 4, the definition of a standard score is

$$z = \frac{X - \overline{X}}{s}$$

Squaring each side, we obtain

$$z^2 = \frac{(X - \overline{X})^2}{s^2}$$

Summing across individuals yields

$$\sum z^2 = \frac{\Sigma(X - \overline{X})^2}{s^2}$$

$$= \frac{1}{s^2} \sum(X - \overline{X})^2$$

$$= \frac{1}{\left(\frac{\Sigma(X - \overline{X})^2}{N}\right)} \cdot \sum(X - \overline{X})^2$$

$$= \frac{N}{\Sigma(X - \overline{X})^2} \cdot \sum(X - \overline{X})^2$$

$$= N$$

$\Sigma z_X z_Y$ by N. This is also advantageous because of a certain property of standard scores. Recall that when the correlation between two variables is positive and perfect, the z scores will be equal to one another. In this case, $\Sigma z_X z_Y$ is equal to $\Sigma z_X^2 = \Sigma z_Y^2$ since $z_X = z_Y$. It turns out that the sum of a set of standard scores squared will always equal N. The proof of this is given in Table 13.2. When the correlation is positive and perfect, dividing by N will always yield a value of $N/N = 1.00$. In the case of a perfect negative relationship, $z_X z_Y$ will equal $-N$. Dividing by N would yield $-N/N = -1.00$ such that a perfect negative correlation will always yield a value of -1.00. Nonperfect linear relationships will always yield a value somewhere between the two extremes of 1.00 and -1.00. This yields the following equation for calculating a correlation coefficient:

$$r = \frac{\Sigma z_X z_Y}{N} \tag{13.3}$$

Although Equation 13.3 yields an intuitive understanding of the correlation coefficient, it is not computationally efficient. We will now present a formula that is easier from a computational standpoint (alternative computational formulas are presented in Appendix 13.1).

Consider a group of ten subjects. Two variables have been measured for each: traditionalism (X) and ideal family size (Y). The scores on X and Y are presented in columns 2 and 5 of Table 13.3. The first step in the computation of the correlation coefficient involves the computation of a sum of squares for the X scores and a sum of squares for the Y scores. This has been done in columns 2 through 7 of Table 13.3 and we find that

$$SS_X = 80$$

$$SS_Y = 92$$

Recall that the formula for a sum of squares is

$$\Sigma(X - \overline{X})^2$$

TABLE 13.3 DATA FOR COMPUTATION OF CORRELATION COEFFICIENT, SLOPE, AND INTERCEPT

Individual	X	$(X - \overline{X})$	$(X - \overline{X})^2$	Y	$(Y - \overline{Y})$	$(Y - \overline{Y})^2$	$(X - \overline{X})(Y - \overline{Y})$
1	9	4	16	10	5	25	20
2	7	2	4	6	1	1	2
3	5	0	0	1	−4	16	0
4	3	−2	4	5	0	0	0
5	1	−4	16	3	−2	4	8
6	1	−4	16	3	−2	4	8
7	3	−2	4	5	0	0	0
8	7	2	4	6	1	1	2
9	5	0	0	1	−4	16	0
10	9	4	16	10	5	25	20
			$SS_X = 80$			$SS_Y = 92$	CP = 60

$$\overline{X} = 5.0$$
$$\overline{Y} = 5.0$$

This can also be written as

$$\Sigma((X - \overline{X})(X - \overline{X}))$$

In other words, we are simply multiplying each deviation score by itself and then summing these products across individuals. In computing the sum of squares for the Y scores, we computed

$$\Sigma((Y - \overline{Y})(Y - \overline{Y}))$$

At this point we introduce a new term called the **cross-product** (CP), which is defined as

$$CP = \Sigma(X - \overline{X})(Y - \overline{Y}) \qquad \qquad [13.4]$$

It is similar to a sum of squares measure but it indicates the extent to which two scores covary. It is computed by first multiplying a subject's deviation score for X by that subject's deviation score for Y, and then summing these products across subjects. This has been done in column 8 of Table 13.3. The first subject's deviation score on X was 4 (column 3) and her deviation score on Y was 5 (column 6). Her score on the cross-products measure is $(4)(5) = 20$. In the example,

$$CP = 60$$

Unlike a sum of squares measure, the CP can be positive, zero, or negative. This is because, unlike the sum of squares, scores are not being squared in its computation.*

The three summary measures, the sum of squares for X, the sum of squares for Y, and the cross-products, allow us to compute the correlation coefficient, slope, and intercept that describe the linear relationship between traditionalism and ideal family size. The formula for the correlation coefficient is

$$r_{XY} = \frac{CP}{\sqrt{SS_X SS_Y}} \qquad \qquad [13.5]$$

In our example,

$$r_{XY} = \frac{60}{\sqrt{(80)(92)}} = .70$$

A correlation coefficient of .70 indicates a relatively strong, positive linear relationship between the two variables. The linear trend can be seen in Figure 13.5,

---

*If we wanted to compute the variance of X and Y, separately, we would divide their respective sum of squares by N. If we divide the cross-products by N, we obtain a statistic called a covariance.

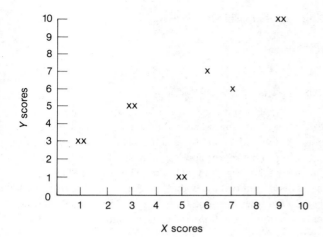

Figure 13.5 Scatterplot for Example on Traditionalism and Ideal Family Size

which presents the scatterplot of the two variables. Note that the relationship is not perfectly linear but it does seem to be approximately linear. Accordingly, we should be able to describe the linear relationship by using the linear model

$$Y = a + bX$$

Since there is not a perfect linear relationship, we must use a special formula for computing the slope and intercept. We will present the formulas and then discuss the logic underlying them. For the slope, the formula is

$$b = \frac{CP}{SS_X} \qquad \textbf{[13.6]}$$

In the example concerning traditionalism and family size,

$$b = \frac{60}{80} = .75$$

The formula for computing the intercept is

$$a = \overline{Y} - b\overline{X} \qquad \textbf{[13.7]}$$

Therefore,

$$a = 5.0 - (.75)(5.0) = 1.25$$

The linear equation describing the relationship between traditionalism (X) and ideal family size (Y) is

$$Y = 1.25 + .75X$$

A line described by the above equation has been drawn in Figure 13.6. This is called the **regression line.** Note that it intersects the Y axis at the value of the intercept (1.25). Further, the slope of the line is such that when X goes up 1 unit, Y goes up .75 unit. In the figure, a line with an arrowhead has been drawn emanating from each data point and extending to the regression line. These illustrate, *visually,* the rationale in defining the values of the slope and intercept. The slope and intercept are defined so as to *minimize* the squared distances that the data points, considered collectively, are from the regression line. The equation $Y = 1.25 + .75X$ accomplishes this. Any other linear equation, by definition, would result in a larger sum of squared distances from the regression line.

The criterion for deriving the values of the slope and intercept is formally called the **least squares criterion.** It can also be illustrated algebraically. Earlier we noted that the linear equation allows us to specify what a person's Y score is, given the person's X score. This was accomplished by substituting the person's score on X into the linear equation. In the present example, the linear equation is

$$Y = 1.25 + .75X$$

Let us take the 10 individuals in the study and substitute each of their X scores into this equation. For the first subject, the score on X was 9. The predicted Y score $(\hat{Y})$ is then $1.25 + (.75)(9) = 8.0$. For the second subject, the score on X was 7. The predicted Y score $(\hat{Y})$ for this subject is $1.25 + (.75)(7) = 6.5$. And so on.

Figure 13.6 Scatterplot and Regression Line for Example on Traditionalism and Ideal Family Size

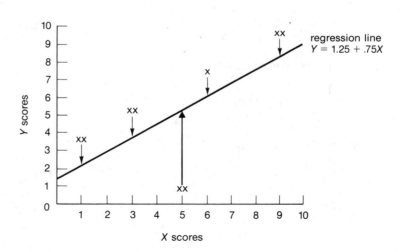

Table 13.4 presents the Y scores for each individual and the predicted scores based upon the linear equation. Inspection of the scores indicates that there is a discrepancy between Y and $\hat{Y}$. This is to be expected, since the linear relationship was not perfect (as shown by the correlation coefficient, $r = .70$). The discrepancies are formally defined in the fourth column of Table 13.4, which reflects the difference between Y and $\hat{Y}$. Note that some of the discrepancies are positive and some are negative. If we were to sum the discrepancy scores, the positive errors would cancel the negative errors and the overall sum would be zero. The least squares criterion concerns itself with the *squares* of the discrepancy scores and, technically, defines the values of the slope and the intercept so as to minimize the sum of these squares; that is,

$$\Sigma(Y - \hat{Y})^2 \text{ is minimized}$$

See Study Exercise 13.2 on page 268.

13.4 / Inference of a Relationship Using the Correlation Coefficient

To this point, we have considered the concepts of a correlation coefficient and a linear equation in order to describe a relationship between two quantitative variables. In practice, we typically want to make an inference about a correlation coefficient in a population based on sample data. As noted in previous chapters, even if a relationship between two variables is observed in a set of sample data, this does not mean that a relationship exists between the variables in the corresponding popu-

TABLE 13.4 SCORES ON X AND Y, PREDICTED SCORES, AND ERROR SCORES

Individual	Y	$\hat{Y}$	$Y - \hat{Y}$
1	10	8.0	2.0
2	6	6.5	− .5
3	1	5.0	−4.0
4	5	3.5	1.5
5	3	2.0	1.0
6	3	2.0	1.0
7	5	3.5	1.5
8	6	6.5	− .5
9	1	5.0	−4.0
10	10	8.0	2.0

Study Exercise 13.2

A director of a political campaign was interested in the relationship between voters' perceptions that his candidate supported labor unions (X) and their willingness to vote for his candidate (Y). Individuals were asked to indicate on a 10 point rating scale the extent to which they thought the candidate supported labor unions, with higher numbers indicating greater perceived support. They also indicated on a similar 10 point scale their willingness to vote for the candidate. The scores on these variables are presented below for nine individuals. Compute the correlation coefficient and the linear equation that describes the relationship between the variables

X	Y
6	5
7	4
8	4
8	5
8	3
9	6
9	7
7	5
10	6

Answer. The first step is to compute the three summary indices, sum of squares X, sum of squares Y, and the cross-products. This has been done in Table 13.5. From this, we find that

$$SS_X = 12$$

$$SS_Y = 12$$

$$CP = 6$$

Using Equations 13.5 to 13.7,

$$r = \frac{CP}{\sqrt{SS_X SS_Y}} = \frac{6}{\sqrt{(12)(12)}} = .50$$

$$b = \frac{CP}{SS_X} = \frac{6}{12} = .50$$

$$a = \overline{Y} - b\overline{X} = 5.0 - (.50)(8.0) = 1.0$$

The correlation coefficient is .70 and the linear equation is $Y = 1.0 + .5X$.

lation. A relationship may exist in a sample even though it doesn't in the population, because of sampling error. The task at hand, then, is to test the viability of a sampling error interpretation.

Null and Alternative Hypotheses.

Recall that a correlation coefficient can range from -1.00 to 1.00. A coefficient of 0 means that there is no linear relationship between the variables. In contrast, nonzero correlation coefficients imply some approximation to a linear relationship. We can therefore state the null and alternative hypotheses as follows

$$H_0: \rho = 0$$

$$H_1: \rho \neq 0$$

where ρ is the Greek letter "r" (rho) and represents the true correlation in the population. The null hypothesis states that there is no linear relationship between the two variables and that the correlation between the two variables is zero. The alternative hypothesis states that the correlation between the two variables is not zero.

Based on the data collected in an experiment, we want to choose between the two hypotheses. We will again use the logic discussed in Chapter 7. First, we will assume that the null hypothesis is true. Then we will translate the value of r in the sample into a t value. This value will be evaluated in terms of whether it falls in the rejection region. If it does, then we will reject the null hypothesis and conclude that there is a linear relationship between the two variables. Otherwise, we will fail to reject the null hypothesis.

TABLE 13.5 CALCULATION OF SS_X, SS_Y, AND CP FOR POLITICAL EXAMPLE

X	$X - \overline{X}$	$(X - \overline{X})^2$	Y	$(Y - \overline{Y})$	$(Y - \overline{Y})^2$	$(X - \overline{X})(Y - \overline{Y})$
6	-2	4	5	0	0	0
7	-1	1	4	-1	1	1
8	0	0	4	-1	1	0
8	0	0	5	0	0	0
8	0	0	3	-2	4	0
9	1	1	6	1	1	1
9	1	1	7	2	4	2
7	-1	1	5	0	0	0
10	2	4	6	1	1	2
$\Sigma X = 72$		$SS_X = 12$	$\Sigma Y = 45$		$SS_Y = 12$	$CP = 6$
		$\overline{X} = 8.0$			$\overline{Y} = 5.0$	

Sampling Distribution of a Correlation Coefficient. Consider the case of a large population in which the correlation between two variables, traditionalism and the ideal number of children, is zero. From the population, we select a random sample of ten individuals and compute the correlation coefficient. We might find that $r = .15$. The fact that it does not equal exactly zero should not be surprising because of sampling error. Now suppose we go to the population again and randomly select ten individuals. For this sample, the correlation might be .03. In principle we could continue this sampling process a large number of times for all possible random samples of size 10. The result would be an entire distribution of correlation coefficients based upon random samples of size 10. Such a distribution is called a **sampling distribution of the correlation coefficient** and it has many of the same properties of sampling distributions as discussed in Chapter 6.

The mean of the sampling distribution of the correlation coefficient is approximately ρ, and the standard deviation is $\sigma_r = (1 - \rho^2)/\sqrt{N - 2}$. The shape of the distribution is nonnormal. When $\rho = 0$, the distribution is symmetrical and nearly normal. When ρ is not equal to zero, the sampling distribution is skewed. Figure 13.7 illustrates the sampling distributions when $\rho = -.8, 0$, and $.8$, respectively, for $N = 10$.

When testing the null hypothesis that $\rho = 0$, the following expression has been shown to be distributed as a t distribution with $N - 2$ degrees of freedom:

$$t = \frac{r - \rho}{\sqrt{\frac{(1 - r^2)}{(N - 2)}}}$$ [13.8]

where r is the observed sample correlation and N is the sample size. Applying this formula to the example on traditionalism and ideal number of children (where $r = .70$ and $N = 10$), we find that

$$t = \frac{.70 - 0}{.252}$$

$$= 2.78$$

Based upon an alpha level of .05, nondirectional test, and with $N - 2 = 10 - 2 = 8$ degrees of freedom, the critical values of t are ± 2.31 (see Appendix E). Since the derived t score is greater than 2.31, we reject the null hypothesis. A sample correlation coefficient of .70 in the present case is just too large to attribute to sampling error. We therefore conclude that the population correlation is nonzero.*

Assumptions of the t Test. The test of the null hypothesis that $\rho = 0$ is based on the assumption that the population distributions of X and Y are such that their joint distribution (that is, their scatterplot) represents a **bivariate normal distribution.** This means that the distribution of Y scores at any value of X is a normal distribution in the population (see Figure 13.8). In addition, the variability of the Y scores at each value of X is assumed to be equal across X scores. Statisticians have shown that the t test is relatively robust to violations of these assumptions, given a reasonable sample size. Table 13.6 presents data from a Monte Carlo study for tests of the hypothesis that $\rho = 0$ for $\alpha = .05$ (Norris and Hjelm, 1961). In this study, five different types of bivariate population distributions were investigated, four of which violated the assumption of bivariate normality. Each population had $\rho = 0$. Using sample sizes of 15, 30, and 90, 2,500 random samples were selected from each

*Procedures for testing hypotheses other than $\rho = 0$ are discussed in Appendix 13.2.

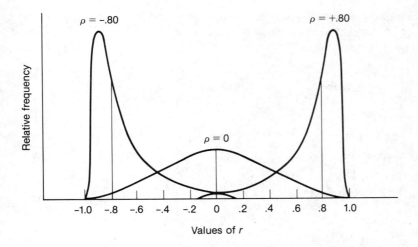

Figure 13.7 Sampling Distributions of r for Three Values of ρ and a Sample Size of 10 (From Minium, 1970)

population, for each of the three sample sizes. As seen in Table 13.6, the actual significance levels (that is, the proportion of times the null hypothesis was rejected) corresponded quite closely to the nominal significance level ($\alpha = .05$) in all cases. For additional perspectives on the robustness of r, see Havilcek and Peterson (1977).

Tabled Values of r. The t statistic computed from Equation 13.8 requires two terms, r and N. It is possible to construct a table of values of r that would lead to rejection of the null hypothesis that $\rho = 0$ for different sample sizes at the .05 and .01 alpha levels. This has been done in Appendix J. Appendix J presents instructions for using this table and allows you to tell at a glance whether the null hypothesis would be rejected given the observed r and N.

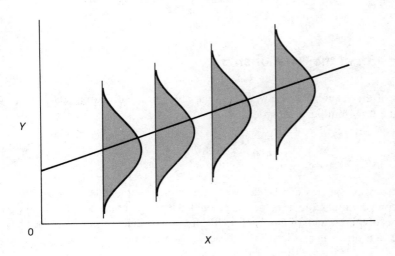

Figure 13.8 Distribution of Y Scores at a Given Value of X for a Bivariate Normal Distribution (Adapted from Glass and Stanley, 1970)

Inference of a Relationship Using the Correlation Coefficient   **271**

TABLE 13.6 APPROXIMATIONS TO ACTUAL PROBABILITIES OF A TYPE I ERROR (α = .05) IN TESTS OF PEARSON CORRELATIONS WHEN ρ = 0 FOR SELECTED SAMPLE SIZES AND SELECTED DISTRIBUTIONS

Population Distribution	Sample Size	Actual Significance Level
Normal	15	.0513
Rectangular	15	.0507
Leptokurtic	15	.0471
Slightly Skewed	15	.0476
Markedly Skewed	15	.0507
Normal	30	.0473
Rectangular	30	.0454
Leptokurtic	30	.0442
Slightly Skewed	30	.0498
Markedly Skewed	30	.0516
Normal	90	.0488
Rectangular	90	.0477
Leptokurtic	90	.0345
Slightly Skewed	90	.0465
Markedly Skewed	90	.0484

Adapted from Norris and Hjelm (1961).

13.5 / Strength of the Relationship

The strength of the relationship between two variables in a correlation analysis is indexed by the now familiar expression

$$1 - \frac{SS_{ERROR}}{SS_{TOTAL}} \qquad [13.9]$$

The relevant sum of squares can be derived with reference to Table 13.4. In this table, the second column represents the scores on the dependent variable (Y) and the sum of squares of this column of numbers equals the sum of squares total. The fourth column represents errors in prediction (the observed score minus the predicted score based upon the linear model). The sum of squares of this column of numbers will yield the sum of squares error. However, it is not necessary to calculate these values since Expression 13.9 has a direct relationship to the correlation coefficient. Specifically,

$$r^2 = 1 - \frac{SS_{ERROR}}{SS_{TOTAL}}$$ [13.10]

In the example on traditionalism and the ideal number of children, we found $r = .70$. Thus,

$$r^2 = .70^2 = .49$$

For this sample, the proportion of variability in the perceived ideal number of children that is associated with variations in traditionalism was .49, as indexed by a linear model.

Study Exercise 13.3

In Study Exercise 13.2, we computed a correlation coefficient between voters' perceptions that a candidate supports labor unions and their willingness to vote for him. We found, based upon a sample size of 9, that the correlation between these two variables was .50. Test whether this correlation can reasonably be attributed to sampling error in the context of a population where the true correlation is zero.

Answer. We begin by formally stating a null and an alternative hypothesis:

$$H_0: \rho = 0$$

$$H_1: \rho \neq 0$$

Using Equation 13.8, we convert the observed correlation coefficient to a t score:

$$t = \frac{r}{\sqrt{\frac{(1 - r^2)}{(N - 2)}}}$$

$$= \frac{.50}{\sqrt{\frac{(1 - .50^2)}{(9 - 2)}}}$$

$$= 1.53$$

For an alpha level of .05, two-tailed test, and a t distribution with $9 - 2 = 7$ degrees of freedom, the critical values of t are ± 2.36. Since 1.53 does not exceed 2.36, nor is it less than -2.36, we fail to reject the null hypothesis. We cannot confidently conclude that ρ does not equal zero.

13.6 / Nature of the Relationship

The sign of the correlation coefficient indicates whether the relationship between X and Y is direct (positive slope) or inverse (negative slope). Aside from this, no additional analyses are performed to discern the nature of the relationship. This is because the nature of the relationship is built into the correlation coefficient. It *assumes* that the two variables are *linearly* related, and hence, the nature of the relationship is specified.

Method of Presentation

When reporting the results of a correlational analysis, most investigators will report only the correlation coefficient and the statistics of the test of the null hypothesis. An example might appear as follows:

The Pearson correlation between traditionalism and ideal family size was .70 ($t = 2.78$, df $= 8$, $p < .05$).

The value of .70 represents the correlation coefficient, r. The square of this correlation is the proportion of variation in Y that is "explained" by its linear relationship to X. Thus, this also constitutes a report of the strength of the relationship. The remaining statistics reported in the parentheses are, by now, self-explanatory. The values of the slope and intercept are seldom reported unless they are central to the theoretical points being addressed.

13.7 / Numerical Example

Computer dating has become increasingly popular. Generally, a participant in a computer dating program will complete a questionnaire indicating his or her general interests and habits. Responses to the questions are then analyzed and the person is matched with a member of the opposite sex who has similar interests. Underlying the matching of individuals is an important assumption: People with similar interests will be attracted to one another. But is this a reasonable assumption?

In order to test the assumption, a researcher interviewed fifteen individuals who had participated in a computer dating program. Each of the individuals had already been on one date with her partner, and each was asked to indicate how much she was attracted to him on a scale of 1 to 10. A score of 1 meant she was not at all attracted to him while a score of 10 meant she was very attracted to him. Before being assigned a partner, each person responded to a ten-item questionnaire about her interests and habits. Their partner responded to the same questionnaire. For purposes of the experiment, the similarity between assigned partners differed. Some individuals were assigned partners who responded the same way on all ten questions, some were assigned partners who responded the same on only nine of the questions, and so on. Thus, similarity between partners could range from 1 to 10 with low scores indicating little similarity and high scores indicating a great deal of similarity. The data for the experiment appear in columns 2 and 5 of Table 13.7.

We are interested in the relationship between two quantitative variables. The independent variable is the similarity with one's partner (X) and the dependent variable is attraction to the partner (Y). The first step of the analysis is to compute

TABLE 13.7 DATA AND CALCULATION OF PRELIMINARY STATISTICS FOR NUMERICAL EXAMPLE ON SIMILARITY AND ATTRACTION

Individual	Similarity (X)	$(X - \bar{X})$	$(X - \bar{X})^2$	Attraction (Y)	$(Y - \bar{Y})$	$(Y - \bar{Y})^2$	$(X - \bar{X})(Y - \bar{Y})$
1	10	4	16	8	3	9	12
2	8	2	4	6	1	1	2
3	6	0	0	4	-1	1	0
4	4	-2	4	2	-3	9	6
5	2	-4	16	3	-2	4	8
6	10	4	16	6	1	1	4
7	8	2	4	8	3	9	6
8	6	0	0	5	0	0	0
9	7	1	1	9	4	16	4
10	4	-2	4	5	0	0	0
11	1	-5	25	3	-2	4	10
12	3	-3	9	1	-4	16	12
13	5	-1	1	3	-2	4	2
14	7	1	1	5	0	0	0
15	9	3	9	7	2	4	6
	$\Sigma = 90$		110	70		78	72

$\bar{X} = 6.0$

$\bar{Y} = 5.0$

$SS_X = 110$
$SS_Y = 78$
$CP = 72$

the three preliminary statistics, sum of squares X, sum of squares Y, and the cross-products. This has been done in Table 13.7. From these we can compute the correlation coefficient:

$$r = \frac{CP}{\sqrt{SS_X SS_Y}} = \frac{72}{\sqrt{(110)(78)}} = .78$$

The null hypothesis of no linear relationship states that the population correlation is zero. The alternative hypothesis states that the population correlation is nonzero:

$$H_0: \rho = 0$$

$$H_1: \rho \neq 0$$

To test the viability of the null hypothesis, we first use Equation 13.8 to convert the correlation coefficient into a t value:

$$t = \frac{r - \rho}{\sqrt{\frac{(1 - r^2)}{(N - 2)}}}$$

$$= \frac{.78}{\sqrt{\frac{(1 - .78^2)}{(15 - 2)}}}$$

$$= 4.48$$

For an alpha level of .05, nondirectional test, and a t distribution with $15 - 2 = 13$ degrees of freedom, the critical values of t are ± 2.16. Since 4.48 is greater than 2.16, we reject the null hypothesis and conclude that ρ does not equal zero.

The strength of the relationship is indexed by the squared correlation coefficient:

$$r^2 = .78^2 = .61$$

For this sample, the proportion of variability in attractiveness ratings that is associated with variations in similarity is .61, as indexed by a linear model.

The nature of the relationship is indicated by the sign of the correlation coefficient. In this case, the correlation was positive, indicating that as the similarity between partners increased, so did the attraction to them. The data might be reported in a research report as follows:

A Pearson correlation was computed between similarity of the partner and attraction to the partner. The observed correlation of .78 was statistically significant ($t = 4.48$, df $= 13$, $p < .05$).

Box 13.1 The Creative Personality

Psychologists have studied extensively the process of creativity and what distinguishes creative individuals from noncreative individuals. Barron (1965) reported the results of an investigation that was designed to illuminate personality differences between certain creative and noncreative persons. One part of this investigation studied groups of professionals, such as women mathematicians. Forty-four female mathematicians attended a three day interview session at the Institute of Personality Assessment and Research (IPAR) at the University of California, Berkeley. During this time period, the mathematicians interacted in both formal and informal settings with the staff of the institute. The particular females studied were invited on the basis of a "nomination" technique and consisted of sixteen women considered by a panel of experts to be "the most original and important women mathematicians in the United States and Canada," and twenty-eight women who were nationally recognized but not distinctively creative in their discipline. During the course of the interviews, the staff was kept unaware of who the creative versus noncreative women were, that is, they were not able to identify the two different groups. The staff rated each woman on a number of personality dimensions and these ratings were correlated with the ratings of creativity of each woman generated by a panel of mathematical experts. The correlations that were the strongest (and which were all statistically significant) were the following:

Positive correlations:

.64	Thinks and associates to ideas in unusual ways; has unconventional thought processes.
.55	Is an interesting, arresting person.
.51	Tends to be rebellious and non-conforming.
.49	Genuinely values intellectual and cognitive matters.
.46	Appears to have a high degree of intellectual capacity.
.42	Is self-dramatizing; histrionic.
.40	Has fluctuating moods.

Negative correlations:

−.62	Judges self and others in conventional terms like "popularity," the "correct thing to do," "social pressures," and so forth.
−.45	Is a genuinely dependable and responsible person.
−.43	Behaves in a sympathetic or considerate manner.
−.40	Favors conservative values in a variety of areas.
−.40	Is moralistic.

Barron summarized the results as follows: "The emphasis is upon genuine unconventionality, high intellectual ability, vividness or even flamboyance of character, moodiness and preoccupation, courage, and self-centeredness. These are people who stand out, and who probably are willing to strike out if impelled to do so."

13.8 / Planning an Investigation Using the Correlation Coefficient

When planning an investigation using the correlation coefficient, Appendix F.3 can facilitate selection of sample sizes in order to achieve desired levels of power for tests of the null hypothesis that $\rho = 0$. Table 13.8 reproduces a portion of this appendix for illustration. The first column indicates different levels of power and the column headings indicate population values of rho squared. As an example, if the desired level of power is .80 and if the researcher suspects that the strength of the relationship in the population corresponds to a ρ^2 of .15, then the number of subjects that should be sampled, for $\alpha = .05$, nondirectional test, is 49. Inspection of Table 13.8 and Appendix F.3 will give the reader a general appreciation for the relationship between power, sample size, alpha levels, and strength of the population relationship for Pearson's correlation coefficient.

13.9 / Alternative Procedures for Analyzing the Relationship Between Two Quantitative Variables

Although most investigators use correlational analyses when testing the relationship between two quantitative variables, some investigators will use the analysis of variance procedures discussed in Chapter 11. Consider two variables, intelligence scores on an intelligence test, X, and academic achievement as indexed by a student's grade point average, Y. One procedure for testing the relationship between these two quantitative variables is the correlational approach already discussed. A second procedure involves classifying the individuals into one of a number of groups based on their intelligence scores (the independent variable). For example, subjects might be divided into three groups: (1) "low" intelligence, (2) "moderate" intelligence and (3) "high" intelligence. Subjects with scores in the lower third of the distribution would be in the low group (for example, IQ = 80–95), those with scores in the middle third of the distribution would be in the moderate group (for example, IQ = 96–110), and those with scores in the upper third of the distribution would be in the high group (for example, IQ = 110–150). Each of these groups is then treated as a category and the relationship between this variable and the quantitative dependent variable analyzed by analysis of variance as discussed in Chapter 11.

There are advantages and disadvantages of the analysis of variance approach relative to the correlational approach. The major disadvantge concerns a loss of precision, with a possible corresponding inflation of SS_{ERROR}. In the above example, the intelligence scores were essentially reduced to a three point scale (low, moderate, and high). Considerable information was lost by doing so, since the original scale makes more than three discriminations. Further, all individuals who are placed into one category (for example, everyone in the low category) are implicitly assumed to be of equal intelligence. If they are not, then the result may be an increase in the within-group variability. This is because the within-group differences in intelligence

TABLE 13.8 SAMPLE SIZES NECESSARY TO ACHIEVE SELECTED LEVELS OF POWER FOR ALPHA = .05, NONDIRECTIONAL TEST, AS A FUNCTION OF ρ^2

Nondirectional Test, Alpha = 0.05

Power	Population Correlation Squared									
	0.01	*0.03*	*0.05*	*0.07*	*0.10*	*0.15*	*0.20*	*0.25*	*0.30*	*0.35*
0.25	166	56	34	25	17	12	9	8	6	6
0.50	384	127	76	54	38	25	19	15	12	10
0.60	489	162	97	69	48	31	23	18	15	13
0.67	570	188	112	80	55	36	27	21	17	14
0.70	616	203	121	86	59	39	29	23	18	15
0.75	692	228	136	96	67	43	32	25	20	17
0.80	783	258	153	109	75	49	36	28	23	19
0.85	895	294	175	124	85	56	41	32	26	21
0.90	1046	344	204	144	100	65	47	37	30	25
0.95	1308	429	255	180	124	80	58	46	37	30
0.99	1828	599	355	251	172	111	81	63	50	42

may cause within-group differences in grade point average. This increased within-group variability will, as discussed in Chapter 11, increase SS_{ERROR} and consequently make the analysis less sensitive. In contrast, the correlational analysis uses the full range of the intelligence scale.

An advantage of the analysis of variance approach relative to the correlational approach concerns the analysis of the nature of the relationship between the two variables. The correlational approach tests for one and only one type of relationship, a linear one. Two variables may be related, but if they are related in a fashion that is nonlinear, the correlation coefficient will not be sensitive to this. In contrast, the analysis of variance approach is sensitive to linear as well as nonlinear relationships.

Consider an example where the relationship between two variables is **curvilinear,** as illustrated in Figure 13.9. A correlation coefficient computed on these data would certainly be near zero. This is because the relationship does not approximate a linear one and the correlational approach is relatively insensitive to **nonlinear relationships.** In contrast, if the subjects had been classified into three groups based on their X scores, the mean Y scores might be as follows:

$$\text{Mean for Low Group} = 10.4$$

$$\text{Mean for Moderate Group} = 15.6$$

$$\text{Mean for High Group} = 10.2$$

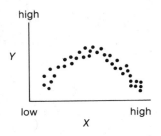

Figure 13.9 Scatterplot of a Curvilinear Relationship Between Two Variables

These means reflect the curvilinear relationship and the analysis of variance approach would be sensitive to this (that is, the null hypothesis of no relationship would be rejected, within the constraints of a type II error). A major difference between the two approaches, then, is the fact that the correlational approach tests only for linear relationships whereas the analysis of variance approach can be used to test for both linear and nonlinear relationships.

It is possible to use correlational procedures to test for nonlinear relationships. This is advantageous since the full range of the scale is used, rather than a collapsed scale. The procedure involves transforming the X scores such that the transformed scores are linearly related to the Y scores. Consider the scatterplots in Figure 13.10. In Figure 13.10a, the relationship between X and Y is clearly linear. However, Figure 13.10b is nonlinear and more closely approximates a logarithmic function. The X scores could therefore be transformed to fit a linear model by using the linear equation

$$Y = a + b \, (\log X)$$

The correlation between Y and log X would then test the hypothesis of a logarithmic relationship between the variables. Discussion of curve-fitting procedures such as these are presented in Hays and Winkler (1971). Although useful curve-fitting approaches have been developed, they are used too infrequently in the social sciences.

Figure 13.10 Scatterplot of a Linear and a Logarithmic Relationship Between Two Variables

(a)

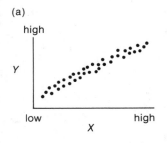

(b)

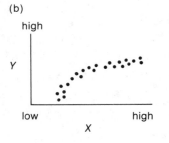

13.10 / The Use of Correlation Analysis for Prediction

Correlation and regression analysis, as discussed in this chapter, are not always used in an inferential context. Sometimes the approach is used in a purely prediction-oriented fashion. Consider the following example: The personnel director of an insurance company has rated every underwriter who has worked for the company during his tenure on a 1 to 10 scale, indicating how successful the underwriter was during his or her stay with the company. The higher the rating, the more successful the underwriter. Across all underwriters, the mean rating was 5.0, with a standard deviation of 2.0. Each of the underwriters was also administered a test designed to measure his or her potential on the job. The test scores could range from 0 to 100, with higher scores indicating greater potential. For these data, it would be possible to do a correlation analysis trying to predict success of the underwriter (Y) from the test score (X). Suppose such an analysis yielded a correlation of .80 and the following regression equation:

$$Y = .5 + .10X$$

The correlation coefficient is relatively large, indicating a fairly substantial linear relationship between test scores and success. The personnel director might therefore use the above equation to help him screen future applicants for underwriting jobs and to *predict* who is going to be successful. Suppose an applicant took the test and had a score of 85. If we put this score into the regression equation, we find a predicted success rating of

$$\hat{Y} = .5 + (.10)(85)$$

$$= .5 + 8.5$$

$$= 9.0$$

A success rating of 9.0 is very high (remember the mean was 5.0 with a standard deviation of 2) and this person would almost certainly be hired.

A useful statistic in the evaluation of linear prediction equations is the **standard error of estimate.** It is an index of predictive error. It can best be understood by reconsidering the conceptual definition of a standard deviation. The formula for a standard deviation is

$$s_Y = \sqrt{\frac{\Sigma(Y - \overline{Y})^2}{N}}$$

It is analogous to the average deviation of Y scores from their mean. The standard error of estimate is defined as

$$s_{YX} = \sqrt{\frac{\Sigma(Y - \hat{Y})^2}{N}} \qquad \textbf{[13.11]}$$

The only difference in the two equations is in the numerator of the right-hand side of the expressions. For the standard deviation, the numerator is the deviation of the Y score from its mean. For the standard error of estimate, the numerator is the deviation of the Y score from its predicted Y score. Thus, the standard error of estimate is analogous to an average error in prediction across subjects, in predicting Y from the linear equation.

There is a simple computational formula for computing the standard error of estimate. It is

$$s_{YX} = s_Y\sqrt{1 - r^2} \qquad \textbf{[13.12]}$$

In the underwriter example, the standard error of estimate is

$$s_{YX} = 2.0\sqrt{1 - .80^2}$$

$$= 1.2$$

On the average, the predicted scores deviated 1.2 units from the observed scores.

There are two perspectives in interpreting the standard error of estimate. First, the absolute magnitude of the statistic is meaningful. In the insurance example, the success scores could range from 1 to 10, and the error in prediction was 1.2 units, on the average. This appears to be a relatively small degree of error. Second, the standard error of estimate can be compared with the standard deviation of Y. The standard deviation of Y indicates what the average error in prediction would be if one were to predict everyone had a score equal to the mean of Y. If X helps to predict Y better, then one would expect the standard error of estimate to be less than this, and the more so, the better. In the underwriter example, the standard deviation of Y was 2.0 and the standard error of estimate was 1.2. This is a reasonable reduction in error.

This fictitious example illustrates how regression equations might be used in a purely predictive orientation. The situation above is rather unrealistic (it is rare that a single test will achieve such high predictive power), and the issues involved in using correlation and regression analysis in a predictive context are quite complex. Interested readers are referred to Wiggins (1973) for an excellent introduction to these issues.

13.11 / Summary

The Pearson product-moment correlation coefficient is typically applied to the analysis of the relationship between two variables when (1) both variables are quantitative in nature and are measured on approximately an interval level and (2) both variables are between-subjects in nature. Correlation analysis is based on the linear model and tests the extent to which two variables are linearly related. The correlation coefficient is an index that reflects the nature (direct or inverse) and strength of the linear relationship between two variables. Least squares estimates of the slope and the intercept describe the line that typifies the relationship.

The test from sample data of the hypothesis that $\rho = 0$ in a population is based on a sampling distribution of the correlation coefficient. This distribution permits an application of basic hypothesis testing principles (Chapter 7). The strength of the linear relationship is indexed by the correlation coefficient squared. The nature of the relationship is indicated by the sign of the correlation coefficient.

An alternative procedure for analyzing the relationship between two quantitative variables is the application of one-way analysis of variance (Chapter 11). This has the disadvantage of losing precision, but has the advantage of flexibility in terms of the nature of the relationship being tested. Finally, correlation and regression analysis will sometimes be used for purely predictive reasons rather than in inferential and theory-building contexts.

APPENDIX 13.1

Computational Formulas

This section presents formulas typically given in textbooks for computing from raw data the statistics reported in this chapter in the most efficient fashion. We will use as an example the following data:

Individual	X	Y	X^2	Y^2	XY
1	10	3	100	9	30
2	8	2	64	4	16
3	6	1	36	1	6
4	5	2	25	4	10
5	7	1	49	1	7
6	9	3	81	9	27
$\Sigma = 45$		12	355	28	96

The value of the correlation coefficient can be computed from the following equation:

$$r = \frac{\sum XY - \frac{(\Sigma X)(\Sigma Y)}{N}}{\sqrt{\left(\Sigma X^2 - \frac{(\Sigma X)^2}{N}\right)\left(\Sigma Y^2 - \frac{(\Sigma Y)^2}{N}\right)}}$$

$$= \frac{96 - \frac{(45)(12)}{6}}{\sqrt{\left(355 - \frac{45^2}{6}\right)\left(28 - \frac{12^2}{6}\right)}} \quad \text{[13.13]}$$

$$= \frac{96 - 90}{\sqrt{(17.5)(4)}}$$

$$= .72$$

The slope is frequently computed from the following equation:

$$b = r\frac{s_Y}{s_X}$$

or

$$b = \frac{N(\Sigma XY) - (\Sigma X)(\Sigma Y)}{N\Sigma X^2 - (\Sigma X)^2} \quad \text{[13.14]}$$

In our example,

$$b = \frac{(6)(96) - (45)(12)}{(6)(355) - (45)^2}$$

$$= \frac{576 - 540}{2130 - 2025}$$

$$= .343$$

APPENDIX 13.2

Testing Null Hypotheses Other Than $\rho = 0$

When the null hypothesis is $\rho = 0$, the t distribution can be used to test the viability of that hypothesis. Occasionally, an investigator will want to test a null hypothesis that ρ is equal to a nonzero number. As noted earlier, when ρ does not equal zero, the sampling distribution of the correlation coefficient is skewed. Fisher (1929) derived a logarithmic function of r, which we will call r', that has two desirable properties: (1) the sampling distribution of r' is approximately normally distributed irrespective of the value of ρ, and (2) the standard error of r' is essentially independent of ρ.* Because of these properties, it is possible to convert a sample r to r' and then use the properties of the normal distribution to test the null hypothesis.

*Fisher referred to his transformation using the symbol Z. Thus, you may see reference in the literature to "Fisher's r to Z transform." We will use the symbol r' to avoid confusing the transformation with a standard score.

The formula for r' is

$$r' = .50 [\log_e (1 + r) - \log_e (1 - r)] \quad \textbf{[13.15]}$$

where $\log_e$ = the natural log of a number and r is the sample correlation coefficient. All terms in Equation 13.15 are constants except r. Thus, the value of r' is completely dependent on r, and r' is simply a rescaling of r. It is not necessary to calculate r' using Equation 13.15 when testing a given null hypothesis. This has been done in Appendix K, which presents a table of values of r' for selected values of r. The standard error of r' is

$$\sigma_{r'} = \frac{1}{\sqrt{N - 3}} \quad \textbf{[13.16]}$$

Figure 13.11 presents sampling distributions of r' when $\rho = -.8, .0,$ and $.8$ when $N = 10$. Compare these with Figure 13.7 illustrating the comparable sampling distributions for r. Unlike r, the sampling distributions of r' are similar in shape and variability. The test of a null hypothesis is based on the following equation:

$$z = \frac{r' - \rho'}{\sigma_{r'}} \quad \textbf{[13.17]}$$

where ρ' = the log transformed value corresponding to the hypothesized value of ρ, and all other terms are as previously defined. The value of z is then compared against a critical value of z from the table of the normal distribution in Appendix C.

As an example, consider the following hypotheses:

$$H_o: \rho = .25$$

$$H_i: \rho \neq .25$$

and the following sample data

$$N = 100$$

$$r = .45$$

From Appendix K, the transformed values of .45 and .25 are $r' = .485$ and $\rho' = .255$. The standard error is

$$\sigma_{r'} = \frac{1}{\sqrt{N - 3}}$$

$$= \frac{1}{\sqrt{100 - 3}}$$

$$= .10$$

The z score is

$$z = \frac{.485 - .255}{.10}$$

$$= 2.30$$

Assuming an alpha level of .05, nondirectional test, the critical values of z are ± 1.96. The observed z exceeds 1.96 and the null hypothesis is therefore rejected.

Figure 13.11 Sampling Distributions of r' for Three Values of ρ Where $N = 10$

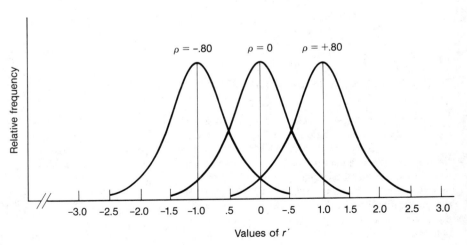

EXERCISES

Exercises to Review Concepts

1. Under what conditions is a Pearson correlation typically used to analyze a bivariate relationship?

2. What are the major assumptions underlying the Pearson correlation in terms of inferring that a relationship exists between two variables?

3. Draw a scatterplot of the following data.

X	Y
10	9
8	7
6	5
9	8
10	8
8	8
5	6

4. What does the *sign* of the slope of the regression line tell you about the relationship between the two variables? The sign of another statistic gives you the same information; which statistic is it?

5. Consider the following set of scores.

Subject	X	Y
1	3	7
2	8	9
3	3	3
4	2	8
5	6	8
6	6	9
7	8	6
8	5	4
9	7	2
10	2	4

(a) Compute SS_X, SS_Y, and CP.

(b) Compute the correlation between X and Y, and compute the slope and intercept of the regression line for predicting Y from X.

(c) Write the equation that describes the regression line for predicting Y from X.

(d) Compute predicted scores and error scores for each subject.

6. Given a perfect linear relationship between two variables, X and Y, and a slope of 3.00 (for predicting Y from X), if X changes 2 units, how many units will Y change?

7. Consider the following set of scores:

Subject	X	Y
1	6	10
2	7	9
3	7	7
4	8	8
5	8	8
6	8	8
7	8	8
8	9	9
9	9	7
10	10	6

(a) Compute SS_X, SS_Y, and CP.

(b) Compute the correlation between X and Y, and the slope and the intercept for predicting Y from X.

(c) Write the equation that describes the regression line for predicting Y from X.

(d) Compute the predicted scores and error scores for each subject.

8. Explain the least squares criterion.

9. Give an example of two variables that you think are positively correlated. Give an example of two variables that you think are negatively correlated.

10. Which correlation coefficient is indicative of a stronger linear relationship, $-.65$ or $.45$?

11. What is the relationship between eta squared and the correlation coefficient?

12. When analyzing the relationship between two quantitative variables, what are the advantages and disadvantages of reducing one of the variables into a qualitative variable and using analysis of variance instead of correlation analysis?

For each of the following experiments, indicate the appropriate statistical test for analyzing the relationship between the variables and state *why* you would use it. Assume that the underlying assumptions of the statistical test have been satisfied.

13. *Experiment I.* An investigator wanted to test the relationship between age and weight in adults. She interviewed one hundred people and asked them their age and measured their weight.

14. *Experiment II.* An investigator wanted to test the relative effects of two drugs on learning. Sixty subjects were randomly assigned to two groups, thirty in each group. Group 1 was given drug A while group 2 was given drug B. The mean learning scores on a paired associated learning task were computed and compared.

15. *Experiment III.* An investigator wanted to test the relationship between preference for brands A, B, or C of aspirin. Two hundred males and females were interviewed. They rated each of the three brands on a preference scale ranging from 1 to 15, with higher scores indicating greater preference.

16. *Experiment IV.* An investigator wanted to test the effects of the nuclear accident at Three Mile Island on attitudes toward nuclear energy. One hundred people had been interviewed prior to the accident and their attitude measured on a 10 point scale, with higher scores indicating a more favorable attitude. Five days after the accident they were reinterviewed and their attitudes measured again.

17. *Experiment V.* An investigator wanted to test the relationship between speed of driving and amount of gas used. One hundred cars were driven, twenty were driven 25 mph, twenty were driven 30 mph, twenty were driven 35 mph, twenty were driven 40 mph, and twenty were driven 45 mph. The amount of gas used over a 50 mile trip was measured.

18. Draw a scatterplot for two variables that are negatively correlated.

19. Draw a scatterplot for two variables that are positively correlated.

20. For Exercise 5, add 3 to each of the X scores and 5 to each of the Y scores. Now recompute the correlation between X and Y. Compare this with your original result. What is the effect of adding a constant to X or Y on the correlation coefficient? Do the same for the slope and intercept. What conclusions can you draw?

Exercises to Apply Concepts

21. A greater number of students are attending graduate school than ever before in academic history. Very few graduate programs can admit all applicants and, accordingly, most have a screening committee whose task is to select those applicants with the most promise. What kinds of criteria discriminate good graduate students from poor ones? Willingham (1974) reviewed forty-three studies from 1952 to 1972 that addressed this question. In his review, Willingham noted that the most commonly used selection criteria were the following: (1) scores on the graduate record exam, which are broken down into three subscores, a quantitative ability score (GRE-Q), a verbal ability score (GRE-V), and an advanced score (GRE-A) that measures mastery and comprehension of materials basic to graduate study in different major fields; (2) the grade point average during undergraduate education; and (3) letters of recommendation. Some of these criteria were found to be reasonable discriminators of the quality of graduate students, while others were quite poor in their prediction of graduate school success. The hypothetical data below examine the relationship between GRE-A scores and graduate students' grade point average after two years of graduate study. The data are representative of those reviewed by Willingham. Analyze the data, draw a conclusion, and write up the results of the analysis using principles discussed in the Method of Presentation section. (Note: In this example, the GRE-A scores have been transformed to a scale ranging from 0 to 80 instead of 0 to 800, as is standardly used.)

$$N = 70$$
$$\overline{X} \text{ for GRE-A} = 55.0$$
$$SS_x = 6250$$
$$\overline{Y} \text{ for GPA} = 3.0$$
$$SS_Y = 10$$
$$CP = 62.5$$

22. Psychologists have recently studied a number of factors concerned with environmental problems. These have included energy conservation, pollution, and maintenance of wildlife, to name a few. Borden (1978) investigated factors that differentiate those who are ecologically concerned versus those who tend to exhibit little concern for the environment. Borden hypothesized that one difference in these individuals may be the extent to which they value technology: People who value technological innovation and who believe that technology can solve most of our world problems should be less likely to perform environmental-conserving behaviors than those who question technology as a panacea to environmental problems. In order to test this hypothesis, Borden administered two assessment scales to a group of individuals. One scale measured the extent to which the individual performed ecological-related behaviors (for example, saving energy), and the other scale measured belief in technology. Both scales were continuous measures ranging from 0 to 20, with progressively higher scores representing the performance of more ecologically related behaviors and belief in technology, respectively. The hypothetical data presented below are representative of the outcome of this investigation. Analyze the data, draw a conclusion, and write up the results of the analysis using principles discussed in the Method of Presentation section.

$$N = 40$$
$$\overline{X} \text{ for belief in technology} = 10.0$$
$$SS_x = 360$$
$$\overline{Y} \text{ for ecological behaviors} = 12.0$$
$$SS_Y = 360$$
$$CP = -126$$

23. Social scientists have studied extensively those factors that determine an effective group leader in small group problem-solving situations. It is generally recognized that the effectiveness of any given leader will depend not only upon his or her leadership style, but also upon the type of problem being addressed and the situation in which the group finds itself. Fiedler (1967) developed a scale designed to measure different leadership styles. The scale involves having individuals think of all the people with whom they have ever worked, and to single out their least-preferred coworker (LPC). They then rate this LPC on a series of scales, such as the extent to which the person was pleasant, friendly, cooperative, etc. A total score is obtained by summing these ratings. According to Fiedler, those who give high ratings on this scale tend to see even a poor coworker in a relatively favorable light. These leaders tend to behave in a manner described as compliant, nondirective, and generally relaxed. In contrast, those who give low ratings on the scale tend to be more demanding, controlling, and managing in their interaction with groups. In one experiment, Fiedler (1964) examined the relationship between a leader's LPC score and group problem-solving performance (time until solution of the problem). In this experiment, the task was relatively unstructured, there were good leader-member relations, and the leader had relatively little control over the rewards given to each group member. The hypothetical data presented below are representative of the results of this experiment. Analyze the data, draw a conclusion, and write up the results of the analysis using principles discussed in the Method of Presentation section.

$$N = 20$$
$$\overline{X} \text{ for LPC score} = 15.0$$
$$SS_x = 320$$
$$\overline{Y} \text{ for time until solution} = 10.0 \text{ minutes}$$
$$SS_Y = 180$$
$$CP = 144$$

Chi Square Test of Independence

14.1 / Use of the Chi Square Test of Independence

The **chi square test of independence** is typically used to analyze the relationship between two variables when

1. both variables are qualitative in nature (that is, measured on a nominal level), and

2. both variables are between-subjects in nature.

Let us consider an example of an investigation that used a chi square analysis. Miller and Swanson (1960) were interested in the relationship between the social class of parents and the kind of discipline they used to influence the behavior of their children. Three types of discipline were characterized: (1) physical (emphasizing physical punishment), (2) psychological (emphasizing psychological punishment such as scolding, withdrawal of affection), and (3) a mixture of physical and psychological. A group of middle class parents and a group of working class parents were studied and the type of discipline each tended to use was noted. It was found that middle class parents were more likely to use psychological discipline whereas working class parents were more likely to use physical discipline.

This investigation concerned the relationship between two qualitative variables: (1) social class membership (middle class versus working class), and (2) type of discipline (physical, psychological, mixed).* Both of the variables are between-subjects in nature. Individuals in the working class category are not the same individuals as those in the middle class category. Similarly, individuals who are classified as using primarily physical discipline are not the same individuals as those who are classified as using primarily psychological discipline who, in turn, are not the same individuals who are classified as using the mixed discipline. In this instance, a chi square test of independence is an appropriate statistical technique for analyzing the relationship between the two variables.

---

*Some social scientists view social class as a quantitative variable. We will consider the issue of using quantitative variables in chi square analysis in a later section of this chapter.

14.2 / Two-Way Frequency Tables

Unlike previous statistical tests, the chi square test of independence uses frequencies to determine the relationship between variables. This is because the variables are measured on nominal scales and it makes no sense to compute means and variances in the manner we have done previously. The chi square test of independence uses a **two-way frequency table** as the basis for analysis (also called a *contingency table*). Such a table is illustrated in Table 14.1, which examines the relationship between a person's gender and his or her political party preference. The table is "two-way" because it examines two variables. The entries within each box (or cell) represent the number of poeple in the sample who were characterized by the corresponding values of the variables. For example, a total of 170 people were studied. Of these, 63 were male Democrats, 17 were male Republicans, 10 were male Independents, 35 were female Democrats, 20 were female Republicans, and 25 were female Independents. The numbers in the last column and bottom row of Table 14.1 indicate how many individuals had each single characteristic. For example, there were 98 Democrats, 37 Republicans, and 35 Independents in the sample. Also, there were 90 males and 80 females. These frequencies are referred to as **marginal frequencies** and are the sum of the frequencies in the corresponding row or column (for example, $63 + 35 = 98$, $17 + 20 = 37$, $10 + 25 = 35$, $63 + 17 + 10 = 90$, $35 + 20 + 25 = 80$). The entire table is referred to as a 2×3 frequency table. The first number indicates the number of rows (2) and the second number indicates the number of columns (3).

TABLE 14.1 EXAMPLE OF A TWO-WAY CONTINGENCY TABLE

	Party Identification			
Gender	Democrat	Republican	Independent	Totals
Male	63	17	10	90
Female	35	20	25	80
Totals	98	37	35	170

14.3 / Inference of a Relationship Using the Chi Square Test of Independence

The logic underlying the chi square test of independence focuses upon the concept of **expected frequency.** The concept is best illustrated using the coin flipping example discussed in Chapter 7. Suppose you are given a coin and asked to determine whether it is fair. If you assume that the coin is fair, and if you were to flip it 100 times, your best guess about the outcome would be 50 heads and 50 tails (since the probability of each is .50). This expected frequency can then be compared with the **observed frequency** when you actually flip the coin. Suppose after 100 flips, you observed 53 heads and 47 tails. Although the observed frequency is not exactly 50/50, you would be hesitant to conclude that the coin is biased because of your knowledge of sampling error. But suppose the result had been 65/35. This is much more discrepant from the expected frequency. At some point the observed frequency will become too discrepant from the expected frequency and you will conclude that the coin is biased. This is analogous to the logic of the chi square analysis whereby you (1) assume a null hypothesis is true; (2) derive a set of expected frequencies based on this assumption; (3) compare the expected frequencies with the frequencies observed in the experiment; and (4) if the discrepancy is large (as defined by a given alpha level), reject the null hypothesis. We will use the example on gender and political party identification to illustrate the chi square test of independence.

Null and Alternative Hypotheses. We begin by stating a null and an alternative hypothesis. The null hypothesis states that there is no relationship between the two categorical variables; that a person's gender and his or her political party identification are independent.* The alternative hypothesis states that there is a relationship between the two variables.

Expected Frequencies and the Chi Square Statistic. The next step of the analysis requires computing the expected frequencies under the assumption of no relationship between the two variables. Consider the Democrat category for the party identification variable. If party identification and gender are unrelated, then one would expect that the proportion of males who are Democrats will be the same as the proportion of females who are Democrats. In our example, 98/170 or 57.65% of the *total* sample were Democrats. This represents an estimate of the percentage of Democrats in the population. If gender and party identification are unrelated, then we would expect 57.65% of the males to be Democrats and 57.65% of the females to be Democrats. In our sample, there were ninety males. 57.65% of 90 is 51.88. This is the expected frequency of Democratic males. There were eighty females in our sample. 57.65% of 80 is 46.12. This is the expected frequency of Democratic females.

Now consider the Republican category for the party identification variable. In the sample, 37/170 or 21.76% of the total sample were Republicans. If party identification and a person's gender are independent, then 21.76% of the ninety males should be Republican (yielding an expected frequency of 19.59) and 21.76% of the eighty females should be Republican (yielding an expected frequency of 17.41). The same procedure is applied for the Independents.

---

*The null hypothesis can be stated formally using probability theory (Chapter 5). If we let A represent a given outcome on sex and B represent a given outcome on political party preference, then H_0: $P(A,B)$ = $P(A)P(B)$ for all values of A and B (see Equation 5.4).

The steps for computing an expected frequency can be summarized as follows:

1. For the group in question (Democratic males), divide the total of the column in which it occurs (98) by the total number of observations (170).

2. Multiply this value by the total of the row in which the group appears (90).

The third column of Table 14.2 presents the expected frequencies for each group. In order to determine how discrepant the observed frequencies are from the expected frequencies, we can compute the difference between the two. This has been done in column 4 of Table 14.2. It is now necessary to combine the discrepancies into an index that reflects the total differences in the observed and expected frequencies. Many indices are possible but one useful index, called the **chi square statistic,** is defined by the following formula:

$$\chi^2 = \sum_{i=1}^{k} \frac{(O_i - E_i)^2}{E_i} \qquad \text{[14.1]}$$

where E is the expected frequency of group i, (male Republicans); O is the observed frequency of group i; k is the number of groups; and χ^2 is the chi square statistic. This index is useful because the sampling distribution of the statistic, under certain conditions, closely approximates a well-known theoretical distribution called the **chi square distribution.** Let us work through the computation of χ^2.

TABLE 14.2 OBSERVED AND EXPECTED FREQUENCIES

Group	Observed Frequency (O)	Expected Frequency (E)	O − E	(O − E)²	(O − E)²/E
Democratic Males	63	51.88	11.12	123.65	2.38
Democratic Females	35	46.12	−11.12	123.65	2.68
Republican Males	17	19.59	− 2.59	6.71	.34
Republican Females	20	17.41	2.59	6.71	.39
Independent Males	10	18.53	− 8.53	72.76	3.93
Independent Females	25	16.47	8.53	72.76	4.42
					14.14

(a) df = 1

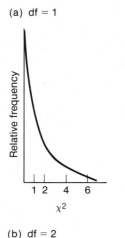

(b) df = 2

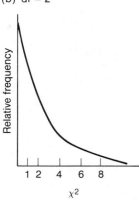

(c) df = 8

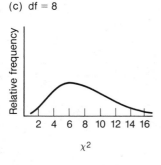

Figure 14.1 Chi Square Distributions for Different Degrees of Freedom

The difference between the observed and expected frequencies (column 4 of Table 14.2) is squared for each group. This has been done in column 5. The squared difference is next divided by the corresponding expected frequency (column 3). For the first group, this would be $123.65/51.88 = 2.38$. This has been done for each group in column 6. Finally, column 6 is summed, yielding the value of the chi square statistic (14.14).*

Sampling Distribution of the Chi Square Statistic. If two categorical variables are truly independent, we would expect the value of χ^2, as computed in the population, to be zero. However, because of sampling error, a χ^2 computed from sample data may be positive even when the null hypothesis is true. Consider a large population where the null hypothesis is true. A sample of 170 individuals is randomly selected and the chi square statistic is computed as before. The value of χ^2 might be .50. Now suppose we repeat this procedure a second time and compute another chi square statistic based on a second random sample. This statistic might differ from the one previously computed because of sampling error. In principle, we could continue to do this a large number of times with the result being a large number of chi square statistics. The sample chi square statistics, if computed on all possible random samples of a given size, would form a **sampling distribution of the chi square statistic.** It would closely approximate, under certain conditions, a well-known theoretical distribution called the chi square distribution. Figure 14.1 presents examples of chi square distributions. There are different chi square distributions depending on the degrees of freedom associated with them. The relevant chi square distribution is one with $(r - 1)(c - 1)$ degrees of freedom, where r is the number of values of the first qualitative variable (or the number of rows) and c is the number of values of the second qualitative variable (or the number of columns). In our example, the degrees of freedom are $(3 - 1)(2 - 1) = 2$.

As with previous sampling distributions, knowledge of the chi square distribution allows us to specify probability levels of obtaining a chi square as large as or larger than the observed chi square statistic under the assumption of the null hypothesis. It is therefore possible to derive a critical value for any given alpha level to test the null hypothesis of no relationship. Appendix L presents a table of critical values for the chi square distribution. In the gender-party identification example, for an alpha level of .05, the critical chi square for 2 degrees of freedom is 5.99. The observed chi square of 14.14 exceeds this value and the null hypothesis is therefore rejected.

Assumptions of the Test of Independence. The preceding inferential test is based on several assumptions. First, in order for the sampling distribution of the chi square statistic to approximate a chi square distribution, the sample size, N, must be

---

*Appendix 14.1 presents a computationally simpler formula for computing χ^2 in a 2×2 contingency table.

relatively large. Second, the *expected* frequency of any group must also be nonzero. Statisticians have recommended that the lowest expected frequency one should observe in order to use the chi square statistic is between 5 (a liberal estimate) and 10 (a more conservative estimate). For applications with df = 1, the expected frequency may be low as 2, according to some statisticians. Third, as noted earlier, the qualitative variables must be between-subjects and each frequency must be based on different individuals.

Study Exercise 14.1

For the experiment by Miller and Swanson discussed earlier, suppose that the frequency table appeared as follows: .

TYPE OF DISCIPLINE

Social Class	Psychological	Mixed	Physical	Totals
Middle Class	49	25	24	98
Working Class	29	48	60	93
Totals	78	73	84	191

Test for the existence of a relationship between these two variables.

Answer. The relevant calculations for the chi square statistics are shown in Table 14.3. We find that $\chi^2 = 30.04$. Based upon $(3 - 1)(2 - 1) = 2$ degrees of freedom, the critical value for χ^2, alpha level = .05, is 5.99. Since the value of 30.04 exceeds 5.99, we reject the null hypothesis and conclude that a relationship exists.

TABLE 14.3 COMPUTATION OF CHI SQUARE FOR MILLER AND SWANSON STUDY

Group	O	E	$O - E$	$(O - E)^2$	$(O - E)^2/E$
Psychological—Middle Class	49	39.98	9.02	81.36	2.04
Psychological—Working Class	29	37.94	− 8.94	79.92	2.11
Mixed—Middle Class	25	37.44	−12.44	154.75	4.13
Mixed—Working Class	48	35.53	12.47	155.50	4.38
Physical—Middle Class	24	43.12	−19.12	365.57	8.48
Physical—Working Class	60	40.92	19.08	364.05	8.90
					$\chi^2 = 30.04$

Yates' Correction for Continuity. When both of the qualitative variables have only two values (that is, when the df $= 1$), then the sampling distribution of the chi square statistic does *not* correspond to a chi square distribution. In this case, a correction factor for the computed chi square statistic has been suggested that supposedly yields a sampling distribution similar to the chi square distribution. This is known as **Yates' correction for continuity** and the formula for computing the chi square statistic is

$$\chi^2 = \sum_{i=1}^{k} \frac{(|O_i - E_i| - .5)^2}{E_i} \qquad [14.2]$$

The correction factor involves subtracting .5 from the *absolute* difference of $O_i - E_i$, then squaring this value, dividing by E_i, and summing across groups. Whether one should apply Yates' correction for continuity is a subject of considerable debate among statisticians (Conover, 1974a, 1974b; Grizzle, 1967; Mantel, 1974; Miettinen, 1974; Plackett, 1964). However, in practice, the correction is applied more often than not.

14.4 / Strength of the Relationship

Our second question concerns assessing the strength of the relationship. A number of different indices for measuring the strength of association between two qualitative variables have been proposed. Among the more common are Pearson's index of mean square contingency, the phi coefficient, Cramer's statistic, gamma, the coefficient of contingency, and the Goodman-Kruskal index of predictive association. Statisticians do not agree on which of these statistics is most appropriate. Probably the most common index of the strength of association is the **fourfold point correlation** (as it is called when applied to the relationship between qualitative variables with two values each) or **Cramer's statistic** (as it is called when df > 1). The strength of association is defined as

$$V = \sqrt{\frac{\chi^2}{N(L-1)}} \qquad [14.3]$$

where χ^2 is the observed chi square statistic, N is the number of subjects in the total sample, and L is the number of values of the qualitative variable that has the fewest values. In the example on sex and party identification, $\chi^2 = 14.14$, $N = 170$, and $L = 2$. L is 2 since the sex variable has two values and party identification has three values. Since 2 is less than 3, $L = 2$.

$$V = \sqrt{\frac{14.14}{170(2-1)}} = .288$$

The fourfold point correlation and Cramer's statistic will always range from 0 to 1.00, where numbers toward 0 reflect no relationship and numbers toward 1.00

reflect a perfect relationship. The statistic can be interpreted *in terms of magnitude* like the correlation coefficient discussed in Chapter 13. To illustrate the mathematical connection between Pearson's correlation coefficient and the present approach, consider the case of a contingency table, with two levels of A and two levels of B. For each subject in the study, a score of 1 is assigned if he or she is characterized by level 1 of A and a score of 2 is assigned if he or she is characterized by level 2. The same is done for the B variable. If the Pearson correlation is computed between the variables, using the formula given in Chapter 13, the absolute value of the result would be equivalent to the fourfold point correlation. Conceptually, a large value for Cramer's statistic indicates a tendency for one category of A to be associated with another category of B.

14.5 / Nature of the Relationship

If the null hypothesis of no relationship is rejected, then additional analyses are required to more fully discern the nature of the relationship. The test of the χ^2 statistic applies only to the data taken as a whole and provides no information as to which cells are responsible for rejecting the null hypothesis. Similar to the F test, the χ^2 test is nondirectional, since all discrepancies from the expected frequencies are reflected in the upper tail of the distribution. Just as we applied the HSD test in analysis of variance to break down the overall relationship, a comparable test can be applied to the chi square test of independence. Procedures have been suggested by Cohen (1967), Ryan (1960), Goodman (1964), and Wike (1971). The procedure recommended by Goodman (1964) is the most popular among statisticians. The mathematics of the approach, however, are complex and interested readers are referred to Appendix 14.2.

From an intuitive perspective, insights into the nature of the relationship can be gained by examining the $(O - E)^2/E$ statistic for each group of individuals. This corresponds to the last column of numbers in Table 14.2. These numbers are summed to yield the overall χ^2 square statistic. The smaller numbers within this column contribute less to the rejection of the null hypothesis than the larger numbers. The two largest numbers in this column were for the Independent females (4.42) and the Independent males (3.93). Examination of the observed minus expected frequencies ($O - E$, column 4) indicates that males were less likely to be Independents than expected (-8.53) and females were more likely to be Independents than expected ($+8.53$), given the assumption of no relationship between sex and party identification. Similarly, males were more likely to be Democrats than expected (11.12) and females were less likely to be Democrats than expected (-11.12). The values of $(O - E)^2/E$ for Republican males (.34) and Republican females (.39) were relatively small and suggest that the differences in these groups contributed little to the rejection of the null hypothesis.

Method of Presentation

The chi square test of independence is typically reported in conjunction with a two-way frequency table of the observed frequencies. Such a table might appear in the format of Table 14.1. The results of the analysis might be written in a research report as follows:

A chi square test of independence was performed on the relationship between a person's sex and party identification. The chi square was statistically significant ($\chi^2 = 14.14$, df = 2, $p < .05$). Cramer's statistic yielded a value of .288. The nature of the relationship was such that males were more likely than females to identify with the Democratic party, whereas the reverse was true with respect to identifying as an Independent. The differences in identification with the Republican party as a function of sex were minimal.

The first sentence identifies the statistical test and the second sentence indicates whether the null hypothesis was rejected. The statistics reported in the parentheses are the observed χ^2 value, the degrees of freedom, and the p value. Cramer's statistic indicates the strength of the relationship. If the analysis had only 1 degree of freedom, this would have been called a fourfold point correlation. The last two sentences report the nature of the relationship. In practice, this should be stated formally using the procedures developed by Goodman. Few social scientists use these procedures and the more common approach is to interpret the nature of the relationship via inspection of the contingency table.

It is sometimes useful to present the observed frequencies in graphic form rather than tabular form. Figure 14.2 illustrates the format that might be used. This format is similar to the frequency polygon discussed in Chapter 2, but two variables are presented. On the abscissa are the values of one variable, in this case, party identification. On the ordinate is the observed frequency. Two different sets of points are plotted, one for males and one for females. A different demarcation is used for each (males are O and females are X). Finally, the points common to males are connected by a solid line whereas the points common to females are connected by a dashed line. Because the abscissa represents a qualitative variable, the lines are not connected with the abscissa, as is traditionally done for frequency polygons.

14.6 / Methodological Considerations

From a methodological standpoint, the data in the investigation on gender and party identification are ambiguous in terms of causality. Even though we observed a relationship between a person's gender and his or her political party identification, this does not necessarily mean there is a causal link between the two variables. Demographic variables such as the sex of a person may be related to political party identification but, if anything, this relationship is probably due to the different

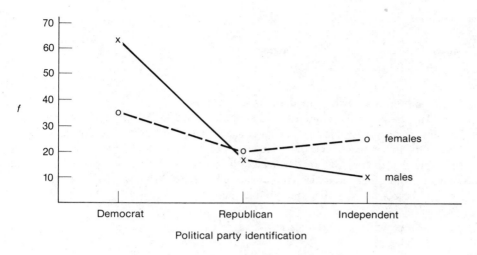

Figure 14.2 Graph of Observed Frequencies for Political Party Identification as a Function of Sex of Subject

experiences that males have relative to females in terms of their socialization, education, and personal development. The data of the present study also suggest that variables other than those that are related to a person's sex have a major impact on political party identification since the strength of the relationship, as indexed by Cramer's statistic, was only .288.

14.7 / Numerical Example

Social psychologists have studied variables that influence altruistic behavior. In one experiment, a researcher investigated the effects of a role model on people's willingness to donate money to a charity. Three hundred people were approached in a shopping center and asked to donate money to a charity. One hundred people were approached just after they had seen a confederate of the experimenter donate money (this represents a *positive model*). Another hundred people were approached just after they had seen a confederate of the experimenter refuse to donate money (this represents a *negative model*). Finally, one hundred people were approached without having observed a model (this represents *no model*). The number of people who donated money and the number of people who refused to donate money in each condition are presented in Table 14.4.

The relevant calculations for the chi square statistic are shown in Table 14.5. We find that $\chi^2 = 16.83$. Based on 2 degrees of freedom, the critical value for χ^2, alpha level = .05, is 5.99 (see Appendix L). Since the value of 16.83 exceeds 5.99, we reject the null hypothesis of no relationship between the variables.

The strength of the relationship is indicated by Cramer's statistic. Using Equation 14.3, we find that

$$V = \sqrt{\frac{16.83}{(300)(2-1)}}$$

$$= .24$$

This represents a weak to moderate relationship.

The nature of the relationship can be discerned intuitively by examination of columns 4 and 6 of Table 14.5. It can be seen that the no model condition contributed very little to the overall chi square ($(O - E)^2/E$ values of .002). The other conditions indicate that the positive model tended to increase donating behavior whereas the negative model tended to decrease donating behavior. What methodological factors should be considered when interpreting these data?

TABLE 14.4 DATA FOR ALTRUISM EXPERIMENT

	Positive Model	Negative Model	No Model	Totals
Donated	63	34	49	146
Did Not Donate	37	66	51	154
Totals	100	100	100	300

TABLE 14.5 COMPUTATION OF CHI SQUARE FOR ALTRUISM EXPERIMENT

Group	O	E	O − E	$(O - E)^2$	$(O - E)^2/E$
Positive—Donated	63	48.67	14.33	205.35	4.22
Positive—Did Not Donate	37	51.33	− 14.33	205.35	4.00
Negative—Donated	34	48.67	− 14.67	215.21	4.42
Negative—Did Not Donate	66	51.33	14.67	215.21	4.19
None—Donated	49	48.67	.33	.11	.002
None—Did Not Donate	51	51.33	−.33	.11	.002

$$\chi^2 = 16.83$$

14.8 / Use of Quantitative Variables in the Test of Independence

The chi square test of independence is typically applied to the analysis of the relationship between two qualitative variables. However, researchers will sometimes group scores on quantitative variables and use the chi square test. An example

Box 14.1 Religion and Intellectualism

Several sociologists have argued that American Catholics have not contributed to the scientific and intellectual development of the United States relative to their numbers in the population. Considerable documentation for this assertion exists. Some social scientists have argued that this is the result of the cultural values of American Catholics, which are said to impede intellectual achievement. Warkov and Greeley (1966), however, have argued that the deficiency may be due to the social conditions surrounding early American Catholic immigrants. According to Warkov and Greeley, Catholic immigrants were too poor and had immigrated too recently to be concerned with much beyond sheer economic survival. They argued that immigrants had little time to concern themselves with matters such as intellectualism and going to college since all of their efforts had to be directed toward establishing an adequate economic base. This would not, however, necessarily be true for more recent generations who have had the opportunity to work from the economic base established by their ancestors. Based on this logic, Warkov and Greeley predicted that there should be a relationship between one's religion (Catholic versus Protestant) and whether a person has attended college for older people but not for younger individuals, since the Catholics in the latter group have had the economic freedom to concern themselves with education while the Catholics in the former group have not. The frequency data collected by Warkov and Greeley are as follows:

YOUNGER GROUP (23–29 YEARS OLD)

	Catholic	Protestant
Attended College	99	265
Did Not Attend College	253	642

OLDER GROUP (50–59 YEARS OLD)

	Catholic	Protestant
Attended College	42	24
Did Not Attend College	310	73

A chi square test of independence for the older group indicates a statistically significant relationship between religion and college attendance ($\chi^2 = 9.57$, df $= 1$, $p < .05$). The strength of the relationship, as indexed by the fourfold point correlation, was .15. The nature of the relationship was such that Protestants were more likely to attend college than Catholics. In contrast, the chi square analysis for the younger group yielded a nonsignificant chi square ($\chi^2 = .16$, df $= 1$, ns) and a phi coefficient of .01. The patterning of the data are more intuitively apparent when expressed as percentages.

PERCENTAGE ATTENDING COLLEGE

	Catholic	Protestant
Younger Group	28%	29%
Older Group	12%	25%

For the older group, the percentage of Catholics attending college was less than Protestants. However, for the younger group, the percentages were roughly the same. These data are consistent with the reasoning of Warkov and Greeley. However, there are other interpretations to the data as well. Can you think of them?

would be the following frequency table for analyzing the relationship between a person's age at which he or she was married and the duration of the marriage (in years):

Age at Marriage	Duration of Marriage			
	<5	5–9	10–14	≥15
<19	42	28	16	18
19–24	33	26	23	23
25–34	12	9	17	15
>34	14	14	10	11

The advantages and disadvantages of reducing a quantitative variable to a small number of groups and then using it in an analysis for qualitative variables have already been discussed in Chapter 13. The same issues apply in the present context. Briefly, when a quantitative variable is collapsed into categories, considerable information is lost by grouping individuals with different scores (an age at marriage of 25 and an age at marriage of 34) into the same group. This potentially can increase the amount of error in the study and make the statistical test less powerful. An advantage of the present technique, however, is that the quantitative variables need only be measured on an ordinal as opposed to an interval level.

14.9 / Planning an Investigation Using the Chi Square Test of Independence

When planning an investigation that will use the chi square test of independence, the selection of sample size in order to achieve desired levels of power can be guided by Appendix F.4. Table 14.6 presents a portion of the appendix for illustration. The first column indicates different levels of power and the column headings indicate population values for Cramer's statistic (or where df $= 1$, the fourfold point correlation). There is a different sample size table required for different contingency tables (2×3, 3×3, 3×4, and so on). Table 14.6 presents the case for a 3×3 contingency table. As an example, if the desired level of power is .80 and if the researcher suspects that the strength of the relationship in the population corresponds to a Cramer's statistic of .30, then the total number of subjects, N, that should be sampled is 66. Inspection of Table 14.6 and Appendix F.4 will give the reader a general appreciation for the relationship between power, sample size, alpha levels, and the strength of the population relationship for the chi square test of independence.

TABLE 14.6 SAMPLE SIZES NECESSARY TO ACHIEVE SELECTED LEVELS OF POWER FOR ALPHA = .05, 3 × 3 TABLE, AS A FUNCTION OF POPULATION VALUES OF CRAMER'S STATISTIC

Type of Table: 3 × 3
Alpha = 0.05

Power	0.1	0.2	0.3	0.4	0.5	0.6	0.7	0.8	0.9
				Population Value of Cramer's Index					
0.25	154	39	17	10	6	4	3	2	2
0.50	321	80	36	20	13	9	7	5	4
0.60	396	99	44	25	16	11	8	6	5
0.70	484	121	54	30	19	13	10	8	6
0.75	536	134	60	34	21	15	11	8	7
0.80	597	149	66	37	24	17	12	9	7
0.85	671	168	75	42	27	19	14	10	8
0.90	770	193	86	48	31	21	16	12	10
0.95	929	232	103	58	37	26	19	15	11
0.99	1262	316	140	79	50	35	26	20	16

14.10 / Summary

The chi square test of independence is typically used when (1) both the independent and dependent variables are qualitative in nature, and (2) both variables are between-subjects in nature. The test is based on the analysis of discrepancies between observed frequencies and expected frequencies from the null hypothesis. The chi square statistic has a sampling distribution that approximates a well-known distribution called the chi square distribution. The test for the existence of a relationship using this distribution requires relatively large sample sizes and expected frequencies greater than 5. When the distribution has 1 degree of freedom, Yates' correction for continuity is typically used. The strength of the relationship between the independent and dependent variables is indicated by the fourfold point correlation or Cramer's statistic. The nature of the relationship can be analyzed by examining the size of the discrepancies between expected and observed frequencies in each group, separately.

APPENDIX 14.1

Computational Formula for a 2 × 2 Contingency Table

In cases where both qualitative variables have two and only two values, an efficient computational formula can be used to derive the chi square statistic. This formula includes Yates' correction for continuity and is

$$\chi^2 = \frac{N\left(|ad - bc| - \frac{N}{2}\right)^2}{(a + b)(c + d)(a + c)(b + d)} \quad \text{[14.4]}$$

where the letters in the following table represent frequencies for the four groups, respectively:

	Variable A	
Variable B	a	b
	c	d

Consider the following frequencies:

	Variable A		Total
Variable B	25	16	41
	15	23	38
Total	40	39	79

Then,

$$\chi^2 = \frac{79\left(|(25)(23) - (16)(15)| - \frac{79}{2}\right)^2}{(25 + 16)(15 + 23)(25 + 15)(16 + 23)}$$

$$= \frac{(79)(295.5^2)}{2430480}$$

$$= 2.84$$

APPENDIX 14.2

Goodman's Simultaneous Confidence Interval Procedure

We will develop the logic of Goodman's procedure using the data in Table 14.1. The technique considers all possible 2 × 2 contingency tables. In the case of a 2 × 3 analysis, there are three such tables:

	Democrat	Republican
Male	63	17
Female	35	20

	Democrat	Independent
Male	63	10
Female	35	25

	Republican	Independent
Male	17	10
Female	20	25

In addition to the fourfold point correlation as a measure of association for 2 × 2 tables, Fisher has suggested the following measure

$$g = \frac{P_{11}P_{22}}{P_{12}P_{21}}$$

where P_{11} is the proportion of subjects relative to the total sample that is in the group designated by the first row and the first column (hence the subscript 11), P_{22} is the proportion of subjects relative to the total sample that is in the group designated by the second row and the second column, P_{12} is the proportion of subjects relative to the total sample that is in the group designated by the first row and the second column, and P_{21} is the proportion of subjects relative to the total sample that is in the group designated by the second row and the first column. The general formula for the P_{ij} is

$$P_{ij} = \frac{X_{ij}}{N} \qquad \textbf{[14.5]}$$

where X_{ij} is the frequency of subjects occurring in the group designated by the ith row and the jth column, and N is the total number of subjects in the 2×2 table.

If the two variables being studied are independent of each other, then g will equal one. Most researchers prefer to have statistical independence associated with a value of zero, and this can be accomplished by taking the natural logarithm of g:

$$y = \log_e g \qquad \textbf{[14.6]}$$

In this case, when the variables are independent, $y = 0$; when the variables are positively related, $y > 0$; and when the variables are negatively related, $y < 0$. Goodman (1964) has shown that the population value of y can be estimated by

$$\hat{y} = (2.3026) \log_{10} \left(\frac{X_{11}X_{22}}{X_{12}X_{21}} \right) \qquad \textbf{[14.7]}$$

with a standard error of

$$\sigma_{\hat{y}}^2 = \frac{1}{X_{11}} + \frac{1}{X_{22}} + \frac{1}{X_{12}} + \frac{1}{X_{21}} \qquad \textbf{[14.8]}$$

Consider the first 2×2 table. We find that

$$\hat{y} = (2.3026) \log_{10} \left(\frac{(63)(20)}{(17)(35)} \right)$$

$$= (2.3026)(.326)$$

$$= .751$$

and

$$\sigma_{\hat{y}}^2 = \frac{1}{63} + \frac{1}{20} + \frac{1}{17} + \frac{1}{35} = .153$$

For large samples, the sampling distribution of y is normally distributed. Under the assumption of the null hypothesis, the observed $\hat{y}$ can be translated into a z score as follows:

$$z = \frac{\hat{y} - 0}{\sigma_{\hat{y}}} \qquad \textbf{[14.9]}$$

Since

$$\sigma_{\hat{y}} = \sqrt{\sigma_{\hat{y}}^2}$$

$$\sigma_{\hat{y}} = \sqrt{.153}$$

$$= .391$$

Then

$$z = \frac{.751 - 0}{.391} = 1.92$$

These procedures can be applied to the remaining 2×2 tables yielding the following:

$$z \text{ for table 1} = 1.92$$

$$z \text{ for table 2} = 3.51$$

$$z \text{ for table 3} = 1.51$$

Each z score is then compared against a critical value, cv, defined as follows:

$$cv = \pm \sqrt{\chi^2_{\text{critical}}} \qquad \textbf{[14.10]}$$

where χ^2_{critical} is the critical value of χ^2 used to reject the original null hypothesis. In this case, it was a χ^2 based on 2 degrees of freedom, and for alpha $= .05$, $\chi^2_{\text{critical}} = 5.99$. Thus

$$cv = \pm \sqrt{5.99}$$

$$= \pm 2.45$$

Only the z score for the second 2×2 table exceeds this value and, consequently, it indicates the source of rejection of the original null hypothesis. Males were more likely to be Democrats and females were more likely to be Independents. This is consistent with the conclusions drawn from the intuitive procedures developed in the main text.

APPENDIX 14.3 CHI SQUARE
GOODNESS-OF-FIT TEST

Another application of the chi square statistic is the *goodness-of-fit test*. This application involves testing whether a distribution of scores in a random sample could, within the constraints of sampling error, come from a population characterized by a type of distribution specified a priori by the investigator. For example, suppose the relative frequencies of marital status for the population of U.S. female adults who are under 40 years of age are as follows:

	rf
Married	.60
Single	.23
Separated	.04
Divorced	.12
Widowed	.01

An investigator conducted a study using a sample of 200 adult females under the age of forty and wanted to know if they were representative of the general population in terms of marital status. The frequency distribution for the sample was:

	f
Married	100
Single	44
Separated	16
Divorced	36
Widowed	4
	200

The null hypothesis is that the sample represents a random selection from a population that does *not* differ in the distribution of marital status from that of the general U.S. population. The alternative hypothesis is that the sample represents a random selection from a population that does differ in the distribution of marital status from that of the general U.S. population. Based on the relative frequencies in the U.S. population previously noted, we can compute the frequencies that would be expected in the sample. The expected frequencies are the respective relative frequencies times the sample size, N. In this case,

	rf	N	$rf \cdot N = E_i$
Married	.60	200	120
Single	.23	200	46
Separated	.04	200	8
Divorced	.12	200	24
Widowed	.01	200	2

The observed and expected frequencies are then used in Equation 14.2 to yield the chi square statistic. The computations are shown in Table 14.7. The resulting χ^2 is 19.42 and is distributed as chi square with $k - 1$ degrees of freedom, where $k =$ the number of groups. Using Appendix L, the critical value for a chi square with $5 - 1 = 4$ degrees of freedom, $\alpha = .05$, is 9.49. The observed chi square exceeds this value, so we reject the null hypothesis that the sample represents a population with a distribution equal to that of the U.S. population.

If we examine the last column in Table 14.7, we can gain an appreciation for which groups were most responsible for the rejection of the null hypothesis. The largest $(O_i - E_i)^2/E_i$ values were for the separated and divorced groups. The discrepancy between the observed frequencies and expected frequencies in these two groups contribute the most to the overall chi square.

The chi square goodness-of-fit test is based on several assumptions. First, random and independent sampling with independent groups is required. Each sampled individual must fall into one and only one category. In addition, the sample must be relatively large. If only two categories are used, the expected frequency should be greater than 10 in any given category. If more than two categories are used, the expected frequency should be greater than 5 in any given category.

TABLE 14.7 COMPUTATION OF CHI SQUARE FOR GOODNESS-OF-FIT TEST

Category	O_i	E_i	$O_i - E_i$	$(O_i - E_i)^2$	$(O_i - E_i)^2/E_i$
Married	100	120	−20	400	3.33
Single	44	46	−2	4	.09
Separated	16	8	8	64	8.00
Divorced	36	24	12	144	6.00
Widowed	4	2	2	4	2.00
					$\chi^2 = 19.42$

EXERCISES

Exercises to Review Concepts

1. Under what conditions is the chi square test of independence typically used to analyze a bivariate relationship?

2. What are the major assumptions underlying the chi square test of independence in terms of inferring that a relationship exists between two variables?

3. Determine the critical value of chi square for each of the following conditions.

  (a) $r = 2, c = 2$
  (b) $r = 3, c = 3$
  (c) $r = 2, c = 3$
  (d) $r = 4, c = 4$

4. Define the concept of expected frequency.

5. In a 3 × 4 frequency table, how many columns are there? How many rows?

6. For the following table, compute the expected frequencies for each group under the assumption of no relationship between A and B.

		B	
	10	20	30
A	15	30	40
	20	40	50

7. When is it appropriate to use Yates' correction for continuity?

8. The relationship between smoking and cancer deaths was studied for a group of one hundred forty cases of people who recently died. The data appear as follows:

	Smoker	Nonsmoker
Cancer Present	36	25
Cancer Absent	34	45

Analyze the relationship between the two variables.

9. When is the fourfold point correlation used as opposed to Cramer's statistic for reporting the strength of a relationship?

10. A researcher investigated the relationship between race and voting behavior in a given city. The following data were obtained for a sample in a presidential election.

	Voted	Did Not Vote
Black	60	40
White	70	30
Oriental	75	25

Analyze the relationship between the two variables.

11. A study was undertaken relating political party preference to religion. The following data for a sample were observed.

	Protestant	Catholic	Jewish
Democrat	40	60	30
Republican	40	20	10
Independent	20	20	10

Analyze the relationship between these two variables.

For each experiment, indicate the type of statistical test you would use to analyze the data *and* explain why (assume the assumptions underlying the test have been satisfied).

12. *Experiment I.* A researcher wanted to test the relationship between college student's scores on an intelligence test and their grade point average.

13. *Experiment II.* A researcher wanted to test the effects of a special summer school learning program on reading skills. A test of reading skills was administered before and after the program and mean performance at the two times compared. Scores on the reading skill test could range from 20 to 100, with higher scores indicating greater reading ability.

14. *Experiment III.* Ninety students were studied in three different classes, a statistics class, a nursing class, and a psychology class. Thirty students were in each class. An anxiety score was derived by administering an anxiety scale (1 = low anxiety through 50 = high anxiety).

15. *Experiment IV.* An investigator wanted to test the relationship between a person's sex and his or her preference for one of three types of cars, Ford, Chevrolet, or Chrysler. Two hundred males and females were interviewed and asked which car they preferred.

16. *Experiment V.* An investigator studied the effects of religion on family size decisions. This investigator interviewed a total of 270 people, 90 of whom were Catholics, 90 of whom were Protestants, and 90 of whom were Jewish. Respondents were asked the number of children they wanted to have in their completed family. The average number of children was compared for the three groups.

17. *Experiment VI.* A survey was conducted that tested the relationship between a person's race and his or her political party preference. Four hundred individuals were interviewed and his or her race and political party preference noted.

Exercises to Apply Concepts

18. Since the inception of television, social scientists have been concerned with the influence of TV on the social development of children. Violence in programming and its effects on aggressive behavior have been the major source of investigation. However, recently, social scientists have studied how television affects sex stereotyping in children. An investigation by McArthur and Eisen (1976) sought to document the amount of sex stereotyping occurring on television. In one analysis, McArthur and Eisen analyzed commercials shown to children on Saturday morning television programs. (Barcus (1971) has estimated that by age 17 the average viewer has seen approximately 350,000 commercials on television.) The commercials were analyzed in terms of whether the central character was male or female and, in turn, whether the central character was portrayed as an authority (that is, an expert about the product) or simply as a user of the product. The data for the investigation are presented below. Analyze the relationship between sex of the central character and the role portrayed by that character and present the results using procedures developed in the Method of Presentation section.

	Authority	Product User
Male	138	114
Female	20	43

19. Presidential debates on television have become more frequent in recent years. It is commonly believed that the "winner" of a debate will benefit greatly at the polls. Research has indicated, however, that the assessment of who wins a debate might not be entirely objective. In one study, a sample of individuals was asked who they thought won a presidential debate, the Democratic or Republican candidate. The individuals were also interviewed prior to the debates to determine the political party they identified with. The data for this study are presented below. Analyze the relationship between political party preference and judged winner of the debate and report your results using the principles discussed in the Method of Presentation section.

Political Party Preference	Winner	
	Democratic Candidate	Republican Candidate
Democrat	71	10
Republican	17	58
Independent	35	32

Nonparametric Statistical Tests

With the exception of the chi square test of independence, all of the statistical tests considered so far require assumptions about the distributions of variables in the populations from which the samples are selected. The most common assumptions concern normal distributions and homogeneity of variances. In most cases, such parametric tests have been shown to be relatively robust to violations of these assumptions. However, when these assumptions are severely violated, nonparametric tests are more appropriate. Nonparametric statistical tests make few assumptions about the nature of the population distribution and hence, are called **distribution-free tests.** In addition, these procedures can be used to analyze quantitative variables that are measured at an ordinal level. In contrast, the parametric statistics using quantitative variables are most appropriate when variables are measured on approximately an interval level.

If nonparametric tests require fewer assumptions about the population distribution and the levels of measurement, then why are parametric tests so much more popular? The major reason is that parametric tests are usually (but not always) more powerful, in a statistical sense, than their nonparametric counterparts. This, plus the fact that parametric tests are relatively robust to violations of assumptions, have contributed greatly to the general preference for parametric procedures. The development of parametric statistics has tended to receive more attention from statisticians relative to nonparametric statistics. This historical bias has yielded a situation whereby certain important statistical questions have been addressed from a parametric perspective but not a nonparametric perspective. As advances are made in this latter respect, the use of nonparametric statistics should increase.

For each of the parametric tests we have considered thus far, there is a nonparametric counterpart. Actually, there are two types of nonparametric counterparts, one type based on ranked scores and the other based on sign tests. We will only consider tests based on ranks, since they are the more popular of the two. The relevant sign tests are discussed by Marascuilo and McSweeney (1977). Table 15.1 presents the names of the parametric tests and their nonparametric counterparts that will be considered in this chapter.

TABLE 15.1 PARAMETRIC TESTS AND THEIR NONPARAMETRIC COUNTERPARTS

Parametric Test	Nonparametric Counterpart
Independent Groups t Test	Mann-Whitney U Test/Wilcoxon Rank Sum Test
Correlated Groups t Test	Wilcoxon Signed-Rank Test
One-way Analysis of Variance	Kruskal-Wallis Test
Repeated Measures Analysis of Variance	Friedman Analysis of Variance by Ranks
Pearson Correlation	Spearman Rank Order Correlation

15.1 / Ranked Scores

Most of the tests discussed in this chapter use ranked scores. To illustrate ranking procedures, consider the following scores on an intelligence test for five individuals:

Individual	IQ Score	Rank
1	138	3
2	139	4
3	142	5
4	130	2
5	126	1

We can rank order the scores from least intelligent to most intelligent by assigning the number 1 to the least intelligent individual and then assigning successive integers to individuals in increasing order of IQ. In some cases, the values of IQ scores will be tied. Consider the following:

Individual	IQ Score	Rank
1	135	3.5
2	135	3.5
3	142	5
4	130	2
5	126	1

In cases where a tie occurs the scores involved are assigned the average of the ranks involved. In the above example, the tie occurred where ranks 3 and 4 were to be assigned. The average of the ranks is $(3 + 4) / 2 = 3.5$. Thus, the rank of 3.5 is assigned to the two scores. The next highest score would be assigned a rank of 5 (not 4) since ranks 3 and 4 have already been "used."

The rank order tests assume that the quantitative variables are continuous in nature. Given this assumption, tie scores are viewed as being the result of imprecision of measurement, since, in principle, more precise measures would eliminate ties. Nevertheless, tied scores often occur in practice and must be dealt with when applying a nonparametric test. The most common approach is to rank the scores using the preceding procedure (assigning the average rank for ties) and then introduce a "correction term" into the formula for the test statistic to adjust for the presence of ties. Unfortunately, statisticians do not agree on the optimal correction formulas. The statistical tests in this chapter will be developed in pure form where no ties exist. As long as the number of ties is minimal, the test statistics will not be affected greatly by tied ranks. Appendix 15.1 presents the most frequently advocated correction formulas. In all of the tests to be described, we will assume that the assumptions of the parametric test have been violated to the extent that non-parametric tests are required. We will consider the issue of selecting parametric versus nonparametric tests in Chapter 16.

15.2 / Rank Test for Two Independent Groups

Use of the Rank Test for Two Independent Groups. The **Mann-Whitney** *U* **test** is the nonparametric counterpart of the independent groups *t* test. It is typically used when (1) the dependent variable is quantitative and continuous in nature, (2) the independent variable is between-subjects in nature, and (3) the independent variable has two and only two values. The test is used in cases where scores on the dependent variable have been converted to ranks. The test is the same as another statistical procedure called the **Wilcoxon rank sum test.**

Analysis of the Relationship. Consider an experiment where ten males and ten females were given an attitude scale (with possible scores from 0 to 50) designed to measure prejudice against women. The scores for each individual are presented in column 3 of Table 15.2. The first step of the analysis involves rank ordering the scores from highest (rank = 1) to lowest (rank = 20) across all individuals. This has been done in column 4 of Table 15.2. Next, the sum of the ranks is computed for each group, separately. Let R_1 equal the sum of the ranks for group 1 (males) and R_2 equal the sum of the ranks for group 2 (females). From Table 15.2 we see that

$$R_1 = 87$$

$$R_2 = 123$$

TABLE 15.2 DATA FOR MANN-WHITNEY U TEST

Individual	Sex	Score	Rank
1	M	48	1
2	M	45	2
3	M	35	5
4	M	33	6
5	M	32	7
6	M	27	12
7	M	31	8
8	M	20	17
9	M	24	15
10	M	25	14
11	F	43	3
12	F	28	11
13	F	29	10
14	F	30	9
15	F	40	4
16	F	19	18
17	F	15	20
18	F	17	19
19	F	26	13
20	F	23	16

$$R_1 = 1 + 2 + 5 + 6 + 7 + 12 + 8 + 17 + 15 + 14 = 87$$

$$R_2 = 3 + 11 + 10 + 9 + 4 + 18 + 20 + 19 + 13 + 16 = 123$$

Given a null hypothesis of no relationship between sex and prejudice against women, we would expect that any differences between R_1 and R_2 would reflect sampling error. There is no reason, under the assumption of the null hypothesis, to expect high scores to concentrate themselves in one group or the other, and we should find that both high- and low-ranked scores are intermingled among the two groups; that is, that $R_1 = R_2$ except for sampling error. In our example, we would expect to find that R_1 and R_2 are both equal to approximately 105 (that is, the sum of the ranks 1 through 20 divided by 2, or $(1 + 2 + 3 + \ldots + 20 / 2 = 105)$. A simple formula for computing the expected R score is

$$\overline{E} = \frac{n_1 (n_1 + n_2 + 1)}{2} \qquad [15.1]$$

Where $\overline{E}$ = the expected rank sum. In the example,

$$\overline{E} = \frac{10\,(10 + 10 + 1)}{2}$$

$$= \frac{210}{2}$$

$$= 105$$

Note that R_1 and R_2 are equidistant from $\overline{E}$ ($87 - 105 = -18$, $123 - 105 = 18$). This will occur whenever $n_1 = n_2$. The test we will develop makes use of the sample statistic R that equals R_1 or R_2, whichever comes from the group with the smaller sample size. The sampling distribution for this statistic has been shown to have a mean equal to $\overline{E}$ and a standard deviation equal to

$$\sigma_R = \sqrt{\frac{n_1\,n_2\,(n_1 + n_2 + 1)}{12}} \qquad \text{[15.2]}$$

where σ_R is the standard deviation of the sampling distribution of the R statistic (or the *standard error of the R statistic*) and all other terms are as previously defined. The shape of the distribution approximates a normal distribution and, in conjunction with a continuity correction suggested by Ferguson (1976), we can translate the R statistic into a z score via the following formula:

$$z = \frac{|R - \overline{E}| - 1}{\sigma_R} \qquad \text{[15.3]}$$

The subtraction of 1 in the numerator represents the continuity correction. In the example,

$$\sigma_R = \sqrt{\frac{(10)\,(10)\,(10 + 10 + 1)}{12}}$$

$$= \sqrt{175}$$

$$= 13.23$$

$$z = \frac{|87 - 105| - 1}{13.23}$$

$$= \frac{17}{13.23}$$

$$= 1.28$$

For an alpha level of .05, nondirectional test, the critical values for z (from Appendix C) are ± 1.96. Since 1.28 does not exceed 1.96 nor is it less than -1.96, we fail to reject the null hypothesis.

The above test is applicable when the sample sizes of both groups are 10 or more. When samples sizes are smaller than 10, the sampling distribution of the R statistic does not approximate a normal distribution. In this case, a new statistic (**the U statistic**) must be computed from the following formulas:

$$U_1 = n_1 \, n_2 + \frac{n_1 \, (n_1 + 1)}{2} - R_1 \qquad [15.4]$$

$$U_2 = n_1 \, n_2 + \frac{n_2 \, (n_2 + 1)}{2} - R_2 \qquad [15.5]$$

U is the smaller of the two values U_1 or U_2. The value of U is compared against critical values of U, which are provided in Appendix M. When the U statistic is used to evaluate the null hypothesis, the statistical test is referred to as the Mann-Whitney U test. U must be *smaller* than the critical value.

The strength of the relationship for the present nonparametric test can be measured using the **Glass rank biserial correlation coefficient.** It is derived using the following formula:

$$r_g = \frac{2 \, (\overline{R}_1 - \overline{R}_2)}{N} \qquad [15.6]$$

where $\overline{R}_1$ is the mean rank for group 1, $\overline{R}_2$ is the mean rank for group 2, and N is the total number of individuals in the experiment. The value of r_g can range from -1 to 1 and its magnitude can be interpreted like that of the Pearson correlation coefficient.

The nature of the relationship tested by the Wilcoxon and Mann-Whitney U tests is ascertained by inspection of R_1 and R_2. If we had rejected the null hypothesis in the present example, the data would indicate that males tend to be more prejudiced against women than females (since $R_1 = 87$ and $R_2 = 123$, and smaller rank scores imply greater degrees of prejudice).

See Method of Presentation on page 314.

15.3 / Rank Test for Two Correlated Groups

Use of the Rank Test for Two Correlated Groups. The **Wilcoxon signed-rank test** is the nonparametric counterpart of the correlated groups t test. It is typically used when (1) the dependent variable is quantitative and continuous in nature, (2) the independent variable is within-subjects in nature, and (3) the independent variable has two and only two values. The test is used in cases where differences in scores on the dependent variable are converted to ranks.

The results of the preceding statistical tests are usually reported as the Wilcoxon rank sum test when the normal distribution is used, and the Mann-Whitney U test when the U statistic is used. Reports typically include information about the value of z or U (whichever is appropriate), the sample sizes, and the p value. A report might appear as follows:

A Wilcoxon rank sum test was applied to the ranked data for males ($n = 10$) and females ($n = 10$). The difference in ranks was statistically nonsignificant ($z = 1.28$, ns).

A report of the Mann-Whitney U test might appear as follows:

A Mann-Whitney U test was applied to the ranked data for blacks ($n = 6$) and whites ($n = 6$). The difference in ranks was statistically significant ($U = 37.0$, $p < .05$). Blacks tended to exhibit a lesser degree of prejudice relative to whites.

Unfortunately, the strength of the relationship is seldom reported. This should be accomplished by reporting the value of the Glass rank biserial correlation coefficient.

Analysis of the Relationship. Consider an investigation in which the attitudes toward nuclear energy for each of ten individuals were measured just before a large nuclear accident and again just after it. The attitude scale was such that scores could range from 0 to 30, with higher numbers indicating more favorable attitudes. The investigator was interested in comparing attitudes before and after the accident. The first three columns of Table 15.3 present the data for the experiment.

The analysis begins by computing the difference between scores in the two conditions (before accident minus after accident). This has been done in column 4 of Table 15.3. The differences are then rank ordered (from smallest to largest), paying no attention to the sign of the difference. Also, differences of zero are ignored. This has been done in column 5 of Table 15.3. We then sum the rank scores for the positive differences (R_p) and sum the rank scores for the negative differences (R_n). This has been done in Table 15.3 and we find that

$$R_p = 29$$

$$R_n = 16$$

If the null hypothesis is true, we would expect the scores for individuals to be about the same in both conditions. Since positive and negative changes would be equally likely to occur and since the average size of the changes in each direction should be about the same, we would expect the sum of the ranks of the positive changes (R_p) to equal the sum of the ranks of the negative changes (R_n), except for sampling error.

TABLE 15.3 DATA FOR NUCLEAR ACCIDENT EXPERIMENT

Individual	Before Accident	After Accident	Difference	Rank
1	27	15	12	9
2	26	16	10	8
3	24	16	8	7
4	20	26	− 6	6
5	19	15	4	4
6	18	20	− 2	2
7	15	14	1	1
8	13	16	− 3	3
9	11	16	− 5	5
10	9	9	0	—

$$R_p = 9 + 8 + 7 + 4 + 1 = 29$$

$$R_n = 6 + 5 + 3 + 2 = 16$$

In the example, we would expect both sums to equal $(29 + 16) / 2$, or 22.5. The symbol $\overline{E}$ will represent the expected value of the sum of ranks. The question then focuses on the following: Given that the null hypothesis is true, how likely is it that we would obtain sums of rank equal to 29 and 16 when we expected to find sums of 22.5? Since both R_p and R_n are equidistant from the expected sum, either can be used in the analysis. Traditionally, whichever sum is smaller is used, and it is called a **T statistic.** In this case, $T = R_n = 16$.

It is possible to construct sampling distributions of the T statistic under the assumption of the null hypothesis and to define critical values of T for rejecting or failing to reject the null hypothesis. This has been done by Wilcoxon (1949) and a table of critical values is presented in Appendix N for $N < 40$.

In the present case, we use the table based on $N = 9$ instead of $N = 10$ since we ignored the scores for one individual whose difference score was zero. For an alpha level of .05, nondirectional test, the critical value of T is 5. For the Wilcoxon signed-rank test, the observed value of T must be *less than* the critical value. The observed T value of 16 exceeds the critical value, so we fail to reject the null hypothesis.

When the sample size is greater than 40, the sampling distribution of T is normally distributed with a standard error equal to the following:

$$\sigma_T = \frac{N (N + 1) (2N + 1)}{24} \qquad [15.7]$$

We can then translate the T statistic into a z score and use the appropriate critical value in Appendix C for testing the null hypothesis. The formula for the z score transformation is

$$z = \frac{T - \overline{E}}{\sigma_T} \qquad [15.8]$$

The Wilcoxon signed-rank test is based on the assumption that the difference scores (column 4 of Table 15.3) reflect ordinal level measurement.

The strength of the relationship in the Wilcoxon signed-rank test can be measured using the **matched-pairs rank biserial correlation coefficient.** It is computed by the following formula:

$$r_c = \frac{4 (T - \overline{E})}{N (N + 1)} \qquad [15.9]$$

where all terms are as previously defined. The magnitude of r_c can be interpreted in the same manner as the Glass rank biserial correlation coefficient.

The nature of the relationship in the Wilcoxon signed-rank test is addressed by simple inspection of the rank sums. In our example, if we had rejected the null hypothesis, we would have concluded that the attitudes toward nuclear energy tended to be less favorable after the nuclear accident than before the nuclear accident.

See Method of Presentation on page 317.

15.4 / Rank Test for Three or More Independent Samples

Use of the Rank Test for Three or More Independent Samples. The statistical test, called the **Kruskal-Wallis test,** is the nonparametric counterpart of one-way analysis of variance. It is typically used when (1) the dependent variable is quantitative and continuous in nature, (2) the independent variable is between-subjects in nature, and (3) the independent variable has three or more groups or values.

Analysis of the Relationship. Consider an experiment studying the effects of the number of roommates on attitudes toward living in dormitories. Eighteen students were assigned dormitory rooms with either one, two, or three roommates. After living with their roommates for one semester, the students were given a questionnaire measuring their attitudes towards living in dormitories. The scale values ranged from 0 to 30, with higher numbers indicating more favorable attitudes. The first two columns of Table 15.4 present the relevant data and column 3 presents the rank order of the data, with lower ranks indicating less favorable attitudes.

The results of the above analysis might appear in a research report as follows:

A Wilcoxon signed-rank test was applied to the differences in attitude scores before versus after the nuclear accident. The test indicated a statistically nonsignificant difference ($T = 16$, $N = 9$, ns).

The statistics reported in the parentheses are the value of the T statistic, the N upon which the

T is based, and the p value if the null hypothesis is rejected. Everything else is self-explanatory. If the sample size had been greater than 40, the parentheses would have contained a report of the derived z score and the p value. When appropriate, the value of the matched-pairs rank biserial correlation coefficient should be reported.

TABLE 15.4 DATA FOR EXPERIMENT ON ATTITUDES TOWARD LIVING IN DORMITORIES

Number of Roommates	Attitude Score	Rank
1	29	17
1	24	12
1	27	15
1	22	10
1	30	18
1	25	13
2	23	11
2	26	14
2	21	9
2	17	5
2	28	16
2	19	7
3	18	6
3	15	3
3	16	4
3	20	8
3	14	2
3	13	1

TABLE 15.5 RANKED DATA FOR EXPERIMENT ON ATTITUDES TOWARD LIVING IN DORMITORIES

One Roommate	Two Roommates	Three Roommates
17	11	6
12	14	3
15	9	4
10	5	8
18	16	2
13	7	1
$R_1 = 85$	$R_2 = 62$	$R_3 = 24$

The rank scores have been retabled in Table 15.5 for purposes of calculation. If the number of roommates does not influence the attitude toward living in the dorm, then we would expect the average rank for the three conditions to be the same, within the constraints of sampling error. To the extent that the average ranks are different, the null hypothesis of no relationship is questionable. The logic underlying the statistic proposed by Kruskal and Wallis to measure the differences in mean ranks can be illustrated by analogy (Friedman, 1972). Consider the following sets of two numbers that each sums to 10 and note what happens when each number is squared and the squares are summed:

$$5^2 + 5^2 = 25 + 25 = 50$$

$$4^2 + 6^2 = 16 + 36 = 52$$

$$3^2 + 7^2 = 9 + 49 = 58$$

$$2^2 + 8^2 = 4 + 64 = 68$$

Note that as the difference between the numbers increases, the sum of the squares of the numbers also increases. This property was used by Kruskal and Wallis (1952) in proposing the following test statistic (**the H statistic**) to reflect the differences between groups:

$$H = \left(\frac{12}{N(N+1)}\right)\left(\sum_{j=1}^{k} \frac{R_j^2}{n_j}\right) - 3(N+1) \qquad \textbf{[15.10]}$$

where R_j is the sum of the ranks for group j, N is the number of subjects in the experiment (in this case, $N = 18$), and n_j is the number of subjects in group j. In this formula, the sum of the ranks in each group is squared and then divided by the number of subjects in that group (to yield an "average" squared rank). The remaining terms in the equation combine with this term to yield a statistic, H, which has a sampling distribution that approximates a chi square distribution with $k - 1$ degrees of freedom. This allows us to specify a critical value for H and test the viability of the null hypothesis. For the data in Table 15.5, the value of H is found to be

$$H = \left(\frac{12}{18\,(18\,+\,1)}\right)\left(\frac{85^2}{6}\,+\,\frac{62^2}{6}\,+\,\frac{24^2}{6}\right)\,-\,3(18\,+\,1)$$

$$=\,10.93$$

For a χ^2 distribution with $k - 1 = 3 - 1 = 2$ degrees of freedom, and an alpha level of .05, the critical value of H is 5.99 (see Appendix L). Since 10.93 exceeds the value of 5.99, we reject the null hypothesis and conclude that the number of roommates and attitude toward living in a dorm are related.

The strength of the relationship in the Kruskal-Wallis test can be calculated using an index called **epsilon squared,** which is directly analogous to eta squared. The formula for epsilon squared is

$$E_R^2 = \frac{H - k + 1}{N - k} \qquad\qquad \textbf{[15.11]}$$

where all terms are as previously defined. In the example,

$$E_R^2 = \frac{10.93 - 3 + 1}{18 - 3}$$

$$=\,.60$$

As with the one-way analysis of variance, rejection of the null hypothesis only tells us that at least one of the three groups differs in ranks, but does not indicate the nature of these differences. Several procedures have been proposed for analyzing the nature of the relationship in a Kruskal-Wallis test (for example, Dunn, 1964; Steel, 1960; Miller, 1966; Wike, 1971; Ryan, 1960). Which one to use, however, is controversial among statisticians. We will use the test recommended by Dunn (1964). The **Dunn procedure** involves applying the Wilcoxon rank sum test or the Mann-Whitney U test to all pairs of rank sums, just as the HSD test was applied to all possible pairs of means. In order to maintain an overall significance level equal to α, however, Dunn recommends that the critical value for rejecting each null hypothesis be set based on a revised alpha equal to $\alpha\,/\,2C$ where C equals the number of contrasts or pairs of ranks tested. In the example, if the original alpha level was .05, then the revised alpha would be $.05/(2)(3)\,=\,.0083$.

Method of Presentation

The results of a Kruskal-Wallis analysis of variance by ranks test might appear in a research report as follows:

A Kruskal-Wallis test was applied to the ranked data relating the number of roommates lived with to a person's attitudes toward living in dormitories. The resulting H statistic was statistically significant (H = 10.9, df = 2, $p < .05$). The strength of the relationship, as indexed by epsilon squared, was .60. A post-hoc test suggested by Dunn (1964) indicated that the attitudes for people with three roommates were less favorable than the attitudes for people with two roommates which, in turn, were less favorable than the attitudes for people with one roommate.

The first sentence identifies the type of test performed. The second sentence states whether the overall null hypothesis was rejected and reports the value of H, the relevant degrees of freedom ($k - 1$), and the alpha level. The last sentence reports the type of follow-up test performed and the results of the test in terms of the nature of the relationship.

15.5 / Rank Test for Three or More Correlated Samples

Use of the Rank Test for Three or More Correlated Samples. A statistical test called the **Friedman analysis of variance by ranks** is the nonparametric counterpart of one-way repeated measures analysis of variance. It is typically used when (1) the dependent variable is quantitative and continuous in nature, (2) the independent variable is within-subjects in nature, and (3) the independent variable has three or more values.

Analysis of the Relationship. Consider an experiment where ten individuals attempted to solve a problem under each of three different conditions: a quiet condition, a slightly noisy condition, and a noisy condition. The dependent measure was the individual's performance on the problem, which was scored from 0 to 20, with higher scores indicating better performance. The three problems had been pretested such that they were of equal difficulty and the order of conditions was counterbalanced. Table 15.6 presents the data for the experiment.

The Friedman test involves first rank ordering the scores *for each individual* across the experimental conditions. This has been done in Table 15.7. For example, the first individual had scores of 10, 6, and 5 and the ranks are 1, 2, and 3, respectively. Individual number 4 had scores of 15, 13, and 17 and the rank order of these is 2, 3, and 1, respectively.

TABLE 15.6 DATA FOR EXPERIMENT ON BACKGROUND NOISE AND PROBLEM SOLVING

Individual	Quiet Condition	Slightly Noisy Condition	Noisy Condition
1	10	6	5
2	15	10	9
3	15	14	8
4	15	13	17
5	8	5	3
6	6	5	2
7	8	7	3
8	9	8	6
9	9	14	15
10	8	14	9

TABLE 15.7 RANKED DATA FOR EXPERIMENT ON BACKGROUND NOISE AND PROBLEM SOLVING

Individual	Quiet Condition	Slightly Noisy Condition	Noisy Condition
1	1	2	3
2	1	2	3
3	1	2	3
4	2	3	1
5	1	2	3
6	1	2	3
7	1	2	3
8	1	2	3
9	3	2	1
10	3	1	2
	$R_1 = 15$	$R_2 = 20$	$R_3 = 25$

If the null hypothesis is true such that background noise does not influence problem solving, then differences in the rank scores are merely a function of chance or sampling error. Overall, we would expect the sum of the ranks in each condition

The method of presentation for the Friedman analysis of variance by ranks is identical to that of the Kruskal-Wallis test discussed previously, except, of course, that the value of χ_r^2 is reported instead of H.

to be the same. Friedman (1973) has suggested a statistic for ascertaining the magnitude of the difference in rank scores. This statistic has a sampling distribution that approximates a chi square distribution with $k - 1$ degrees of freedom. The statistic is

$$\chi_r^2 = \left(\frac{12\Sigma R_j^2}{Nk(k + 1)}\right) - 3N(k + 1) \qquad \textbf{[15.12]}$$

where R_j is the sum of the rank scores in condition j, N is the total number of individuals in the experiment, and k is the number of conditions in the experiment. For the example,

$$\chi_r^2 = \frac{(12)(15^2 + 20^2 + 25^2)}{(10)(3)(3 + 1)} - (3)(10)(3 + 1)$$

$$= 125 - 120$$

$$= 5.0$$

For a χ^2 distribution with $k - 1$ or 2 degrees of freedom, the critical value of χ_r^2 is 5.99. Since 5.0 does not exceed the critical value, we fail to reject the null hypothesis.

The strength of the relationship in the Friedman test can be estimated using an index of epsilon squared:

$$E_R^2 = \frac{\chi_r^2 - (k + 1)}{Nk} \qquad \textbf{[15.13]}$$

where all terms are as previously defined.

If we had rejected the null hypothesis, then it would be necessary to do additional analyses to discern the nature of the relationship. The procedure recommended by Dunn (1964) could be used in this instance. This would involve comparing all possible pairs of ranks using the Wilcoxon signed-rank test, but adjusting the alpha level for defining the critical value to $\alpha/2C$, as discussed in the section on the Kruskal-Wallis test.

See Method of Presentation on page 322.

15.6 / Spearman's Rank Order Correlation

Use of Spearman's Rank Order Correlation. **Spearman's rank order correlation** is typically used when both the independent and dependent variables are quantitative and continuous in nature and are in the form of ranks. It is the counterpart to the Pearson correlation coefficient and examines the degree to which rank scores on two variables are linearly related.

Analysis of the Relationship. Consider a study where an investigator wanted to study the relationship between pollution and cancer mortality rates. A pollution index was developed that could range from 0 to 4, with higher numbers indicating greater pollution. This index was applied to twenty cities in the United States. For each city, a measure of cancer mortality per 100,000 people was also obtained. The relevant data are presented in the first three columns of Table 15.8.

The first step of the analysis is to rank order the scores for each variable, separately. This has been done in columns 4 and 5 of Table 15.8. The Spearman rank order correlation procedure involves computing a correlation coefficient with respect to the ranked data. The correlation can range from -1.00 to 0 to 1.00 and is interpreted the same way as the Pearson correlation. The derivation of Spearman's rank order correlation involves computing the difference, d, between the ranked scores for each individual. This has been done in column 6 of Table 15.8. The differences are then squared, as in column 7. The following formula is used to compute Spearman's correlation:

$$r_s = 1 - \left(\frac{6\Sigma d^2}{N\,(N^2 - 1)} \right) \qquad \textbf{[15.14]}$$

For our data,

$$r_s = \left(\frac{(6)\,(670)}{(20)\,(20^2 - 1)} \right)$$

$$= 1 - \frac{4020}{7980}$$

$$= .50$$

The rank order correlation between pollution and cancer mortality was .50.

TABLE 15.8 DATA FOR EXPERIMENT ON POLLUTION AND CANCER MORTALITY

City	Pollution Index	Cancer Mortality	R_1	R_2	d	d^2
1	1.0	124	1	1	0	0
2	1.5	131	4	7	− 3	9
3	2.0	135	8	10	− 2	4
4	2.5	139	11	14	− 3	9
5	3.0	145	15	18	− 3	9
6	1.6	125	5	2	3	9
7	2.8	130	14	6	8	64
8	1.2	136	2	11	− 9	81
9	1.7	140	6	15	− 9	81
10	2.2	146	9	19	− 10	100
11	2.6	126	12	3	9	81
12	3.2	129	16	5	11	121
13	3.7	137	18	12	6	36
14	4.0	141	20	16	4	16
15	1.3	133	3	9	− 6	36
16	1.9	127	7	4	3	9
17	2.4	132	10	8	2	4
18	2.7	138	13	13	0	0
19	3.4	142	17	17	0	0
20	3.8	148	19	20	− 1	1
						$\Sigma = 670$

When the sample size is greater than 10, Spearman's correlation can be converted to a score in a t distribution with $N - 2$ degrees of freedom. The following formula estimates the standard error for the t test:

$$\hat{s}_{r_s} = \frac{1}{\sqrt{\dfrac{N - 2}{1 - r_s^2}}} \qquad [15.15]$$

In the example,

$$\hat{s}_{r_s} = \frac{1}{\sqrt{\dfrac{20 - 2}{1 - .50^2}}}$$

$$= \frac{1}{4.90}$$

$$= .204$$

Under the assumption that the null hypothesis is true (that is, $\rho_s = 0$), the sample r_s can be converted into a t score as follows:

$$t = \frac{r_s - \rho_s}{\hat{s}_{r_s}} \qquad \text{[15.16]}$$

$$= \frac{.50 - 0}{.204}$$

$$= 2.45$$

The critical values for a t score with $N - 2 = 20 - 2 = 18$ degrees of freedom are ± 2.10. Since 2.45 exceeds 2.10, we reject the null hypothesis and conclude that the linear relationship between the pollution index and cancer mortality is nonzero and positive.

Given that we rejected the null hypothesis, it is meaningful to interpret the strength and nature of the relationship. As with the Pearson correlation, the strength of the relationship is indicated by r_s^2 and the nature of the relationship is indicated by the sign of the correlation coefficient.

When N is less than 10, the application of the t test is not appropriate. Appendix O presents critical values for r_s for small Ns that can be used to infer the existence of a relationship when $N < 10$. An alternative to Spearman's index of association between ranked scores is an index presented by Kendall called *tau*. The computation of tau is considerably more complex than that of r_s, but it has certain advantageous mathematical properties. For a comparison of the two techniques, see Glass and Stanley (1970).

Method of Presentation

The method of presentation for Spearman's rank order correlation is identical to that of the Pearson correlation. The results of the above experiment might appear in a research report as follows:

A Spearman rank order correlation was performed on the ranked scores for the pollution index and cancer mortality. The observed correlation ($r_s = .50$) was statistically significant ($t = 2.45$, df $= 18$, $p < .05$). As pollution increased in cities, so did cancer mortality rates.

15.7 / Summary

When the assumptions of a parametric test are markedly violated, then a nonparametric statistic may be used to analyze the relationship between two variables. Nonparametric statistics tend to make few assumptions about population distributions and are well-suited to analyzing data that are not normally distributed. The major rank-based nonparametric tests for analyzing bivariate relationships are the Mann-Whitney U test, Wilcoxon's rank sum test, Wilcoxon's signed-rank test, the Kruskal-Wallis test, Friedman's analysis of variance by ranks, and Spearman's rank order correlation. Each of these is a counterpart to a parametric test, as indicated in Table 15.1.

APPENDIX 15.1

Corrections for Ties in Nonparametric Rank Order Statistics

All of the tests we have considered in this chapter deal with ties by assigning the average rank that the tied observations occupy. However, some of the approaches require modifications in the derivation of statistics in order to maintain a close approximation to the relevant sampling distribution when a large number of ties occur. The two tests where this is the case are the Wilcoxon rank sum test and the Kruskal-Wallis test.

For the Wilcoxon rank sum test, the formula for z when numerous ties occur is

$$z = \frac{(R_1 - \overline{R}) - 1}{\hat{s}_R}$$

where

$$\hat{s}_R = \sqrt{\left(\frac{n_1 n_2}{N(N-1)}\right)\left(\frac{N^3 - N}{12} - \Sigma T\right)}$$

where N equals $n_1 + n_2$, and T equals $(t^3 - t) / 12$, where t is the number of values tied at a particular rank. The summation of T extends over all groups of ties.

For the Kruskal-Wallis test, when numerous ties occur the value of H is computed by the following formula:

$$H = \frac{\left(\frac{12}{N(N+1)}\right)\left(\sum_{j=1}^{k} \frac{R_j}{n_j}\right) - 3(N+1)}{1 - \frac{\Sigma T}{N^3 - N}}$$

where T equals $(t^3 - t) / 12$, where t is the number of values tied at a particular rank. As before, the summation of T extends over all groups of ties.

EXERCISES

1. Rank order each of the following sets of scores, with the lowest number receiving a rank of 1.

Set I	Set II	Set III	Set IV
135	131	130	104
132	131	136	102
136	136	135	102
131	135	130	102
134	132	135	105
134	134	134	106

For each of the following problems, assume that the standard assumptions of the parametric tests have been violated to such an extent that a nonparametric test is required. You are to decide which nonparametric test should be applied and then analyze the data. Draw a conclusion and report your results using procedures discussed in the appropriate Method of Presentation section.

2. A high school counselor wanted to test the relationship between owning a car and performance in school. A random sample of fifteen students who own cars was selected as well as fifteen students who did not own cars. The grade point averages of these students were obtained. The data for the two groups are presented below. Use the appropriate nonparametric statistical test to analyze these data.

Owns Car			Does Not Own Car		
3.10	2.10	2.90	2.20	3.00	2.00
2.50	2.30	3.90	2.40	2.19	3.40
2.75	2.15	2.70	2.60	2.98	3.60
2.96	3.30	2.64	2.80	2.77	4.00
3.50	3.70	1.90	3.20	2.66	3.80

3. A researcher was interested in the effects of marijuana on recognition of auditory stimuli. A group of eight subjects was presented an auditory stimulus that was increased in intensity until it could be heard (as signaled by the subject). The amount of time, in seconds, it took the subject to recognize the stimulus was recorded. This task was performed in each of two conditions: (1) while under the influence of marijuana, and (2) while not under the influence of marijuana. The data are presented below. Analyze them using the appropriate nonparametric statistical test.

Subject	Under Influence of Marijuana	Normal
1	.56	.63
2	.54	.65
3	.58	.64
4	.67	.41
5	.72	.71
6	.43	.44
7	.84	.83
8	1.02	.86

4. A social psychologist was interested in comparing the attitudes of males and females toward a newly developed male birth control pill. A group of eight males and eight females was presented a description of the characteristics of the pill and asked how favorable they would feel about its use (either by themselves or their partner) on a 0 to 100 scale. Higher numbers indicated more favorable attitudes. The data for this study are presented below. Analyze these data using the appropriate nonparametric statistical test.

Males		Females	
50	59	52	61
48	65	50	66
74	67	76	86
56	82	63	84

5. A psychologist tested the effects of a group encounter session on individuals' self-esteem. A scale measuring self-esteem (with scores ranging from 10 to 80 and higher scores implying higher levels of self-esteem) was administered to a group of ten individuals just before they participated in the session and again just after. The data are presented below. Analyze these data using the appropriate nonparametric statistical test.

Subject	Before Session	After Session
1	40	42
2	38	39
3	36	37
4	39	41
5	28	26
6	30	32
7	65	63
8	69	70
9	72	71
10	71	82

6. A researcher wanted to compare the relative amounts of violence on three major television networks during prime time. During a week, two independent judges rated all television shows shown on the networks in terms of their violence. The ratings were made on a scale from 0 (no violence) to 50 (considerable violence). The average of these ratings for the two judges was computed for each of eight randomly selected shows from each of the three networks. The data are presented below. Use the appropriate nonparametric statistical test to analyze these data.

Network A		Network B		Network C	
20	38	21	39	17.5	43
28	40	29	41	27	26
32	42	33	45	31	50
35	15	34	13	30	16

7. A researcher was interested in the relationship between crime rates in cities and the size of a city's police force. For eighteen large cities, the rank order of the crime rates (1 = lowest, 18 = highest) and the size of the police force (1 = smallest, 18 = largest) were obtained. The data are below. Analyze these data using the appropriate nonparametric statistical technique.

City	Crime Rate	Size of Police Force
1	3	4
2	15	16
3	4	5
4	1	2
5	8	12
6	14	15
7	13	6
8	2	1
9	7	11
10	12	7
11	16	18
12	6	16
13	11	9
14	17	14
15	10	8
16	18	17
17	9	13
18	5	3

8. A researcher wanted to compare the quality ratings for picture tubes of three brands of television, A, B, C. A group of ten repairmen rated each brand on a scale from 0 to 50, with higher scores indicating higher qual-

ity. The data are presented below. Analyze the data using the appropriate nonparametric statistical technique.

Repairman	Brand A	Brand B	Brand C
1	15	17	19
2	27	28	29
3	31	33	32
4	18	20	22
5	41	39	36
6	44	35	37
7	13	14	16
8	43	44	45
9	24	26	30
10	20	23	25

9. A psychologist was interested in the effects of the affective content of words on learning. Fifteen subjects were read a list of twenty-five words and asked to recall as many as they could. For five subjects, the words were all positive (for example, "intelligent"), for another five subjects the words were all negative (for example, "conceited"), and for the remaining five subjects, the words were all neutral (for example, "door"). The words comprising lists were equated on potentially relevant dimensions such as the frequency with which they occurred in literature, etc. Learning scores were derived by scoring the number of words correctly recalled. The data are presented below. Analyze the data using the appropriate nonparametric technique.

Positive Words	Negative Words	Neutral Words
20	15	16
18	10	21
22	9	14
17	7	13
19	11	12

10. An educator wanted to test the effects of a film on students' knowledge about health. A health knowledge test (with scores ranging from 0 to 100, with higher scores indicating more knowledge) was administered to ten students before the film, just after the film, and again one week later. The data are presented below. Analyze these data using the appropriate nonparametric statistical technique.

Student	Before Film	Immediately After Film	One Week After Film
1	28	31	29
2	32	39	37
3	14	23	21
4	20	17	15
5	36	35	33
6	40	43	41
7	24	27	25
8	16	19	18
9	10	13	11
10	44	47	45

11. A researcher wanted to test the relationship between conservatism and support for gun control. A group of twenty individuals was given a scale measuring degree of political conservatism (0 to 100, with higher scores indicating greater degrees of conservatism) and another scale measuring attitudes toward gun control (0 to 50, with higher scores indicating more positive attitudes). The data are presented below. Analyze the data using the appropriate nonparametric statistical technique.

Subject	Conservatism Score	Attitude Toward Gun Control
1	50	10
2	89	48
3	75	28
4	56	14
5	64	24
6	66	26
7	85	47
8	79	35
9	87	46
10	60	23
11	68	32
12	52	11
13	54	15
14	58	16
15	83	44
16	81	45
17	77	34
18	73	30
19	62	22
20	76	31

16

Selecting the Appropriate Statistical Technique to Analyze Bivariate Relationships

To this point, we have considered statistical procedures for analyzing the relationship between two variables. Three questions have guided our analysis: (1) Given sample data, can we infer that a relationship exists between two variables in the population? (2) What is the strength of the relationship? (3) What is the nature of the relationship? We have considered eleven types of analyses for studying bivariate relationships. These analyses and the specific statistics used to address each question are summarized in Table 16.1.

The present chapter is concerned with issues in selecting one of the eleven types of tests to analyze one's data. The problem of selecting a statistical test is a matter that should be considered both before and after data collection occurs. Prior to data collection, the researcher should explicitly specify the statistical techniques that will be used to analyze the data. After the data have been collected, the researcher must ensure that the assumptions underlying the relevant statistical test have not been violated to an extent that the planned analyses are inappropriate. If the violations are too marked, then an alternative statistical test may be necessary. This possibility should also be considered prior to data collection. The procedures for testing assumptions underlying a statistical test are generally complex and can't be considered in an introductory text. Interested readers are referred to Hays (1963) and Winer (1971) for a description of these procedures. In the discussion that follows, we will have to be somewhat vague about tests of assumptions underlying various statistical procedures. It must be emphasized that the following guidelines for selecting a statistical test are only rules-of-thumb. There are potential exceptions to every one of them. Nevertheless, the guidelines should prove to be of some heuristic value in the vast majority of cases.

The first step in deciding the appropriate statistical test to use is to identify the independent and dependent variables and note whether each is qualitative or quantitative in nature. Four different situations can arise and are illustrated as follows:

| | Dependent Variable | |
Independent Variable	Qualitative	Quantitative
Qualitative	Case I	Case II
Quantitative	Case III	Case IV

TABLE 16.1 STATISTICAL TESTS FOR ANALYZING BIVARIATE RELATIONSHIPS

Independent Groups t Test

Inference of a Relationship:	t statistic
Strength of the Relationship:	Eta squared
Nature of the Relationship:	Inspection of group means

Correlated Groups t Test

Inference of a Relationship:	t statistic
Strength of the Relationship:	Eta squared
Nature of the Relationship:	Inspection of group means

One-Way Between-Subjects Analysis of Variance

Inference of a Relationship:	F ratio
Strength of the Relationship:	Eta squared
Nature of the Relationship:	HSD test

One-Way Repeated Measures Analysis of Variance

Inference of a Relationship:	F ratio
Strength of the Relationship:	Eta squared
Nature of the Relationship:	HSD test

Pearson Correlation

Inference of a Relationship:	t statistic
Strength of the Relationship:	r^2
Nature of the Relationship:	Sign of r

Chi Square Test of Independence

Inference of a Relationship:	Chi square statistic
Strength of the Relationship:	Fourfold point correlation or Cramer's statistic
Nature of the Relationship:	Goodman's Simultaneous Confidence Interval Procedure

Mann-Whitney U Test/Wilcoxon Rank Sum Test

Inference of a Relationship:	U statistic/z statistic
Strength of the Relationship:	Glass rank biserial correlation coefficient
Nature of the Relationship:	Inspection of rank sums

Wilcoxon Signed-Rank Test

Inference of a Relationship:	T statistic/z statistic
Strength of the Relationship:	Matched-pairs rank biserial correlation coefficient
Nature of the Relationship:	Inspection of rank sums

Inference of a Relationship:	H statistic
Strength of the Relationship:	Epsilon squared
Nature of the Relationship:	Dunn's procedure

Friedman Analysis of Variance by Ranks

Inference of a Relationship:	χ_r^2
Strength of the Relationship:	Epsilon squared
Nature of the Relationship:	Dunn's procedure

Spearman's Rank Order Correlation

Inference of a Relationship:	t statistic derived from rank correlation
Strength of the Relationship:	Rank correlation coefficient
Nature of the Relationship:	Sign of rank correlation

Consider the following description of an experiment:

A researcher wanted to test the psychological component of acupuncture treatments for headaches. The experiment used two groups of subjects, all of whom had a history of frequent and severe headaches. One group of subjects was given an acupuncture treatment for one week, in which needles were placed in certain critical nerve points. They were told that this process should reduce the pain they feel when a headache occurs. In fact, the needles were placed such that they should have no effect at all. The second group served as a control group and received no treatment during the "acupuncture week." Each subject then rated their next headache on a 100 point rating scale for pain. The results indicated that people who *thought* they were receiving treatment to reduce pain reported less pain, on the average, than those who were not told to expect a reduction in pain.

In this experiment the independent variable was the expectation of a reduction in pain (expected versus not expected) and it represents a qualitative variable with two values. The dependent variable was the self-report of pain of the next headache as measured by the rating scale. It constitutes a quantitative variable. In this instance, we would be concerned with a case II situation (a qualitative independent variable and a quantitative dependent variable).

16.1 / Case I: The Relationship Between Two Qualitative Variables

When both of the variables involved in the analysis are qualitative, and both of the qualitative variables are between-subjects in nature, the most likely candidate for data analysis is the chi square test of independence (Chapter 14). This would occur, for example, when examining the relationship between the political party preference of fathers and the political party preference of their sons, as indicated by the following frequency table:

Father's Preference	Democrat	Son's Preference Republican	Independent
Democrat	150	28	30
Republican	26	180	33
Independent	38	45	140

As noted in Chapter 14, the chi square test of independence requires a relatively large sample size, and if the expected frequency in any given cell is less than 5, the test may not be appropriate. If one or both of the qualitative variables are within-subjects in nature, then none of the procedures we have considered can be used to analyze the relationship between the variables. The appropriate statistical procedures in this instance are discussed in McNemar (1962) and Marascuilo and McSweeney (1977). To summarize, the most typical statistic used to analyze the relationship between two qualitative variables (for example, political party preference and a person's gender) is the chi square test of independence.

16.2 / Case II: The Relationship Between a Qualitative Independent Variable and a Quantitative Dependent Variable

In cases where the independent variable is qualitative and the dependent variable is quantitative, the first decision point is whether or not to use a parametric (Chapters 9–13) or nonparametric (Chapter 15) test. Several factors should be weighed in making this decision, which we will now consider.

Number of Values on the Dependent Variable. As noted in Chapter 8, the choice of a nonparametric versus a parametric statistical technique for Case II situations is a controversial matter. One characteristic that is relevant is the number of values of the quantitative dependent variable. Consider the case where the dependent variable is an attitude toward nuclear energy and the independent variable is a person's political party preference (Democrat, Republican, Independent). The relevant attitude measure could conceivably have only two values (unfavorable and favorable) or it could have 50 values (scores from 0 to 50, with higher scores indicating more favorable attitudes). In cases where the dependent measure has only two values, most researchers will *not* use a parametric test, such as the *t* test or analysis of variance, to analyze the data. As noted above, an assumption of these tests is that the population scores are normally distributed. A dichotomous dependent measure (that is, one with two values) cannot be normally distributed and thus, this assumption of the parametric tests is violated. When the dependent measure has a small number of values (for example, three or fewer), most researchers will apply the chi square test of independence. This would involve formulating a 3 × 2 frequency table as follows:

	Unfavorable	*Favorable*
Democrat		
Republican		
Independent		

The procedures developed in Chapter 14 and summarized in Table 16.1 would then be applied.

There is some evidence to suggest that the number of values of the dependent measure may not be a limiting factor in the application of a parametric test. Hsu and Feldt (1969) investigated the effect of scale limitations on type I error rates involving independent groups. Four scale lengths were studied: five, four, three, and two points. Generally speaking, nominal and actual significance levels were relatively close. Lunney (1970) has shown that with equal ns and with $n > 10$ per group, a dichotomous dependent variable can be subjected to a parametric test (such as the t test or one-way analysis of variance) and the conclusions drawn will be robust to violations of the normality assumption. Thus, even in the case of dichotomous measures, some researchers will prefer a parametric test over the chi square test of independence.*

In sum, when the dependent variable has a small number of values, most researchers will either use the chi square test of independence or, given equal ns and $n > 10$ per group, a parametric test.

Interval versus Ordinal Data and Distributional Analyses. Given a dependent variable with a sufficient range of values, additional characteristics of the dependent measure must be considered. If the dependent variable is measured on an ordinal level that seriously departs from interval level characteristics, a nonparametric statistic will typically be used. Similarly, if the distributional assumptions of the parametric tests are *markedly* violated and one has relatively small sample sizes, then nonparametric tests should be used. Otherwise a parametric test can be used to good effect.

*With a small number of values on the dependent measure, the rank order techniques discussed in Chapter 15 are of questionable value. This is because there will usually be a large number of ties in ranks, and as noted in that chapter, this can create serious problems for those approaches.

A Decision Tree. Table 16.2 presents a decision tree that can be used to guide the selection of the appropriate statistical test in Case II situations. The three critical facets are (1) use of a parametric versus a nonparametric procedure, (2) whether the independent variable is between-subjects or within-subjects in nature, and (3) the number of values or groups characterizing the independent variable. Let us consider an example where an investigator studied the relationship between religion (Catholic, Protestant, Jewish) and the ideal number of children people wanted to have in their completed family. In this case, the independent variable (religion) is qualitative and the dependent variable (ideal number of children) is quantitative. Hence, this a case II situation. The first step involves deciding whether to use a parametric or nonparametric statistic. The investigator would begin by examining the range of scores obtained on the ideal number of children variable. Let us suppose that the range was from 0 to 9, with most of the scores occurring in the 1, 2, 3, and 4 range. This is probably sufficient variability to permit a parametric test. Next, the experimenter would examine the distribution of scores for each religious group separately. The investigator might note a slight degree of negative skewness in each group, suggesting a departure from the normality assumption of parametric tests. In addition, the investigator would examine the variances of scores within each religious group to insure that the assumption of equal variances is reasonable. Suppose in this

TABLE 16.2 CASE II: SELECTION OF STATISTICAL TESTS

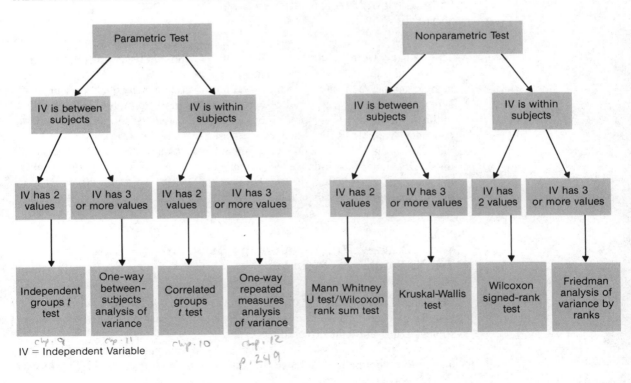

IV = Independent Variable

case it was found that the variances for Catholics and Protestants tended to be the same, but the variance for Jews tended to be somewhat smaller. This would suggest another deviation from the parametric assumption of equal variances.

Suppose the investigator interviewed a total of fifty Catholics, fifty Protestants, and fifty Jews. Given the large sample sizes, a parametric test would probably be robust to the minor violations of the assumptions noted above. This, coupled with the fact that the ideal number of children is an interval measure, would lead to the decision to use a parametric test.

Looking at Table 16.2, the next decision requires classifying the independent variable as between-subjects or within-subjects in nature. In this case, the independent variable is between-subjects since the fifty Catholics are different individuals than the fifty Protestants who, in turn, are different individuals than the fifty Jews.

Finally, one needs to note the number of values on the independent variable. In this case there are three (Catholic, Protestant, Jewish). This would dictate a one-way analysis of variance as the data analytic technique. Referring to Table 16.1, we would analyze the relationship between the variables by using the F ratio to infer the existence of a relationship, eta squared to examine the strength of the relationship, and an HSD test to determine the nature of the relationship.

16.3 / Case III: The Relationship Between a Quantitative Independent Variable and a Qualitative Dependent Variable

The statistics we have considered for analyzing the relationship between a quantitative and a qualitative variable have all been presented with the dependent variable being quantitative in nature and the independent variable being qualitative in nature. When the reverse is true, most researchers will simply apply the decision criteria outlined for Case II in the preceding section, but for statistical purposes, treat the dependent variable as the "independent variable" and the independent variable as the "dependent variable." Of course, the interpretation of the tests would maintain the original independent-dependent variable distinction and it is only for the sake of the statistical analysis that the two are reversed. All of the issues discussed in the previous section would then apply.

16.4 / Case IV: The Relationship Between Two Quantitative Variables

Considerations for analyzing the relationship between two quantitative variables are complex, since so many options are available to the researcher. We will discuss the strategies that are typically used.

Given two variables that are measured on approximately an interval level, and that are each normally distributed, the most common analysis used is that of Pearson's correlation coefficient (Chapter 13). As noted in Chapter 13, this technique, as it is usually applied, tests for a *linear* relationship between the variables. If the expected relationship is nonlinear, then researchers will typically collapse the independent variable into three to five groups and analyze the data via analysis of variance techniques (see Section 13.8 for an elaboration and discussion of the disadvantages of this approach). Although it is not recommended, some researchers will collapse both variables into a small number of groups and then perform a chi square test of independence on the data.

Any time one (and only one) of the variables has a small number of values associated with it (for example, five or fewer), Table 16.2 and the statistics elaborated in Case II could be applied. Essentially, the variable with the small number of values is treated as the independent variable and all of the considerations discussed in Section 16.2 then apply. As an example, a researcher might investigate the relationship between age and scores on an intelligence test. The test is administered to thirty five-year-old children, thirty seven-year-old children, and thirty nine-year-old children. In this case, both variables are quantitative in nature, but age has only a few values (5, 7, and 9). Further, the study was structured so as to yield these three well-defined groups on the independent variable. Under these circumstances, most investigators would use Table 16.2 to decide the appropriate statistical test to use, treating age as the independent variable and the intelligence scores as the dependent variable. The resulting mode of analysis would most likely be the one-way between-subjects analysis of variance.

When the two variables are clearly measured at an ordinal level, then the Spearman rank order correlation will be applied (Chapter 15). As with the Pearson correlation, there is always the option of collapsing the independent variable into three to five groups and then analyuzing the data using the Kruskal-Wallis test. The relative advantages and disadvantages of doing so parallel closely those discussed in Section 13.8 on the Pearson correlation.

16.5 / Independent-Dependent Variables versus Predictor-Criterion Variables

In Chapter 1, an independent variable was defined as the "presumed cause" of the dependent variable. In Chapter 8, factors that influence our ability to draw causal inferences were discussed, including the problem of confounding variables and disturbance variables. Although a large majority of research in the social sciences is concerned with making causal inferences, some research areas are not. For example, one branch of personnel selection is concerned with predicting potential success of job applicants. The goal of this research is purely prediction and the personnel tests are selected accordingly. In essence, the researcher doesn't care whether a causal relationship exists; he or she is only interested in *predicting* job success. In these instances, the terminology of "independent variable" and "dependent variable" is not used. Rather, the terms "predictor variable" and "criterion variable" are used. This is to emphasize the fact that no causal relationship is assumed. The statistical analysis of bivariate relationships is the same for predictor and criterion variables as for independent and dependent variables. As noted in Chapter 8, the ability to make causal inferences is a function of one's research design, not the statistics used to analyze the data generated by that design. In order to select the appropriate statistical technique for analyzing a bivariate "predictor-criterion" relationship, you can just substitute the word "predictor" for "independent" and "criterion" for "dependent" throughout the text.

Another situation where the terms "independent variable" and "dependent variable" do not apply is in certain measurement situations where an investigator is trying to determine if two tests measure the same variable. For example, a researcher might test the relationship between scores on the Stanford-Binet intelligence test with those of the Wechsler intelligence test. No causal relationship is assumed. Nor can one of the tests be designated a "predictor variable" and the other a "criterion variable." The question of interest is whether two measures, each supposedly measuring the same thing, are related to each other. The selection of the appropriate statistical test for this case can be accomplished in the present scheme by arbitrarily treating one of the variables as the independent variable and the other as the dependent variable. Of course, the interpretation of one's results would not make such a distinction, since it is arbitrary and not appropriate in this instance.

16.6 / Summary

When deciding which type of statistic to use to analyze a bivariate relationship, it is useful to distinguish between four cases: (1) analyzing the relationship between two qualitative variables, (2) analyzing the relationship between a qualitative independent variable and a quantitative dependent variable, (3) analyzing the relationship between a quantitative independent variable and a qualitative dependent variable, and (4) analyzing the relationship between two quantitative variables. Given between-subjects variables, case I is typically analyzed by the chi square statistic. Case II situations are usually analyzed by the rules outlined in Table 16.2. Case III situations typically involve reversing the role of the independent and dependent variable for statistical purposes only and then applying Table 16.2. Case IV situations can use either correlational or analysis of variance based techniques.

EXERCISES

For each of the following studies, assume that the assumptions necessary for a parametric test are satisfied. Indicate what type of statistical technique you would use to analyze the data yielded by the study and state why you would select that technique.

1. An investigator wanted to test if changes in mood were associated with certain times of the year. Specifically, he wondered if people would tend to be more

depressed in the winter than in the spring. Two hundred individuals were administered the Marlowe depression scale in December (winter) and again in May (spring). Scores on the Marlow scale range from 0 to 50, with higher scores indicating greater degrees of depression.

2. A consumer psychologist was interested in the effects of color of ice cream on taste perceptions. Four hundred individuals tasted each of three different "types" of vanilla ice cream. Each type was actually identical to each other, but differed in the particular shade of yellow used to color it. The order in which individuals tasted each of the three types of ice cream was randomized. After tasting a given sample of ice cream, the individual rated it in terms of the quality of taste on a scale from 1 to 10. Higher ratings indicated higher quality.

3. A researcher was interested in whether the noise level of rock and roll music could affect how houseplants grow. Forty seeds were randomly assigned to one of two conditions. In one condition, plants were grown with a steady background of rock and roll music playing at an average volume level. In the other condition, plants were grown under identical conditions, but without any music playing. After six months, the growth in inches was measured for each plant.

4. A psychologist was interested in the relationship between imagination as a child and creativity in the adult years. A sample of one hundred people classified as creative and a sample of one hundred people classified as noncreative were each asked whether they had an imaginary friend during childhood. Answers to this question were in terms of a Yes-No response.

5. A researcher wanted to test the relationship between social class and how dogmatic an individual is. Dogmatism refers to closed-mindedness and the tendency to be inflexible in thought and intolerant of other viewpoints. A sample of 500 people were given a Dogmatism scale. Scores could range from 1 to 70, with higher scores indicating higher degrees of dogmatism. The social class of each individual was measured using Duncan's occupational index in which scores can range from 1 to 100, with higher numbers indicating higher social class.

6. An investigator wanted to test the effects of social influence on drinking behavior. Sixty subjects were randomly assigned to one of three groups, yielding twenty subjects per group. Subjects came to an experiment and were given a soft drink to consume while waiting. A confederate of the experimenter, posing as another subject, was also given a drink. In one group the confederate sipped the drink at a much faster rate than the subject, in another group the confederate sipped the drink at about the same rate as the subject, and in the last group the confederate sipped the drink at a slower rate than the subject. The time it took for each subject to consume his or her drink was then measured.

7. An investigator wanted to test if hypnosis could influence responses to a biofeedback task. Fifty subjects were randomly assigned to one of two conditions. In one condition, subjects were hypnotized and then told to try to make one of their hands warmer by just thinking about it. The other group of subjects was given the same task but were not placed under hypnosis. Temperature was measured in centigrade units to the nearest hundredth of a degree using a special physiological device.

8. A researcher was interested in examining the relationship between physiological arousal and time before taking a test. A group of fifty individuals were scheduled to take an exam on a Thursday. Just prior to a given individual's bedtime, the individual was instructed to administer the Palmer Sweat Index, a physiological measure of arousal with scores ranging from 0 to 100. This was done on Tuesday (two days before the exam), Wednesday (one day before the exam), Thursday (day of the exam), and Friday (day after the exam). The relationship between days before/after the exam and physiological arousal was then examined.

9. In general, our bodies feel weak in the morning when we first get up. A researcher was interested in testing whether people are actually weaker when they first wake up in the morning as compared to a few hours later. A group of fifty subjects were instructed to squeeze a dynamometer (which measures grip strength) when they first woke up in the morning and again three hours later. Scores on the dynamometer could range from 0 to 30.

10. An investigator wanted to test for possible race discrimination in loan officers at banks. A sample of 120 loan officers was given background information on an applicant and asked how much money they would loan the individual. Forty of the loan officers were provided with the information that the applicant was black, forty were told he was white, and the remaining forty were told he was Chicano. Aside from this, the descriptions of the applicant were identical. The relationship between race of the applicant and the amount of money the loan officers were willing to give was assessed.

11. An investigator wanted to test if male college students were more highly likely to use marijuana than female college students. A sample of 200 male and 200 female college students were asked whether they had ever used marijuana.

12. A researcher was interested in testing the relationship between performance in college and a person's income five years later. For a group of 800 individuals, data were obtained on their grade point average while in college and their salary after being out of college for five years, measured in dollars.

For each of the following problems, assume that the standard assumptions of the relevant parametric tests have been violated to such an extent that a nonparametric test is required. Indicate what type of statistical technique you would use to analyze the data yielded by the study and state why you would use that technique.

13. A professor wanted to test the relationship between how quickly a sutdent finishes an exam and his or her performance on the exam. During the class period in which an exam was given, the professor kept track of the order in which students turned in their tests. Specifically, she divided the exams into three groups: the first third of the class to turn in the exam, the middle third of the class to turn in the exam, and the last third of the class to turn in the exam. Scores on the exam could range from 0 to 100 and performance was related to which group the individuals were in.

14. A consumer psychologist was interested in people's impressions of individuals who buy generic foods. A sample of 200 individuals was given a hypothetical shopping list of a person who was planning a party. For half the individuals, the list contained some generic brands whereas for the other half, the list contained only national brands. All products (for example, potato chips, beer, and so one) on the two lists were identical and the only difference was the brand (generic versus national). After reading the list, each subject rated the shopper on a scale from 1 to 100, with higher scores indicating the shopper was perceived as more discriminating in her food preferences.

15. An educational psychologist was interested in the relationship between a psychology department's national reputation and the quantity of scientific articles published by members of its faculty. A national ranking of the top one hundred American psychology departments was obtained (where 1 = the best department, 100 = the worst department) and for each department the number of publications generated by its faculty was tabulated. The relationship between these indices of quality and quantity was then ascertained.

Additional Topics

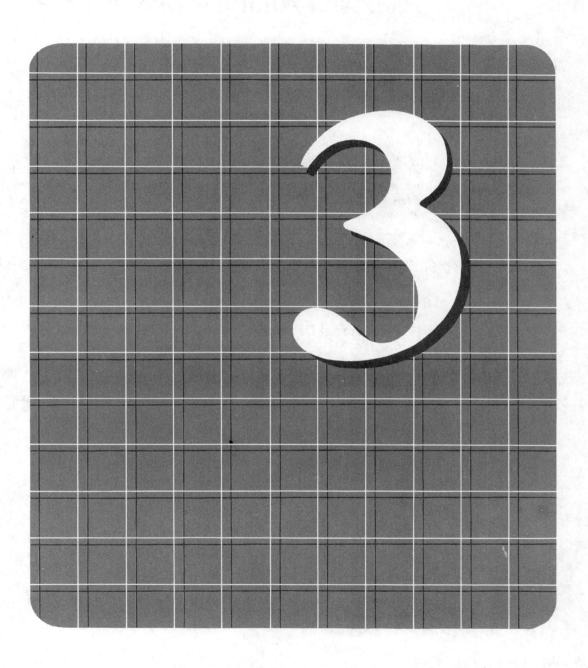

Two-Way Between-Subjects Factorial Analysis of Variance

To this point, we have considered statistical techniques for analyzing the relationship between two variables, an independent variable and a dependent variable. In real life, it is rare that a given dependent variable is determined by one and only one independent variable. A person's attitude toward abortion might be influenced by the religion she was raised in, how religious she is, the type of upbringing she had, and so on. Statistical techniques have been developed that analyze the relationship between more than one independent variable and a given dependent variable. This chapter will consider one such technique, called **two-way between-subjects factorial analysis of variance.**

17.1 / Factorial Designs

Suppose an investigator was interested in studying factors that influence the number of children people want to have in their completed family. He posits that one important factor is an individual's religion. The investigator might conduct a study in which fifty Catholics and fifty Protestants are asked what they think is the ideal number of children. Suppose this is done, that the data are analyzed using the independent groups t test discussed in Chapter 9, and that the results indicate that Catholics, on the average, tend to want more children than Protestants.

Now suppose the investigator was interested in studying the effects of a second factor, how religious an individual is, on ideal family size. Suppose further that the investigator has a measure of religiosity that reliably categorizes individuals into one of two categories, religious versus not religious. Again, a study could be conducted in which fifty religious and fifty nonreligious individuals are asked what they think is the ideal number of children. Suppose this is done, the independent groups t test is applied, and the results show that religious individuals, on the average, tend to want more children than nonreligious individuals.

From the two studies, we would conclude that both religion and religiosity are related to an individual's ideal family size. However, we do not know anything about how the two variables act *in conjunction with one another* to influence ideal family size. The joint effects of two independent variables, such as religion and religiosity, can be studied using a **factorial design.** Table 17.1 illustrates a factorial design for the present example. There are four groups of individuals in the study: (1) religious Catholics, (2) nonreligious Catholics, (3) religious Protestants, and (4) nonreligious Protestants. The four groups are defined by combining the two levels of each of the two independent variables, or **factors,** as they are also called. That is, the first factor, religion, has two levels (or values), Catholic and Protestant. The second factor, religiosity, also has two levels, religious and not religious. If we combine the two levels of religion with the two levels of religiosity, we have a total of four groups.*

TABLE 17.1 EXAMPLE OF A FACTORIAL DESIGN

	Religious Denomination	
Religiosity	*Catholic*	*Protestant*
Religious	Religious Catholics	Religious Protestants
Not Religious	Not Religious Catholics	Not Religious Protestants

The above factorial design is called a 2 × 2 (read "two by two") factorial design. In the "2 × 2" notation, the first number refers to the number of levels of the first factor and the second number refers to the number of levels of the second factor. If the investigation had used three levels of religion instead of two (Catholics, Protestants, and Jews), then the design would be a 3 × 2 factorial design. This would involve six groups of subjects instead of four: (1) religious Catholics, (2) nonreligious Catholics, (3) religious Protestants, (4) nonreligious Protestants, (5) religious Jews, and (6) nonreligious Jews. The number of independent groups required in a between-subjects factorial design is simply the product of the levels of each factor. In a 2 × 2 design it is two multiplied by two, or four. In a 4 × 2 design, the number of independent groups required would be eight.

The present chapter is restricted to the case where the relationship between two independent variables and one dependent variable is studied. However, there are factorial designs that examine the relationship between three or more independent variables and a dependent variable. For example, in addition to religion and religiosity an investigator might also study the relationship of a person's sex (male

*The designation of religion as the first factor and religiosity as the second factor is completely arbitrary. We could have specified religiosity as the first factor and religion as the second factor.

versus female) to ideal family size. This would involve analyzing three factors (or independent variables) in the context of a $2 \times 2 \times 2$ factorial design. The first factor, consisting of two levels, would be religion; the second factor, also consisting of two levels, would be religiosity; and the third factor, also consisting of two levels, would be a person's sex. This $2 \times 2 \times 2$ factorial design would involve studying eight different groups of individuals (two multiplied by two multiplied by two again is eight).

17.2 / Use of Two-Way Between-Subjects Factorial Analysis of Variance

Two-way between-subjects analysis of variance is typically used under the same conditions as either the independent groups *t* test (discussed in Chapter 9) or the one-way between-subjects analysis of variance (discussed in Chapter 11) except that two independent variables rather than one independent variable are studied. Thus, two-way between-subjects analysis of variance would be used when

1. the dependent variable is quantitative in nature and is measured on approximately an interval level;

2. the independent variables are both between-subjects in nature (they can be either qualitative or quantitative);

3. the independent variables have two or more values (that is, two or more levels); and

4. the independent variables are combined so as to form a factorial design.

Consider the example studying the effects of religion and religiosity on ideal family size. Ideal family size is the dependent variable and it is quantitative in nature. Religion is one of the independent variables and it is between-subjects in nature. Religiosity is the other independent variable and it is also between-subjects in nature. Finally, the two factors are combined so as to yield a 2×2 factorial design. In this instance, a two-way between-subjects analysis of variance would be used to analyze the data.

17.3 / The Concept of Main Effects and Interactions

A major feature of a factorial design is that it allows us to study the relationship between *two* independent variables and a dependent variable. Specifically, it allows us to address three issues, phrased here in terms of the example on ideal family size:

1. Is there a relationship between religion, considered alone, and ideal family size?

2. Is there a relationship between religiosity, considered alone, and ideal family size?

3. Is there a relationship between some specific *combination* of religion and religiosity and ideal family size, independent of the effects of religion alone and religiosity alone?

The first two questions are addressed in terms of **main effects.** The "main effect" for religion is the comparison of the mean ideal family size for all Catholics in the study as compared with the mean ideal family size for all Protestants in the study. The null hypothesis for the main effect of religion is

$$H_0: \mu_C = \mu_P$$

where μ_C is the population mean for Catholics and μ_P is the population mean for Protestants. The "main effect" for religiosity is the comparison of the mean ideal family size for all religious individuals in the study as compared with the mean ideal family size for all nonreligious individuals in the study. The null hypothesis for the main effect of religiosity is

$$H_0: \mu_R = \mu_{NR}$$

where μ_R is the population mean for religious individuals and μ_{NR} is the population mean for nonreligious individuals.

Table 17.2 presents hypothetical *population* means for the four groups in question as well as the appropriate main effect means. In this case, there is a relationship between religion and ideal family size since $\mu_C \neq \mu_P$. The nature of this relationship is such that Catholics, on the average, want more children than Protestants ($\mu_C = 3.0$ as compared with $\mu_P = 2.0$). There is also a relationship between religiosity and ideal family size as indicated by the main effect of religiosity. On the average, religious individuals want more children than nonreligious individuals ($\mu_R = 3.0$ as compared with $\mu_{NR} = 2.0$).

The third question addressed in factorial designs concerns **interaction effects.** An interaction refers to the case where the *nature* of the relationship between one of the independent variables and the dependent variables changes as a function of the other independent variable. Consider the relationship between religion and ideal family size. Examine this relationship just for religious individuals in Table 17.2. *Considering just religious individuals,* there is a relationship between religion and ideal family size: Catholics, on the average, want more children than Protestants (4.0 versus 2.0). Now examine the same relationship for nonreligious individuals. For these subjects, religion is unrelated to ideal family size. On the average, Catholics want the same number of children as Protestants (2.0 versus 2.0). This illustrates an interaction effect. The nature of the relationship between religion and ideal family size differs depending on one's religiosity. For religious individuals, Catholics tend to want more children than Protestants. For nonreligious individuals, however, Catholics tend to want the same number of children as Protestants. A powerful feature of factorial designs is that they allow us to test for such effects.

	Religion		
Religiosity	Catholics	Protestants	Total
Religious	4.0	2.0	3.0
Nonreligious	2.0	2.0	2.0
Total	3.0	2.0	

Main Effect for Religion

Population mean for Catholics = 3.0

Population mean for Protestants = 2.0

Main Effect for Religiosity

Population mean for Religious people = 3.0

Population mean for Nonreligious people = 2.0

Although we have stated that the relationship between religion and ideal family size depends on one's religiosity, we could just as easily state the reverse: that the nature of the relationship between religiosity and ideal family size differs as a function of religion. Consider just Catholics. Inspection of Table 17.2 shows that religious Catholics, on the average, want more children than nonreligious Catholics (4.0 versus 2.0). Now consider just Protestants. Table 17.2 reveals that religious Protestants, on the average, want the same number of children as nonreligious Protestants (2.0 versus 2.0). The nature of the relationship between religiosity and ideal family size depends upon one's religion. For Catholics, religious individuals tend to desire more children than nonreligious individuals. For Protestants, religious individuals tend to want the same number of children as nonreligious individuals. The manner in which we choose to state the interaction effect is arbitrary and is usually guided by the research questions being investigated. The important point is that religion and religiosity are interacting such that their joint effects uniquely influence the dependent variable.

Identifying Main Effects and Interactions. In statistical jargon, a given main effect is said to "occur" if the investigator rejects the null hypothesis concerning that main effect. An investigator might conclude in a study that there is "no main effect of religion" and that there "is a main effect" of religiosity. This would mean that the null hypothesis was rejected for religiosity but not religion. The "presence" of an interaction effect means that the null hypothesis of no interaction effect was rejected.

Figure 17.1 presents examples of different population means for the four groups defined by the two levels of religion and two levels of religiosity.* The examples have also been depicted in graphs. On the abscissa of the graph are two demarcations representing the two levels of religion, Catholic and Protestant. On the ordinate of the graph are the units of the dependent variable, in this case, ideal family size. Directly above the demarcation for Protestants on the abscissa are two points that correspond to the mean ideal family size for religious and nonreligious Protestants, respectively. Directly above the demarcation for Catholics on the abscissa are also two points that correspond to the mean ideal family size for religious and non-religious Catholics, respectively. Finally, the two points for religious individuals are connected by a straight line and the two points for nonreligious individuals are connected by a straight line.

Figure 17.1a indicates a case where there are no main effects nor is there an interaction effect. The means for all groups are 4.0. In Figure 17.1b, there is a main effect of religion (Protestants want fewer children, $\mu_P = 2.0$, than Catholics, $\mu_C = 8.0$), but no main effect of religiosity ($\mu_R = 5.0$ and $\mu_{NR} = 5.0$). There is also no interaction effect since for both religious and nonreligious individuals, Catholics, on the average, want six ($8 - 2$) more children than Protestants. Graphically, a lack of interaction is indicated by two parallel lines. Note that this is the case in Figure 17.1b. Each line represents the relationship between religion and ideal family size. One of the lines is the relationship between the two variables for religious individuals and the other line represents this relationship for nonreligious individuals. If the lines are parallel, this means the relationships are the same for both religious and nonreligious individuals. Thus, there is no interaction effect. If the lines were *not* parallel, then the nature of the relationship between religion and ideal family size would depend upon religiosity, and an interaction effect would be present.

Figure 17.1c presents the case where there is no main effect of religion ($\mu_P = 5.0$ and $\mu_C = 5.0$), but there is a main effect of religiosity ($\mu_R = 7.0$, $\mu_{NR} = 3.0$). Note the differential elevation in the lines for religious and nonreligious individuals in the graph. This reflects the main effect of religiosity. Again, the lines are parallel, indicating no interaction effect: The nature of the relationship between religion and ideal family size is the same for both religious and nonreligious individuals.

Figure 17.1d presents a case where there is a main effect of religion ($\mu_C = 6.5$ and $\mu_P = 3.5$) as well as a main effect of religiosity ($\mu_R = 6.5$ and $\mu_{NR} = 3.5$). Again, however, there is no interaction effect as indicated by the parallel lines: For both Protestants and Catholics, religious individuals, on the average, wanted three more children than nonreligious individuals (for Catholics, $\mu_R - \mu_{NR} = 8 - 5 = 3$ and for Protestants, $\mu_R - \mu_{NR} = 5 - 2 = 3$).

---

*Note that population means and not sample means are used in these examples. Population means are used so we can develop the theoretical concepts of main effects and interactions without regard to sampling error.

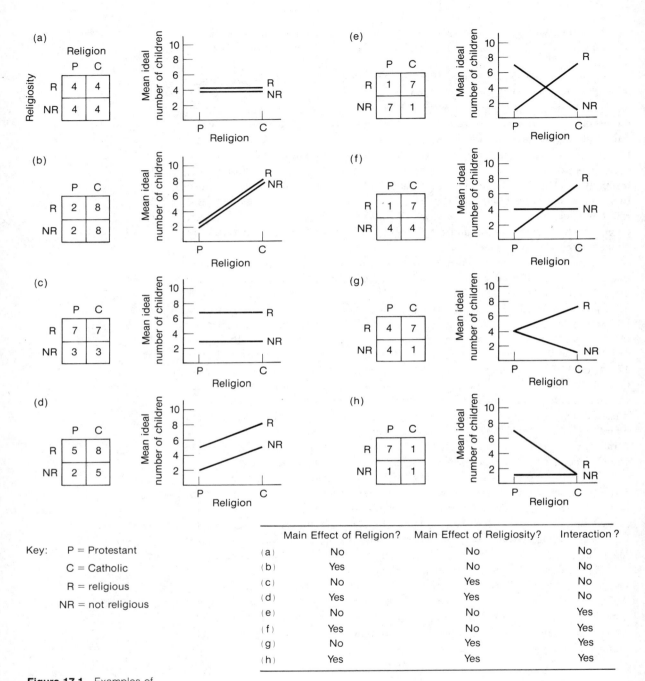

	Main Effect of Religion?	Main Effect of Religiosity?	Interaction?
(a)	No	No	No
(b)	Yes	No	No
(c)	No	Yes	No
(d)	Yes	Yes	No
(e)	No	No	Yes
(f)	Yes	No	Yes
(g)	No	Yes	Yes
(h)	Yes	Yes	Yes

Key: P = Protestant
      C = Catholic
      R = religious
      NR = not religious

Figure 17.1 Examples of Main Effects and Interaction Effects in a 2 × 2 Factorial Design (Adapted from Johnson and Liebert, 1977)

Figures 17.1e through 17.1h represent cases where interaction effects occur. Note that in each graph, the lines are not parallel. This signifies that the relationship between religion and ideal family size differs, depending upon one's religiosity. Examine the examples carefully, noting the nature of the interaction effects and also whether main effects occur.

Main Effects and Interaction Effects in Designs with More Than Two Levels per Factor. All of the above examples were from a 2 × 2 factorial design. The principles for detecting main effects and interaction effects are readily generalizable to designs with more levels. Figure 17.2 presents examples of a 3 × 2 factorial design (using three levels of religion rather than two: Catholic, Protestant, and Jewish) and Figure 17.3 presents examples of a 3 × 3 factorial design (using three levels of religion and three levels of religiosity). Whenever the lines are parallel, a lack of interaction is indicated. When the lines are not parallel (that is, do not follow the same pattern), the presence of an interaction is indicated. Note especially Figures 17.2c and d and Figure 17.3c and d. None of these graphs indicates an interaction effect because the lines all follow the same pattern. The nature of the relationship between one independent variable and the dependent variable is the same irrespective of the levels of the other independent variable.

17.4 / Hypothesis Testing in Factorial Analysis of Variance

Breakdown of SS_BETWEEN. In Chapter 11, the relationship between an independent variable and a dependent variable was analyzed by defining the total variability in terms of two components, between-group variability and within-group variability. Recall that

$$SS_{TOTAL} = SS_{BETWEEN} + SS_{WITHIN} \qquad [17.1]$$

The $SS_{BETWEEN}$ and SS_{WITHIN} were each divided by their respective degrees of freedom and an F ratio formed by dividing the resulting $MS_{BETWEEN}$ by MS_{WITHIN}. In factorial analysis of variance, the same steps are taken. However, the overall $SS_{BETWEEN}$ is further "broken down" into three components: (1) variability due to factor A (religion), (2) variability due to factor B (religiosity), and (3) variability due to the interaction of A and B (abbreviated as "A × B"). Thus,

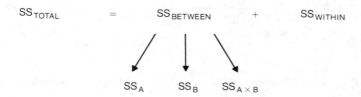

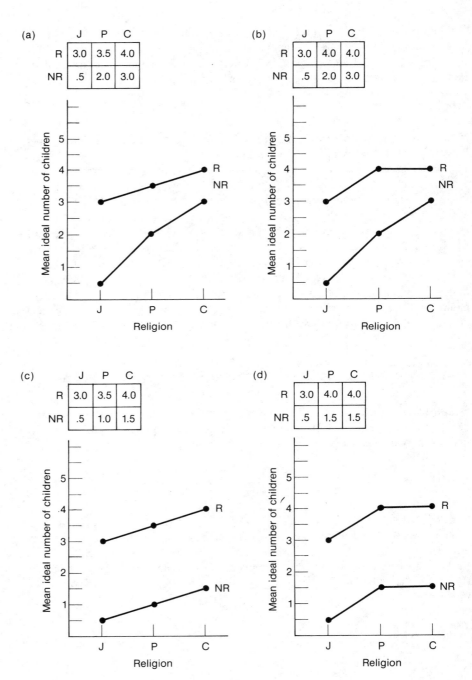

Figure 17.2 Examples of Main Effects and Interaction Effects in a 3 × 2 Factorial Design

Key: J = Jewish, P = Protestant, C = Catholic, R = religious, NR = not religious

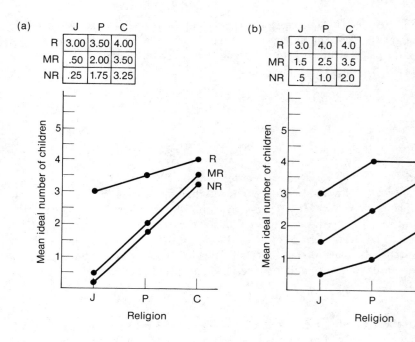

(a)

	J	P	C
R	3.00	3.50	4.00
MR	.50	2.00	3.50
NR	.25	1.75	3.25

(b)

	J	P	C
R	3.0	4.0	4.0
MR	1.5	2.5	3.5
NR	.5	1.0	2.0

Figure 17.3 Examples of Main Effects and Interaction Effects in a 3 × 3 Factorial Design

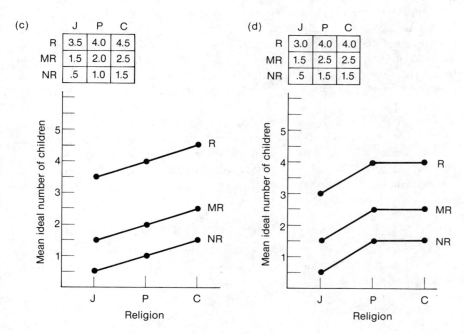

(c)

	J	P	C
R	3.5	4.0	4.5
MR	1.5	2.0	2.5
NR	.5	1.0	1.5

(d)

	J	P	C
R	3.0	4.0	4.0
MR	1.5	2.5	2.5
NR	.5	1.5	1.5

Key: J = Jewish, P = Protestant, C = Catholic,
R = religious, MR = moderately religious, NR = not religious

where SS_A is the sum of squares defining between-group variability for factor A, SS_B is the sum of squares defining between-group variability for factor B, and $SS_{A \times B}$ is the sum of squares defining between-group variability for the interaction of factor A and factor B. Stated more formally,

$$SS_{BETWEEN} = SS_A + SS_B + SS_{A \times B} \qquad [17.2]$$

or

$$SS_{TOTAL} = SS_A + SS_B + SS_{A \times B} + SS_{WITHIN} \qquad [17.3]$$

In factorial analysis of variance, each of the components of the $SS_{BETWEEN}$ is divided by its respective degrees of freedoms. The resulting mean squares, MS_A, MS_B, and $MS_{A \times B}$, are then divided by the MS_{WITHIN}, yielding three F ratios. The F ratio formed by MS_A/MS_{WITHIN} is used to test the null hypothesis with regard to the main effect of factor A. The F ratio formed by MS_B/MS_{WITHIN} is used to test the null hypothesis with respect to the main effect of factor B. Finally, the F ratio formed by $MS_{A \times B}/MS_{WITHIN}$ is used to test the null hypothesis with respect to the interaction of factors A and B.

Computation of SS_A, SS_B, and $SS_{A \times B}$. Table 17.3 presents data for a 2×2 factorial design that will be used as a numerical example. Unlike previous sections of this chapter, we are now working with sample data. The first factor, A, is religion, the second factor, B, is religiosity, and the dependent variable is the ideal number of children. Table 17.3 introduces some new notation. The main effect means for factor A are indicated by $\overline{Y}_{A_1}$ and $\overline{Y}_{A_2}$ (corresponding to level 1 and level 2, respectively). The main effect means for factor B are indicated by $\overline{Y}_{B_1}$ and $\overline{Y}_{B_2}$. The mean scores for each of the four groups comprising the 2×2 design are indicated by $\overline{Y}_1$, $\overline{Y}_2$, $\overline{Y}_3$, and $\overline{Y}_4$.

The sum of squares total is computed using the procedures discussed in Chapter 11. A grand mean is first computed by summing all scores and dividing by N, the total number of subjects in the experiment. In the example, the sum of all scores is 60, hence $G = 60/20 = 3.0$. The grand mean is then subtracted from each score, and the resulting deviation scores are squared and then summed. This has been done in Table 17.4. We find that

$$SS_{TOTAL} = 66.0$$

The sum of squares for between group variability due to factor A (religion) is computed using procedures identical to those in Chapter 11. This involves defining the treatment effect for each level:

$$T_{A_1} = \text{Treatment effect for Catholics} = \overline{Y}_{A_1} - G = 4.0 - 3.0 = 1.0$$

$$T_{A_2} = \text{Treatment effect for Protestants} = \overline{Y}_{A_2} - G = 2.0 - 3.0 = -1.0$$

The treatment effects are squared, multiplied by the sample size for the group in question, and these are summed:

TABLE 17.3 DATA FOR 2 × 2 FACTORIAL EXAMPLE

Religious Catholics	Nonreligious Catholics	Religious Protestants	Nonreligious Protestants
7	5	0	2
4	2	4	4
3	1	2	0
5	3	2	2
6	4	2	2
$\Sigma Y = 25.0$	15.0	10.0	10.0
$\overline{Y}_1 = 5.0$	$\overline{Y}_2 = 3.0$	$\overline{Y}_3 = 2.0$	$\overline{Y}_4 = 2.0$

Main Effects for Religion

Catholics: $\Sigma Y = 40$ $\overline{Y}_{A_1} = \dfrac{40}{10} = 4.0$

Protestants: $\Sigma Y = 20$ $\overline{Y}_{A_2} = \dfrac{20}{10} = 2.0$

Main Effects for Religiosity

Religious: $\Sigma Y = 35$ $\overline{Y}_{B_1} = \dfrac{35}{10} = 3.5$

Nonreligious: $\Sigma Y = 25$ $\overline{Y}_{B_2} = \dfrac{25}{10} = 2.5$

$$SS_A = n_{A_1}T_{A_1}^2 + n_{A_2}T_{A_2}^2$$

$$= (10)(1.0^2) + (10)(-1.0^2)$$

$$= 10 + 10$$

$$= 20.0$$

In general notation

$$SS_A = \sum_{j=1}^{k} n_{A_j}T_{A_j}^2 \qquad [17.4]$$

where k is the number of levels of factor A, n_{A_j} is the number of subjects in level A_j, and all other terms are as previously defined.

The sum of squares for between-group variability due to factor B (religiosity) is computed using identical procedures but using factor B instead of A. The treatment effects for B are:

TABLE 17.4 CALCULATION OF SS_{TOTAL}

Y	$(Y - G)$	$(Y - G)^2$
7	4	16
4	1	1
3	0	0
5	2	4
6	3	9
5	2	4
2	-1	1
1	-2	4
3	0	0
4	1	1
0	-3	9
4	1	1
2	-1	1
2	-1	1
2	-1	1
2	-1	1
4	1	1
0	-3	9
2	-1	1
2	-1	1
$\Sigma Y = 60$		$SS_{TOTAL} = 66.0$

$$G = \frac{60}{20} = 3.0$$

T_{B_1} = Treatment effect for religious subjects = $\overline{Y}_{B_1} - G = 3.5 - 3.0 = .5$

T_{B_2} = Treatment effect for nonreligious subjects = $\overline{Y}_{B_2} - G = 2.5 - 3.0 = -.5$

Then,

$$SS_B = n_{B_1} T_{B_1}^2 + n_{B_2} T_{B_2}^2$$

$$= (10)(.5^2) + (10)(-.5^2)$$

$$= 2.5 + 2.5$$

$$= 5.0$$

In general notation,

$$\text{SS}_\text{B} = \sum_{j=1}^{m} n_{\text{B}_j} T_{\text{B}_j}^2 \qquad \text{[17.5]}$$

where m is the number of levels of factor B, n_{B_j} is the number of subjects in level B_j, and all other terms are as previously defined.

The sum of squares for between-group variability due to the interaction of A and B is computed by first computing treatment effects for the four groups comprising the 2 × 2 design:*

T_1 = Treatment effect for religious Catholics = $\overline{Y}_1 - G = 5.0 - 3.0 = 2.0$

T_2 = Treatment effect for nonreligious Catholics = $\overline{Y}_2 - G = 3.0 - 3.0 = 0$

T_3 = Treatment effect for religious Protestants = $\overline{Y}_3 - G = 2.0 - 3.0 = -1.0$

T_4 = Treatment effect for nonreligious Protestants = $\overline{Y}_4 - G = 2.0 - 3.0 = -1.0$

The treatment effects are squared, multiplied by the n for each group, and then summed, that is:

$$n_1 T_1^2 + n_2 T_2^2 + n_3 T_3^2 + n_4 T_4^2$$

$$= (5)(2.0^2) + (5)(0^2) + (5)(-1.0^2) + (5)(-1.0^2)$$

$$= 20 + 0 + 5 + 5$$

$$= 30.0$$

In general notation, this is represented by the following term:

$$\sum_{j=1}^{k \times m} n_j T_j^2$$

*Technically, these are not treatment effects for the A × B interaction, but rather they represent the treatment effects for all sources of between-group variability, combined.

The $SS_{A \times B}$ is defined as

$$SS_{A \times B} = \sum_{j=1}^{k \times m} n_j T_j^2 - SS_A - SS_B \qquad \text{[17.6]}$$

$$= 30.0 - 20.0 - 5.0$$

$$= 5.0$$

where all terms are as previously defined. The term with the summation sign on the right-hand side of Equation 17.6 represents all of the between-group variability. The sum of squares for the interaction effect is the between-group variability after the variability due to the main effects is removed. This is why SS_A and SS_B are subtracted in Equation 17.6.

To summarize, we have found that

$$SS_{TOTAL} = 66.0$$

$$SS_A = 20.0$$

$$SS_B = 5.0$$

$$SS_{A \times B} = 5.0$$

Recall from Equation 17.3 that

$$SS_{TOTAL} = SS_A + SS_B + SS_{A \times B} + SS_{WITHIN}$$

We can compute SS_{WITHIN} from the above by simple algebraic manipulation:

$$SS_{WITHIN} = SS_{TOTAL} - SS_A - SS_B - SS_{A \times B} \qquad \text{[17.7]}$$

$$= 66.0 - 20.0 - 5.0 - 5.0$$

$$= 36.0$$

The Summary Table. Each of the above sums of squares has a certain number of degrees of freedom associated with it. They are:

$$df_{TOTAL} = N - 1 = 20 - 1 = 19 \qquad \text{[17.8]}$$

$$df_A = k - 1 = 2 - 1 = 1 \qquad \text{[17.9]}$$

$$df_B = m - 1 = 2 - 1 = 1 \qquad \text{[17.10]}$$

$$df_{A \times B} = (k - 1)(m - 1) = (2 - 1)(2 - 1) = 1 \qquad \text{[17.11]}$$

$$df_{\text{WITHIN}} = \sum_{j=1}^{k \times m} (n_j - 1) = (5 - 1) + (5 - 1) + (5 - 1) + (5 - 1)$$

$$= 16 \qquad \text{[17.12]}$$

The relevant mean squares are then defined as

$$MS_A = \frac{SS_A}{df_A} = \frac{20}{1} = 20.0 \qquad \text{[17.13]}$$

$$MS_B = \frac{SS_B}{df_B} = \frac{5}{1} = 5.0 \qquad \text{[17.14]}$$

$$MS_{A \times B} = \frac{SS_{A \times B}}{df_{A \times B}} = \frac{5}{1} = 5.0 \qquad \text{[17.15]}$$

$$MS_{\text{WITHIN}} = \frac{SS_{\text{WITHIN}}}{df_{\text{WITHIN}}} = \frac{36}{16} = 2.25 \qquad \text{[17.16]}$$

and the F ratios to test the relevant null hypotheses are

$$F_A = \frac{MS_A}{MS_{\text{WITHIN}}} = \frac{20.0}{2.25} = 8.89 \qquad \text{[17.17]}$$

$$F_B = \frac{MS_B}{MS_{\text{WITHIN}}} = \frac{5.0}{2.25} = 2.22 \qquad \text{[17.18]}$$

$$F_{A \times B} = \frac{MS_{A \times B}}{MS_{\text{WITHIN}}} = \frac{5.0}{2.25} = 2.22 \qquad \text{[17.19]}$$

All of the above calculations can be summarized in a table as follows:

Source	SS	df	MS	F
Religion (A)	20.0	1	20.00	8.89
Religiosity (B)	5.0	1	5.00	2.22
A × B	5.0	1	5.00	2.22
Within	36.0	16	2.25	
Total	66.0	19		

This table is similar in format to the summary table in Chapter 11. However, in this table, the sum of squares between is represented by its three components, the main effect for A, the main effect for B, and the A × B interaction. We can now address the three questions about a relationship for each of these sources of variability.

Inference of a Relationship. The null and alternative hypotheses for the main effect of religion are

$$H_0: \mu_C = \mu_P$$

$$H_1: \mu_C \neq \mu_P$$

The test of the null hypothesis is made with reference to the F ratio derived from MS_A/MS_{WITHIN}. The F equals 8.89 with 1 and 16 degrees of freedom. Appendix G indicates that the critical value for F, $\alpha = .05$, is 4.49. Since the observed F exceeds this value, the null hypothesis is rejected. There *is* a relationship between religion and ideal family size.

The null and alternative hypotheses for the main effect of religiosity are

$$H_0: \mu_R = \mu_{NR}$$

$$H_1: \mu_R \neq \mu_{NR}$$

The test of the null hypothesis is made with reference to the F ratio derived from MS_B/MS_{WITHIN}. The F equals 2.22 with 1 and 16 degrees of freedom. This does not exceed the critical value of 4.49 and hence, we fail to reject the null hypothesis. We cannot confidently conclude that there is a relationship between religiosity and ideal family size.

The null hypothesis for the interaction states that the relationship between religion and ideal family size is the same for both religious and nonreligious individuals. The alternative hypothesis is that the relationship between religion and ideal family size will differ as a function of religiosity. This is tested by examination of the F ratio derived from $MS_{A \times B}/MS_{WITHIN}$. The F equals 2.22 with 1 and 16 degrees of freedom. This does not exceed the critical value of 4.49 and hence, we fail to reject the null hypothesis. We cannot confidently conclude that an interaction effect is present.

Assumptions of the F Tests. The preceding F tests are based on the same assumptions underlying the one-way between-subjects analysis of variance discussed in Chapter 11. These are:

1. Within-group variances are homogeneous.

2. Population scores within groups are normally distributed.

3. Subjects are independently and randomly selected from independent populations.

As in the one-way analysis of variance, statisticians have shown that the F tests are robust to violations of the first two assumptions when sample sizes are equal and greater than 15 in each group. Interested readers are referred to Glass, Peckham, and Sanders (1974).

Strength of the Relationship. The strength of the relationship for the three sources of between-group variability is computed using the familiar eta squared:

$$\text{Eta}_A^2 = \frac{SS_A}{SS_{TOTAL}} \qquad\qquad [17.20]$$

$$\text{Eta}_B^2 = \frac{SS_B}{SS_{TOTAL}} \qquad\qquad [17.21]$$

$$\text{Eta}_{A \times B}^2 = \frac{SS_{A \times B}}{SS_{TOTAL}} \qquad\qquad [17.22]$$

where Eta_A^2 is eta squared for the main effect of A, Eta_B^2 is eta squared for the main effect of B, and $\text{Eta}_{A \times B}^2$ is eta squared for the interaction of A $\times$ B. We find that

$$\text{Eta}_A^2 = \frac{20.0}{66.0} = .30$$

$$\text{Eta}_B^2 = \frac{5.0}{66.0} = .08$$

$$\text{Eta}_{A \times B}^2 = \frac{5.0}{66.0} = .08$$

Thus, the proportion of variability in ideal family size that is associated with the main effect of religion was .30. This represents a relationship of moderate strength. The proportion of variability associated with religiosity and the religion by religiosity interaction was relatively small.

Nature of the Relationship. When a significant main effect has only two levels, as in the present case, the nature of the relationship is determined in the same fashion as the independent groups t test. This involves examination of the two means involved. In the present example, the nature of the relationship between religion and ideal family size was such that the Catholics, on the average, wanted more children ($\overline{Y} = 4.0$) than Protestants ($\overline{Y} = 2.0$).

If a statistically significant main effect had three or more levels, then the nature of the relationship would be determined using the HSD procedure discussed in Chapter 11. This involves computing the absolute difference between all possible pairs of means comprising the main effect and then comparing each of these against a "critical difference," defined in Equation 11.13 as

$$CD = q \sqrt{\frac{MS_{WITHIN}}{n}}$$

where q is the studentized range value (see Appendix I), n is the number of subjects per group, and MS_{WITHIN} is as previously defined. If the absolute difference between a given pair of means equals or exceeds CD, then the null hypothesis regarding that pair of means is rejected. Otherwise, we fail to reject it. We will consider a numerical example using a factor with three levels shortly.

Simple Main Effects. When an interaction effect is statistically significant, the nature of the interaction is tested using a statistical procedure called **simple main effects analysis.** This involves evaluating the viability of the null hypothesis $\mu_C = \mu_P$ at each level of religiosity, that is,

For level 1 (religious) of factor B: H_0: $\mu_C = \mu_P$

                                            H_1: $\mu_C \neq \mu_P$

For level 2 (nonreligious) of factor B: H_0: $\mu_C = \mu_P$

                                            H_1: $\mu_C \neq \mu_P$

and evaluating the null hypothesis that $\mu_R = \mu_{NR}$ at each level of religion,

For level 1 (Catholics) of factor A: H_0: $\mu_R = \mu_{NR}$

                                            H_1: $\mu_R \neq \mu_{NR}$

For level 2 (Protestants) of factor A: H_0: $\mu_R = \mu_{NR}$

                                            H_1: $\mu_R \neq \mu_{NR}$

Since a significant interaction effect implies that the relationship between an independent variable and a dependent variable differs depending upon the level of the other independent variable, the above tests will often reveal the nature of the interaction effect. In the example with religion and religiosity, the interaction effect was nonsignificant. Therefore, a simple main effects analysis would not be performed. The computational procedures for simple main effects analysis are provided in Appendix 17.2 for the numerical example discussed later.

Method of Presentation

Factorial analysis of variance is reported in much the same way as one-way analysis of variance. A summary table will usually be provided like the one in Section 17.3. In addition, a table of mean scores for each individual group will be given if these are relevant to the interpretation of results. In the previous example, this probably would not be done since only the main effect for religion was statistically significant. However, if the investigator wanted to highlight the similarities as well as differences between means when interpreting the results, a table of all the means would be given. The results of the example might be reported as follows:

A 2 × 2 factorial analysis of variance was performed analyzing ideal family size as a function of religion (Catholic versus Protestant) and religiosity (religious versus not religious). The summary of this analysis is reported in Table 1 and the relevant mean scores are reported in Table 2. The only statistically significant effect was the main effect for religion ($F = 8.89$, df = 1, 16, $p < .05$). On the average, Catholics wanted more children ($\bar{Y} = 4.0$) than Protestants ($\bar{Y} = 2.0$). The strength of this effect, as indexed by eta squared, was .30.

This report would be accompanied by the following two tables:

TABLE 1 SUMMARY TABLE

Source	df	MS	F
Religion (A)	1	20.00	8.89*
Religiosity (B)	1	5.00	2.22
A × B	1	5.00	2.22
Within	16	2.25	
Total	19		

*$p < .05$.

TABLE 2 MEAN IDEAL FAMILY SIZE AS A FUNCTION OF RELIGION AND RELIGIOSITY[a]

	Religiosity	
	Religious	Not Religious
Catholic	5.0	3.0
Protestant	2.0	2.0

[a]$n = 5$ per group.

The first sentence of the report indicates the statistical test used and explicitly defines the independent and dependent variables. The term "between-subjects" is not used since it is assumed that this analysis used between-subjects analysis of variance. The remainder of the text highlights the statistically significant findings. The summary table and table of means "fill out" the rest of the results.

17.5 / Methodological Considerations

Some methodological considerations are worth pointing out in the context of the study relating religion and religiosity to ideal family size. These help to place the results of the statistical test in proper context. First, there are the usual problems of nonrandom assignment to groups associated with independent variables such as religion and religiosity. In addition, there are the problems of uncontrolled disturbance variables and generalizability of the data. Beyond this, however, we can use the investigation to illustrate an important methodological strategy. In Chapter 11, on one-way between-subjects analysis of variance, one of the numerical examples was concerned with the relationship between religion and ideal family size. In that chapter, it was pointed out that religiosity of the individual was acting as a disturbance variable and creating within-group variability (that is, it was increasing the size of the sum of squares within). In the present investigation, religiosity and religion were combined to form four groups and the within-group variability was based on variation of scores within each of the four groups separately. As such, the disturbance effects of religiosity did not enter into the computation of within-group variability since, like religion, it was held constant within groups. This highlights an important advantage of factorial designs. Not only does it allow us to assess the relationship between two independent variables and their interaction with a dependent variable, but it also "removes" the effects of these variables on the within-group variance. This potentially makes the test of either main effect more sensitive than if one of the variables was left uncontrolled and took on the role of a disturbance variable. Thus, another strategy for dealing with disturbance variables, over and above those discussed in Chapter 8, is to bring the variable into the experimental design by including it as a factor.

17.6 / Numerical Example

Social psychologists have studied extensively factors that influence the ability of a speaker to persuade an audience to take the speaker's position on an issue. One important factor influencing the amount of attitude change a speaker can generate on a given issue is the discrepancy between the position advocated by the speaker relative to the position of the audience. Up to a point, the more extreme the speaker's position, the greater the attitude change that will result. However, if the position becomes too extreme, then the speaker loses credibility and the persuasiveness of the speaker lessens. Aronson, Turner, and Carlsmith (1963) tested the effects of message discrepancy and initial expertise of the speaker on attitude change. These researchers hypothesized that the relationship between message discrepancy and attitude change would differ depending upon the initial expertise of the source. Specifically, they argued that speakers with high initial expertise could take much more discrepant positions than speakers with low initial expertise, and still obtain large degrees of attitude change.

In an experiment, college students evaluated the quality of a passage of poetry on a 20 point scale (0 = low quality to 20 = high quality), and then received a written "communication" concerning these verses authored by either an expert (a famous poetry critic) or a nonexpert (an undergraduate student taking beginning English). The communications were identical for the two sources and the only difference between them was the source they were attributed to (either the expert or nonexpert). In addition, the written communications were constructed so as to be either slightly discrepant, moderately discrepant, or very discrepant from the subjects' initial ratings of quality. For example, in the large discrepancy condition, if the subject rated the passage as being relatively high in quality, the written communication argued that the passage was low in quality. After reading the communication, subjects re-rated the passages. The resulting design was s 2 × 3 factorial with two levels of expertise (high versus low) and three levels of discrepancy (small, medium, large). The dependent variable was the amount of change in the quality ratings after reading the communications. The data for this experiment are presented in Table 17.5.

TABLE 17.5 DATA FOR 2 × 3 FACTORIAL DESIGN: ATTITUDE CHANGE AS A FUNCTION OF SPEAKER EXPERTISE AND MESSAGE DISCREPANCY

High Expertise Small Discrepancy	Low Expertise Small Discrepancy	High Expertise Medium Discrepancy	Low Expertise Medium Discrepancy	High Expertise Large Discrepancy	Low Expertise Large Discrepancy
3	1	6	3	5	0
4	0	5	2	4	1
2	2	5	3	6	1
3	1	5	4	5	2
3	1	4	3	5	1
$\Sigma Y_1 = 15$	$\Sigma Y_2 = 5$	$\Sigma Y_3 = 25$	$\Sigma Y_4 = 15$	$\Sigma Y_5 = 25$	$\Sigma Y_6 = 5$
$\overline{Y}_1 = 3.00$	$\overline{Y}_2 = 1.00$	$\overline{Y}_3 = 5.00$	$\overline{Y}_4 = 3.00$	$\overline{Y}_5 = 5.00$	$\overline{Y}_6 = 1.00$

Main Effect for Expertise of Source

High Expertise: $\Sigma Y_{A_1} = 65$ $\overline{Y}_{A_1} = \dfrac{65}{15} = 4.33$

Low Expertise: $\Sigma Y_{A_2} = 25$ $\overline{Y}_{A_2} = \dfrac{25}{15} = 1.67$

Main Effect for Message Discrepancy

Small Discrepancy: $\Sigma Y_{B_1} = 20$ $\overline{Y}_{B_1} = \dfrac{20}{10} = 2.00$

Medium Discrepancy: $\Sigma Y_{B_2} = 40$ $\overline{Y}_{B_2} = \dfrac{40}{10} = 4.00$

Large Discrepancy: $\Sigma Y_{B_3} = 30$ $\overline{Y}_{B_3} = \dfrac{30}{10} = 3.00$

We begin by computing the sum of squares total. This has been done in Table 17.6. The grand mean, G, is equal to $90/30 = 3.0$. The grand mean is subtracted from each score, the resulting deviation scores are squared and then summed. We find that

$$SS_{TOTAL} = 95.0$$

The sum of squares for between-group variability due to factor A (expertise) is computed by calculating the relevant treatment effects:

$$T_{A_1} = \text{Treatment effect for high expertise source} = \overline{Y}_{A_1} - G$$

$$= 4.33 - 3.00 = 1.33$$

$$T_{A_2} = \text{Treatment effect for low expertise source} = \overline{Y}_{A_2} - G$$

$$= 1.67 - 3.00 = -1.33$$

The treatment effects are squared, multiplied by the relevant n, and then summed: summed:

$$SS_A = n_{A_1}T_{A_1}^2 + n_{A_2}T_{A_2}^2$$

$$= (15)(1.33^2) + (15)(-1.33^2)$$

$$= 26.533 + 26.533$$

$$= 53.07$$

The sum of squares for between-group variability due to factor B (discrepancy) is computed similarly. We first define the relevant treatment effects:

$$T_{B_1} = \text{Treatment effect for small discrepancy} = \overline{Y}_{B_1} - G$$

$$= 2.00 - 3.00 = -1.00$$

$$T_{B_2} = \text{Treatment effect for medium discrepancy} = \overline{Y}_{B_2} - G$$

$$= 4.00 - 3.00 = 1.00$$

$$T_{B_3} = \text{Treatment effect for large discrepancy} = \overline{Y}_{B_3} - G$$

$$= 3.00 - 3.00 = .00$$

TABLE 17.6 CALCULATION OF SS$_{TOTAL}$

Y	$(Y - G)$	$(Y - G)^2$
3	0	0
4	1	1
2	-1	1
3	0	0
3	0	0
1	-2	4
0	-3	9
2	-1	1
1	-2	4
1	-2	4
6	3	9
5	2	4
5	2	4
5	2	4
4	1	1
3	0	0
2	-1	1
3	0	0
4	1	1
3	0	0
5	2	4
4	1	1
6	3	9
5	2	4
5	2	4
0	-3	9
1	-2	4
1	-2	4
2	-2	4
1	-2	4

$Y = 90$ SS$_{TOTAL}$ = 95.0

$$G = \frac{90}{30} = 3.0$$

The treatment effects are then squared, multiplied by the relevant n, and the products summed:

$$SS_B = n_{B_1}T_{B_1}^2 + n_{B_2}T_{B_2}^2 + n_{B_3}T_{B_3}^2$$

$$= (10)(-1.00^2) + 10(1.00)^2 + 10(.00^2)$$

$$= 10 + 10$$

$$= 20.00$$

The sum of squares for the interaction of factors A and B is computed by first defining the treatment effects for each of the groups comprising the factorial design:

T_1 = Treatment effect for high expert, small discrepancy = $\overline{Y}_1 - G$

$\quad = 3.00 - 3.00 = .00$

T_2 = Treatment effect for low expert, small discrepancy = $\overline{Y}_2 - G$

$\quad = 1.00 - 3.00 = -2.00$

T_3 = Treatment effect for high expert, medium discrepancy = $\overline{Y}_3 - G$

$\quad = 5.00 - 3.00 = 2.00$

T_4 = Treatment effect for low expert, medium discrepancy = $\overline{Y}_4 - G$

$\quad = 3.00 - 3.00 = .00$

T_5 = Treatment effect for high expert, large discrepancy = $\overline{Y}_5 - G$

$\quad = 5.00 - 3.00 = 2.00$

T_6 = Treatment effect for low expert, large discrepancy = $\overline{Y}_6 - G$

$\quad = 1.00 - 3.00 = -2.00$

The treatment effects are each squared, multiplied by the relevant n, and summed:

$$\sum_{j=1}^{6} n_j T_j^2 = n_1 T_1^2 + n_2 T_2^2 + n_3 T_3^2 + n_4 T_4^2 + n_5 T_5^2 + n_6 T_6^2$$

$$= (5)(.00^2) + (5)(-2.00^2) + (5)(2.00^2) + (5)(.00^2)$$

$$+ (5)(2.00^2) + (5)(-2.00^2)$$

$$= 0 + 20 + 20 + 0 + 20 + 20$$

$$= 80.00$$

The $SS_{A \times B}$ is then computed by subtracting SS_A and SS_B from this result:

$$SS_{A \times B} = \sum_{j=1}^{6} n_j T_j^2 - SS_A - SS_B$$

$$= 80.00 - 20.00 - 53.07$$

$$= 6.93$$

The sum of squares within groups is computed by

$$SS_{WITHIN} = SS_{TOTAL} - SS_A - SS_B - SS_{A \times B}$$

$$= 95.00 - 53.07 - 20.00 - 6.93$$

$$= 15.00$$

The relevant degrees of freedom are

$$df_A = k - 1 = 2 - 1 = 1$$

$$df_B = m - 1 = 3 - 1 = 2$$

$$df_{A \times B} = (k - 1)(m - 1) = (2 - 1)(3 - 1) = 2$$

$$df_{WITHIN} = \sum_{j=1}^{6}(n_j - 1) = 24$$

$$df_{TOTAL} = N - 1 = 29$$

The relevant mean squares are

$$MS_A = \frac{SS_A}{df_A} = \frac{53.07}{1} = 53.07$$

$$MS_B = \frac{SS_B}{df_B} = \frac{20.00}{2} = 10.00$$

$$MS_{A \times B} = \frac{SS_{A \times B}}{df_{A \times B}} = \frac{6.93}{2} = 3.47$$

$$MS_{WITHIN} = \frac{SS_{WITHIN}}{df_{WITHIN}} = \frac{15}{24} = .62$$

The relevant F ratios are

$$F_A = \frac{MS_A}{MS_{WITHIN}} = \frac{53.07}{.62} = 85.60$$

$$F_B = \frac{MS_B}{MS_{WITHIN}} = \frac{10.00}{.62} = 16.13$$

$$F_{A \times B} = \frac{MS_{A \times B}}{MS_{WITHIN}} = \frac{3.47}{.62} = 5.60$$

This yields the following summary table:

Source	SS	df	MS	F
Expertise (A)	53.07	1	53.07	85.60
Discrepancy (B)	20.00	2	10.00	16.13
A × B	6.93	2	3.47	5.60
Within	12.00	24	.62	
Total	92.00	29		

Inference of a Relationship. The null and alternative hypotheses for the main effect of expertise are

$$H_0: \mu_{A_1} = \mu_{A_2}$$

$$H_1: \mu_{A_1} \neq \mu_{A_2}$$

where μ_{A_1} is the population mean for the high expertise condition and μ_{A_2} is the population mean for the low expertise condition. The critical value for F with 1 and 24 degrees of freedom, $\alpha = .05$, is 4.26 (see Appendix G). Since 85.60 exceeds this value, we reject the null hypothesis. There is a relationship between expertise of the source and attitude change.

The null and alternative hypotheses for the main effect of discrepancy are:

$$H_0: \mu_{B_1} = \mu_{B_2} = \mu_{B_3}$$

$$H_1: \text{The three population means are not all equal}$$

where μ_{B_1} is the population mean for the small discrepancy condition, μ_{B_2} is the population mean for the medium discrepancy condition, and μ_{B_3} is the population mean for the large discrepancy condition. The critical value for F with 2 and 24 degrees of freedom, $\alpha = .05$, is 3.40 (see Appendix G). Since 16.13 exceeds this value, we reject the null hypothesis. There is a relationship between message discrepancy and attitude change.

The null hypothesis for the interaction effect is that the relationship between message discrepancy and attitude change is the same for both high expertise and low expertise communicators. The alternative hypothesis is that the nature of the relationship between message discrepancy and attitude change depends on the expertise of the communicator. The critical value for F with 2 and 24 degrees of freedom, $\alpha = .05$, is 3.40. Since 5.60 exceeds this value, we reject the null hypothesis and conclude that an interaction is present.

Strength of the Relationship. The strengths of the various relationships are computed using Equations 17.20 to 17.22:

$$\text{Eta}_A^2 = \frac{\text{SS}_A}{\text{SS}_{\text{TOTAL}}} \quad \frac{53.07}{95.00} = .56$$

$$\text{Eta}_B^2 = \frac{\text{SS}_B}{\text{SS}_{\text{TOTAL}}} = \frac{20.00}{95.00} = .21$$

$$\text{Eta}_{A \times B}^2 = \frac{\text{SS}_{A \times B}}{\text{SS}_{\text{TOTAL}}} = \frac{6.93}{95.00} = .07$$

These are interpreted as the proportion of variability in the dependent variable that is associated with the particular source of between-group variability. The strongest effect was for the main effect of communicator expertise ($\text{Eta}^2 = .56$). Although the interaction effect was statistically significant, the strength of the effect was relatively weak ($\text{Eta}^2 = .07$).

Nature of the Relationship. For the main effect of expertise, the nature of the relationship is determined by examination of the means, since it has only two levels. Messages attributed to a communicator with high expertise resulted in, on the average, more attitude change ($\overline{Y} = 4.33$) than messages attributed to a communicator with low expertise ($\overline{Y} = 1.67$).

The main effect of message discrepancy has three levels and hence, it is necessary to apply the HSD procedure to determine the nature of the relationship between message discrepancy and attitude change. The value of the critical difference is

$$\text{CD} = q\sqrt{\frac{\text{MS}_{\text{WITHIN}}}{n}}$$

$$= 3.53\sqrt{\frac{.62}{10}}$$

$$= .88$$

Ten is used as n in the above formula, since each main effect mean was based on ten subjects. The comparison of the relevant pairs of means, as developed in Chapter 11, is summarized as follows, in which the absolute difference between a pair of means is compared with CD:

Null Hypothesis	Absolute Mean Difference	CD	Is Null Hypothesis Rejected?
$\mu_{B_1} = \mu_{B_2}$	$\lvert 2.0 - 4.0 \rvert = 2.0$	.88	Yes
$\mu_{B_1} = \mu_{B_3}$	$\lvert 2.0 - 3.0 \rvert = 1.0$	.88	Yes
$\mu_{B_2} = \mu_{B_3}$	$\lvert 3.0 - 4.0 \rvert = 1.0$	.88	Yes

The nature of the relationship between message discrepancy and attitude change is such that small amounts of message discrepancy lead to the smallest amount of attitude change, followed by messages with large discrepancies. The most effective message for producing attitude change is one of moderate discrepancy.

Since the interaction effect was statistically significant, it is necessary to perform a simple main effects analysis to determine the nature of the interaction. Because of the complexity of the analyses, this is presented in Appendix 17.2. Inspection of the means and the results of the analysis indicate that high expertise communicator was more persuasive than the low expertise communicator in the large discrepancy condition as opposed to the other discrepancy conditions.

The results of the analysis might be reported in a journal as follows:

A 2 × 3 factorial analysis of variance was performed analyzing attitude change as a function of communicator expertise (high versus low) and message discrepancy (small versus medium versus large). The summary table is presented in Table 1 and relevant mean scores are presented in Table 2. All effects in the analysis were statistically significant. The main effect of communicator expertise ($F = 85.60$, df = 1, 24, $p < .05$) was such that high expertise communicators produced more attitude change, on the average, than low expertise communicators ($\overline{Y} = 4.33$ versus $\overline{Y} = 1.67$, respectively). The strength of this effect, as indexed by eta squared, was .56. The nature of the main effect of message discrepancy ($F = 16.13$, df = 2, 24, $p < .05$) was analyzed using Tukey's honest significance difference (HSD) method. This analysis indicated that all pairwise differences between the small ($\overline{Y} = 2.00$), medium ($\overline{Y} = 4.00$), and large ($\overline{Y} = 3.00$) discrepancy messages were statistically significant. The strength of the relationship between message discrepancy and attitude change, as indexed by eta squared, was .21. The significant interaction effect ($F = 5.60$, df = 2, 24, $p < .05$) was analyzed using simple main effects analysis and, where appropriate, Tukey's HSD procedure. The results of this analysis indicate that under conditions of large message discrepancy, high expertise communicators lead to more attitude change than low expertise communicators and do so more in this condition than under conditions of small or medium discrepancy. The strength of the overall interaction effect, as indexed by eta squared, was .07.

Box 17.1 Luck versus Skill

When someone is successful at a task, we will sometimes attribute that success to the person's ability to perform well. Alternatively, we may simply think that the person "got lucky" and that ability has relatively little to do with success. Johnson (1976) replicated the results of Deaux and Emswiller (1974) who were interested in the extent to which these two different attributes would be used to explain success on a task by males as opposed to females. Subjects listened to tape recordings that indicated that a man or woman had succeeded either at a traditionally masculine task (for example, identifying wrenches, screwdrivers, and so on) or at a traditionally feminine task (for example, identifying various cooking utensils). Subjects were then asked to indicate on a 13 point scale the extent to which performance on the task was due to luck (scores near 1) as opposed to ability (scores near 13). The experimental design was thus a 2 × 2 between-subjects factorial design, with sex of the performer (male versus female) and type of task (masculine task versus feminine task) as the independent variables and ratings of luck-ability as the dependent variable. Deaux and Emswiller expected that for male performers, success would tend to be attributed to ability on masculine tasks but luck on feminine tasks. For female performers, the reverse was expected. Thus, the investigators predicted an interaction effect, with the relationship between ability-luck attributions and sex of the performer depending on the nature of the task being performed. Although a statistically significant interaction effect was observed, the nature of the interaction was not what the experimenters expected. The relevant summary table and mean scores were as follows:

Source	SS	df	MS	F
Sex of Performer (A)	17.64	1	17.64	4.38*
Type of Task (B)	9.92	1	9.92	2.46
A × B	36.61	1	36.61	9.08*
Within	386.88	96	4.03	
Total	451.05	99		

*$p < .01$.

	Masculine Task	Feminine Task	
Male Performer	9.53	8.95	9.24
Female Performer	7.48	9.32	8.40
	8.50	9.14	

The main effect of the sex of performer indicated that, on the average, the success of male performers was more likely to be attributed to ability than that of female performers. The strength of this effect, as indexed by eta squared, was only .04, however. The main effect of type of task was not statistically significant. The significant interaction effect was analyzed using simple main effects analysis. These analyses indicated that on masculine tasks, males' performances were, on the average, attributed more to ability than luck ($F = 13.04$, df = 1, 96, $p < .05$) as compared with females' performances on that task. However, for feminine tasks, the differences between males and females were not statistically significant ($F = .42$, df = 1, 96, ns). The strength of the overall interaction effect, as indexed by eta squared, was .08. What type of interpretation might you give to these data?

This report would be accompanied by the following tables:

TABLE 1 SUMMARY TABLE OF ANALYSIS OF VARIANCE

Source	df	MS	F
Expertise (A)	1	53.07	85.60*
Discrepancy (B)	2	10.00	16.13*
A × B	2	3.47	5.60*
Within	24	.62	
Total	29		

*$p < .01$.

TABLE 2 MEAN ATTITUDE CHANGE SCORES AS A FUNCTION OF COMMUNICATOR EXPERTISE AND MESSAGE DISCREPANCY

	Small Discrepancy	Medium Discrepancy	Large Discrepancy
High Expertise	3.00	5.00	5.00
Low Expertise	1.00	3.00	1.00

$n = 5$ per group

17.7 / Unequal Sample Sizes

In the examples thus far, the individual group sample sizes, n_j, have been equivalent. This is not always the case in psychological research. For example, in animal experimentation, a subject in an experiment might be lost due to accidental death or temporary sickness. Or a subject scheduled to participate in a particular condition of a learning experiment may never show up. Unequal sample sizes in groups raise additional issues for the analysis of variance techniques discussed in this chapter.

When all groups are of equal sample size, the two independent variables, A and B, are unrelated to each other; that is, there is no relationship between the independent variables. Consider the following sample sizes in a 2 × 2 factorial design:

	Religion	
Sex	Catholic	Protestant
Male	10	10
Female	10	10

In this experiment, there are equal sample sizes in each group, and the two independent variables, *in the context of this experiment,* are unrelated to one another. This can be thought of in terms of conditional probabilities: The probability of being a Catholic is the same for both males and females as is the probability of being a Protestant. The variable of religion is independent of a person's sex. Now consider the case of unequal sample sizes:

	Catholic	Protestant
Male	6	10
Female	10	8

In this case, a relationship between a person's sex and religion exists. For example, if you know that a subject is male, you know it is more likely that he is a Protestant than a Catholic. If you know a subject is a female, it is more likely that she is a Catholic than a Protestant. Thus, there is a relationship between the two independent variables.*

The introduction of a relationship between independent variables creates a number of statistical and conceptual issues for testing the main effects of A, B and the A × B interaction. Consider factor A in a design with unequal sample sizes. On the basis of an analysis of variance, we might conclude that A is related to the dependent variable and accounts for 20% of the variability in it. However, because factor B is related to factor A, some of the variability in the dependent variable due to A may actually be due to factor B. Factor B is acting like a confounding variable in that it is related to factor A and, hence, could explain some of the between-group variability resulting from A. The problem faced by statisticians is what to do about this explained variance that is common to both factor A and factor B.

*Another way of thinking about this is that if one were to do a χ^2 analysis on the cell frequencies (as discussed in Chapter 14), the χ^2 would equal 0 in the case of equal ns (indicating no relationship) whereas it would be nonzero and positive in the case of unequal ns (indicating the presence of a relationship). Unequal sample sizes do *not* introduce a relationship between the two independent variables when the sample sizes for the groups comprising one factor are proportional across levels of the other factor (for example, the ratio of males to females is 60 to 10 for both Catholics and Protestants).

One consideration that determines how to proceed with the analysis is whether the correlation that has been introduced between factor A and factor B as a result of unequal ns is theoretically meaningful or not. Suppose one selects a random sample of 200 college students at a university and finds the following sample sizes when subjects are categorized in a 2 × 4 design:

	Catholic	Protestant	Jewish	Other
Male	40	33	17	10
Female	31	42	15	12

The unequal ns produce a relationship between sex and religion. The relationship is theoretically meaningful because it probably exists in the population of interest. That is, for the population of college students at this university, there *is* a relationship between a person's sex and his or her religion. The observed relationship in the sample is not an artifact of the method of data collection or of subject selection procedures. Rather, it reflects a relationship that truly exists in the real world, a relationship that should be taken into consideration.

In contrast, an experiment might test the relative effects of two different memory aids and a person's gender on ability to recall materials. Twenty subjects are scheduled in each group, but for some reason, four subjects don't show up, yielding the following ns:

	Memory Aid 1	Memory Aid 2
Males	18	20
Females	20	18

A relationship between the two independent variables has been introduced because of the unequal ns. However, the relationship is not theoretically meaningful. It is simply a function of four subjects not showing up, for whatever reason. In this case, the relationship introduced by the unequal ns should not be considered in the analysis. When the unequal ns are *not* theoretically meaningful, the analysis of variance described in this chapter is altered slightly and an **unweighted means factorial analysis of variance** is performed. Procedures for doing so are discussed in Winer (1971). When the unequal ns *are* theoretically meaningful, then a **least squares factorial analysis of variance** is performed. Procedures for this technique are also discussed in Winer (1971).

17.8 / Planning an Investigation Using Two-Way Factorial Analysis of Variance

When planning an investigation that will use two-way between-subjects factorial analysis of variance, the selection of sample size in order to achieve desired levels of power can be facilitated by the use of Appendix F.2. Table 17.7 presents a portion of this appendix for purposes of exposition. The first column indicates different levels of power and the column headings represent population values of eta squared. There are different tables for different numbers of degrees of freedom. Table 17.7 represents the case where df = 1. In two-way factorial analysis of variance there are three relevant population eta squares, that for the main effect of A, that for the main effect of B, and that for the interaction effect. The determination of sample size for a given cell requires that you calculate the necessary cell size for each individual effect and then use the *largest* sample size of the three. This will ensure that the minimum power requirements are met (for a discussion of possible compromises, see Cohen 1977). In addition, the sample size given in Appendix F.2 must be adjusted to take into account the nature of the factorial design (2×3, 3×3, and so on). The tabled value should be adjusted as follows

$$n' = \frac{(n - 1)(df_e + 1)}{(k)(m)}$$

where n' is the adjusted cell sample size, n is the tabled value indicating the sample size per cell, df_e is the degrees of freedom for the effect in question [either $k - 1$, $m - 1$, or $(k - 1)(m - 1)$], k is the number of levels of factor A, and m is the number of levels of factor B. The value of n' is rounded up to the nearest integer.

TABLE 17.7 CELL SAMPLE SIZES NECESSARY TO ACHIEVE SELECTED LEVELS OF POWER FOR ALPHA = .05, DF = 1 AS A FUNCTION OF POPULATION VALUES OF ETA SQUARED

Degrees of Freedom = 1
Alpha = 0.05

Power	Population Eta Squared										
	0.01	0.03	0.05	0.07	0.10	0.15	0.20	0.25	0.30	0.35	0.40
0.10	22	8	5	4	3	2	2	2	—	—	—
0.50	193	63	38	27	18	12	9	7	5	5	4
0.70	310	101	60	42	29	18	13	10	8	7	6
0.80	393	128	76	53	36	23	17	13	10	8	7
0.90	526	171	101	71	48	31	22	17	13	11	9
0.95	651	211	125	87	60	38	27	21	16	13	11
0.99	920	298	176	123	84	53	38	29	22	18	15

As an example, consider a 2×2 design in which the investigator suspects that the population values of eta squared for A, B, and A $\times$ B are .15, .15, and .10, respectively. For power of .80 for each effect, alpha = .05, we determine n' for A, B, and A $\times$ B:

n' for A. Tabled value of n for population eta squared = .15, df = $k - 1 = 1$ is 23:

$$n' = \frac{(23 - 1)(1 + 1)}{(2)(2)} = 11$$

n' for B. Tabled value of n for population eta squared = .15, df = $m - 1 = 1$ is 23:

$$n' = \frac{(23 - 1)(1 + 1)}{(2)(2)} = 11$$

n' for A $\times$ B. Tabled value of n for population eta squared = .10, df = $(k - 1)(m - 1) = 1$ is 36:

$$n' = \frac{(36 - 1)(1 + 1)}{(2)(2)} = 17.5; \text{ rounded to } 18$$

The largest cell size dictates the number of subjects per group. In this experiment, the investigator would use eighteen subjects in each of the four groups. Note that this will increase the power of the test beyond the .80 level for the two main effects.

17.9 / Summary

Two-way between-subjects factorial analysis of variance allows an investigator to study the separate and joint effects of two independent variables on a dependent variable. This is accomplished by the analysis of main effects and interaction effects in a factorial design. A factorial design is one in which the k levels of one independent variable are combined with the m levels of another independent variable, yielding $k \times m$ independent groups. A test of a main effect refers to the test of the relationship between one of the independent variables and the dependent variable. A test of an interaction effect refers to the test of whether the nature of the relationship between one of the independent variables and the dependent variable depends on the value of the other independent variable. The statistical analysis of main and interaction effects is similar to one-way analysis of variance except that the overall between-group variability is broken down into three components, between-group variability due to factor A, between-group variability due to factor B, and between-group variability due to the interaction of A and B. Hypothesis testing procedures are then applied to these three sources of between-group variability. When a statistically significant main effect is obtained, the nature of the effect is determined using the HSD test. When a statistically significant interaction effect is obtained, the nature of the effect is determined using simple main effect analysis. The computational procedures discussed in the chapter are only applicable to cases of equal sample sizes in the individual groups. When unequal sample sizes result, either an unweighted means or least squares analysis of variance should be used.

APPENDIX 17.1

Computational Formulas

This appendix presents computationally efficient formulas for computing the relevant sums of squares in a two-way between-subjects factorial analysis of variance. The relevant mean squares, degrees of freedom, and F ratios are computed using procedures discussed in the text. Consider the following 2×4 factorial design consisting of factors A and B with four subjects per group (n). The scores on the dependent variable are:

	b_1	b_2	b_3	b_4
	3	4	7	7
	6	5	8	8
a_1	3	4	7	9
	3	3	6	8
	1	2	5	10
	2	3	6	10
a_2	2	4	5	9
	2	3	6	11

k = number of levels of factor A = 2
m = number of levels of factor B = 4

We begin by calculating the sum of all scores and the sum of all scores squared:

$$\Sigma Y = 3 + 6 + 3 + \ldots + 11 = 172$$

$$\Sigma Y^2 = 3^2 + 6^2 + 3^2 + \ldots + 11^2 = 1,160$$

Next, compute the sum of Y scores in each group separately, and generate the following table with the sums as entries:

	b_1	b_2	b_3	b_4	Total
a_1	15	16	28	32	91 = A_1
a_2	7	12	22	40	81 = A_2
Total	22	28	50	72	
	B_1	B_2	B_3	B_4	

Let A_1 be the sum of all scores at level a_1, and A_2 be the sum of all scores at level a_2. Similarly, B_1 is the sum of all scores at level b_1, and so on through B_4. Thus, A_1 = 91, A_2 = 81, B_1 = 22, B_2 = 28, B_3 = 50, and B_4 = 72. For each entry in the 2×4 table of sums, square each sum and add these squared sums together.

$$(I) = 15^2 + 16^2 + 28^2 + 32^2 + 7^2 + 12^2 + 22^2 + 40^2 = 4566.00$$

Then divide this result by n:

$$(II) = \frac{4566.000}{4} = 1141.500$$

Then compute

$$(III) = \frac{(\Sigma Y)^2}{(n)(k)(m)} = \frac{172^2}{(4)(2)(4)} = 924.500$$

$$(\text{IV}) = \sum_{j=1}^{k} \frac{A_j^2}{(n)(m)} = \frac{91^2}{(4)(4)} + \frac{81^2}{(4)(4)} = 927.625$$

$$(\text{V}) = \sum_{j=1}^{l} \frac{B_j^2}{(n)(k)} = \frac{22^2}{(4)(2)} + \frac{28^2}{(4)(2)} + \frac{50^2}{(4)(2)}$$

$$+ \frac{72^2}{(4)(2)} = 1,119.000$$

Then,

$$SS_{\text{TOTAL}} = \Sigma Y^2 - (\text{III}) = 1,160.00 - 924.500$$
$$= 235.500$$

$$SS_A = (\text{IV}) - (\text{III}) = 927.625 - 924.500$$
$$= 3.125$$

$$SS_B = (\text{V}) - (\text{III}) = 1,119.000 - 924.500$$
$$= 194.500$$

$$SS_{A \times B} = (\text{II}) - (\text{IV}) - (\text{V}) + (\text{III})$$
$$= 1,141.500 - 927.625 - 1,119.000$$
$$+ 924.500 = 19.375$$

$$SS_{\text{WITHIN}} = \Sigma Y^2 - (\text{II}) = 1,160.000 - 1,141.500$$
$$= 18.500$$

APPENDIX 17.2

Computation of Simple Main Effects

This appendix describes the computation of simple main effects for the expertise-message discrepancy experiment. The first step involves testing the nature of the relationship between message discrepancy and attitude change at each of the levels of expertise, respectively:

For high expertise sources: H_0: $\mu_{B_1} = \mu_{B_2} = \mu_{B_3}$

H_1: The three means are not all equal

For low expertise sources: H_0: $\mu_{B_1} = \mu_{B_2} = \mu_{B_3}$

H_1: The three means are not all equal

The null and alternative hypotheses, because they involve more than two means, are stated in the format of one-way analysis of variance. We will first discuss procedures for testing the "simple main effect" of message discrepancy on attitude change for the high expertise source condition. This is accomplished by computing a sum of squares between for message discrepancy for subjects in the high expert condition only. In essence, a sum of squares between is computed as if a one-way analysis of variance was being performed on just these subjects. The grand mean, considering only these subjects, is $65/15 = 4.33 = G_{A_1}$. The treatment effects are

$T_{B_1 \text{ at } A_1}$ = Treatment effect for small discrepancy
$= 3.00 - 4.33 = -1.33$

$T_{B_2 \text{ at } A_1}$ = Treatment effect for medium discrepancy
$= 5.00 - 4.33 = .67$

$T_{B_3 \text{ at } A_1}$ = Treatment effect for large discrepancy
$= 5.00 - 4.33 = .67$

and

$$SS_{B \text{ at } A_1} = (5)(-1.33^2) + (5)(.67^2) + (5)(.67^2)$$

$$= 13.32$$

This sum of squares has $(k - 1) = 3 - 1 = 2$ degrees of freedom. The resulting mean square is

$$MS_{B \text{ at } A_1} = \frac{SS_{B \text{ at } A_1}}{df} = \frac{13.32}{2} = 6.66$$

An F ratio is formed by dividing $MS_{B \text{ at } A_1}$ by the original MS_{WITHIN} from the original summary table. This MS_{WITHIN} is used in preference to a mean square within based only on high expertise subjects since it has more degrees of freedom (24 as opposed to 12) and hence is a better estimate of the population within-group variance. Recall that an assumption of analysis of variance is that within-group population variances are equal across groups. The MS_{WITHIN} based on all groups will thus be the best estimate of within-group variability. The F ratio is

$$F_{B \text{ at } A_1} = \frac{6.66}{.62} = 10.74$$

For an F with 2 and 24 degrees of freedom, the critical value is 3.40. Since 13.32 exceeds this value, the null hypothesis that $\mu_{B_1} = \mu_{B_2} = \mu_{B_3}$ for subjects in the high expert condition is rejected. An HSD test is therefore required to discern the nature of the differences between means. We define the value of CD as follows:

$$CD = 3.53 \sqrt{\frac{.62}{5}}$$

$$= 1.24$$

Then we compute the appropriate significance tests:

Null Hypothesis	Absolute Mean Difference	CD	Is Null Hypothesis Rejected?
$\mu_{B_1} = \mu_{B_2}$	$\lvert 3.0 - 5.0 \rvert = 2.0$	1.24	Yes
$\mu_{B_1} = \mu_{B_3}$	$\lvert 3.0 - 5.0 \rvert = 2.0$	1.24	Yes
$\mu_{B_2} = \mu_{B_3}$	$\lvert 5.0 - 5.0 \rvert = 0.0$	1.24	No

For subjects in the high expertise condition, both medium discrepant and large discrepant messages led to more attitude change than small discrepancy messages. However, there were no statistically significant differences between medium discrepant and large discrepant messages.

The same procedures are applied to the analysis of message discrepancy and attitude change for subjects in the low expertise condition. This has been done in Table 17.8. The results of this analysis show that the relationship between message discrepancy and attitude change for subjects in the low expertise condition is different than that for subjects in the high expertise condition. For low expertise subjects, both small discrepancy and large discrepancy messages were about equally persuasive and both were significantly less persuasive than medium discrepant messages. This was not the case for subjects in the high expertise condition. The fact that the nature of

the relationship between message discrepancy and attitude change differs depending on the expertise of the communicator is what produced the significant interaction effect. The simple main effects analysis has made the nature of the interaction explicit.

It is also possible to do simple main effects analysis examining the relationship between communicator expertise and attitude change at each level of message discrepancy. This has been done in Table 17.9. In this case, the analysis locates the interaction in the fact that the relationship between communicator expertise and attitude change was different in the large discrepancy condition as opposed to the small discrepancy and medium discrepancy conditions. Graphs of the individual group means are presented in Figure 17.4. In Figure 17.4a, message discrepancy is plotted on the abscissa whereas in Figure 17.4b, the communicator expertise is plotted on the abscissa. In both cases, the lines are not parallel, suggesting an interaction effect.

TABLE 17.8 SIMPLE MAIN EFFECTS ANALYSIS FOR SUBJECTS IN LOW EXPERTISE CONDITION

H_0: $\mu_{B_1} = \mu_{B_2} = \mu_{B_3}$

H_1: The three population means are not all equal

$$G_{A_2} = \frac{25}{15} = 1.67 \qquad T_{B_1 \text{ at } A_2} = 1.00 - 1.67 = -.67$$

$$T_{B_2 \text{ at } A_2} = 3.00 - 1.67 = 1.33$$

$$T_{B_3 \text{ at } A_2} = 1.00 - 1.67 = -.67$$

$$SS_{B \text{ at } A_2} = (5)(-.67^2) + (5)(1.33^2) + (5)(-.67^2)$$
$$= 13.32$$

$$df = k - 1 = 3 - 1 = 2$$

$$MS_{B \text{ at } A_2} = \frac{13.32}{2} = 6.66$$

$$F = \frac{MS_{B \text{ at } A_2}}{MS_{WITHIN}} = \frac{6.66}{.62} = 10.74*$$

$*p < .05.$

Figure 17.4 Mean Attitude Change as a Function of Communicator Expertise and Message Discrepancy

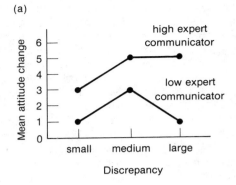

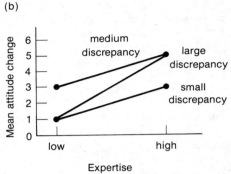

(TABLE 17.8)

HSD test

$$CD = 3.53 \sqrt{\frac{.62}{5}} = 1.24$$

Null Hypothesis	Absolute Mean Difference	CD	Is Null Hypothesis Rejected?
$\mu_{B_1} = \mu_{B_2}$	$|1.00 - 3.00| = 2.00$	1.24	Yes
$\mu_{B_1} = \mu_{B_3}$	$|1.00 - 1.00| = \ .00$	1.24	No
$\mu_{B_2} = \mu_{B_3}$	$|3.00 - 1.00| = 2.00$	1.24	Yes

TABLE 17.9 SIMPLE MAIN EFFECTS ANALYSIS OF EACH LEVEL OF MESSAGE DISCREPANCY

Small Discrepancy

$$\overline{Y}_{A_1} = 3.00 \qquad \overline{Y}_{A_2} = 1.00$$

$$G_{B_1} = \frac{20}{10} = 2.00$$

$$T_{A_1 \text{ at } B_1} = 3.00 - 2.00 = 1.00$$
$$T_{A_2 \text{ at } B_1} = 1.00 - 2.00 = -1.00$$
$$SS_{A \text{ at } B_1} = (5)(1.00^2) + (5)(-1.00^2) = 10$$
$$df = M - 1 = 2$$

$$MS_{A \text{ at } B_1} = \frac{10}{1} = 10.00$$

$$F = \frac{MS_{A \text{ at } B_1}}{MS_{\text{WITHIN}}} = \frac{10.00}{.62} = 16.13^*$$

Medium Discrepancy

$$\overline{Y}_{A_1} = 5.00 \qquad \overline{Y}_{A_2} = 3.00$$

$$G_{B_2} = \frac{40}{10} = 4.00$$

$$T_{A_1 \text{ at } B_2} = 5.00 - 4.00 = 1.00$$
$$T_{A_2 \text{ at } B_1} = 3.00 - 4.00 = -1.00$$
$$SS_{A \text{ at } B_2} = (5)(1.00^2) + (5)(-1.00^2) = 10$$
$$df = M - 1 = 1$$

$$MS_{A \text{ at } B_2} = \frac{10}{1} = 10.00$$

$$F = \frac{MS_{A \text{ at } B_2}}{MS_{\text{WITHIN}}} = \frac{10.00}{.62} = 16.13^*$$

Large Discrepancy

$$\overline{Y}_{A_1} = 5.00 \qquad \overline{Y}_{A_2} = 1.00$$

$$G_{B_3} = \frac{30}{10} = 3.00$$

$$T_{A_2 \text{ at } B_3} = 5.00 - 3.00 = 2.00$$
$$T_{A_2 \text{ at } B_3} = 1.00 - 3.00 = -2.00$$
$$SS_{A \text{ at } B_3} = (5)(2.00^2) + (5)(-2.00^2) = 20$$
$$df = M - 1 = 1$$

$$MS_{A \text{ at } B_3} = \frac{20}{1} = 20.00$$

$$F = \frac{MS_{A \text{ at } B_3}}{MS_{\text{WITHIN}}} = \frac{20.00}{.62} = 32.26^*$$

$^*p < .05.$

The analysis of simple main effects makes explicit the nature of an interaction effect. You may wonder why the social scientists bother computing the statistical significance of the overall interaction effect since one could move directly into the computation of simple main effects analysis. The reason we don't do this is similar to the reason we do an overall F test in one-way between-subjects analysis of variance and then follow this up with Tukey's HSD procedure when the F is statistically significant. Simple main effects analysis involves a large number of statistical tests and the overall test of the interaction effect maintains an alpha level of .05 in testing possible differences in relationships. Thus, if the overall interaction effect is not statistically significant, one should *not* apply simple main effects analysis, since this would violate the hypothesis testing logic of specifying alpha levels.

EXERCISES

Exercises to Review Concepts

1. How many independent variables are there in a 3×3 factorial design? How many independent variables are there in a $2 \times 2 \times 2$ factorial design?

2. For each of the following sets of population means, indicate if there is a main effect of factor A, a main effect of factor B, and/or an interaction effect.

(a)

	B_1	B_2
A_1	4	5
A_2	4	5

(b)

	B_1	B_2
A_1	6	6
A_2	4	4

(c)

	B_1	B_2
A_1	5	10
A_2	10	5

(d)

	B_1	B_2	B_3
A_1	1	2	3
A_2	5	6	7
A_3	8	9	10

(e)

	B_1	B_2	B_3
A_1	4	7	9
A_2	5	6	7
A_3	6	8	10

(f)

	B_1	B_2	B_3
A_1	5	6	6
A_2	7	8	4
A_3	9	10	2

3. Give an example of an experiment that would use a 2×3 factorial design.

4. How many independent groups of subjects are there in a 3×3 between-subjects factorial design? How many in a 2×3 between-subjects factorial design? In a 4×3 between-subjects factorial design?

5. Complete the following summary table from a between-subjects factorial analysis of variance:

Source	SS	df	MS	F
A		2		
B	45	3		
A × B	60			
Within	216			
Total	341	119		

Consider the following analysis of variance summary table from a between-subjects factorial design:

Source	SS	df	MS	F
A	50	1	50	10
B	40	2	20	4
A × B	40	2	20	4
Within	270	54	5	
Total	400	59		

6. How many levels of factor A were there in this experiment? How many levels of factor B?

7. What was the total number of subjects in this experiment? How many independent groups were there in the factorial design? If there were an equal number of subjects in each independent group, how many subjects were in each group?

8. What was the total amount of between-group variability (that is, what is the value of sum of squares between)?

9. What proportion of variance in the dependent variable is associated with factor A? What proportion of variance in the dependent variable is associated with factor B? What proportion of variance in the dependent variable is associated with the A × B interaction? What proportion of variance in the dependent variable is unexplained in this study?

10. Test the viability of the null hypotheses with respect to the main effect of A, the main effect of B, and the interaction effect. In doing so, explicitly state the respective null and alternative hypotheses.

11. For the following 2 × 3 factorial design, generate a set of population means that would reflect a main effect of A, no main effect of B, and an A × B interaction effect:

	B_1	B_2	B_3
A_1			
A_2			

12. For the following 2 × 2 factorial design, generate a set of population means that would reflect no main effect of A, no main effect of B, but an interaction effect of A × B:

	B_1	B_2
A_1		
A_2		

13. Suppose you are told that an experiment was performed in the context of a 2 × 2 factorial design with five subjects per group. You are also given the relevant group means:

	B_1	B_2
A_1	10.0	12.0
A_2	15.0	19.0

You are only told that the main effect of B was statistically significant ($F = 5.44$, df = 1, 17, $p < .05$). From just this information, it is possible to derive the entire summary table for the analysis. Derive the summary table.

14. What are the major assumptions underlying the F test for inferring a relationship in factorial analysis of variance?

15. What is the difference between an unweighted means analysis of variance and a least squares analysis of variance?

16. When should one use two-way between-subjects factorial analysis of variance to analyze one's data?

17. Given the following means and summary table, analyze as completely as possible the relationship of factors A, B, and their interaction with the dependent variable. There are ten subjects per group.

	B_1	B_2	B_3
A_1	10.0	15.0	20.0
A_2	14.0	20.0	23.0

$\overline{Y}_{A_1} = 15.0$ $\overline{Y}_{B_1} = 12.0$

$\overline{Y}_{A_2} = 19.0$ $\overline{Y}_{B_2} = 17.5$

$\overline{Y}_{B_3} = 21.5$

Source	SS	df	MS	F
A	240	1	240	24.0
B	910	2	455	45.5
A × B	10	2	5	.5
Within	540	54	10	
Total	1700	59		

18. Given the following means and summary table, analyze as completely as possible the relationship of factors A, B, and their interaction with the dependent variable.

There are five subjects per group.

	B_1	B_2	B_3
A_1	10.0	12.0	14.0
A_2	12.0	8.0	4.0

$\overline{Y}_{A_1} = 12.0 \qquad \overline{Y}_{B_1} = 11.0$

$\overline{Y}_{A_2} = 8.0 \qquad \overline{Y}_{B_2} = 10.0$

$\qquad\qquad\qquad \overline{Y}_{B_3} = 9.0$

Source	SS	df	MS	F
A	120	1	120	10.00
B	20	2	10	.83
A × B	380	2	190	15.83
Within	288	24	12	
Total	808	29		

Exercises to Apply Concepts

19. Psychologists have studied extensively factors that contribute to weight gain in individuals. It is currently believed that individuals use at least two sources of "cues" to decide that they are hungry and should eat. One source is internal cues or changes in one's physiology that create or suggest hunger. A second source is external cues, which occur in the environment and suggest to the individual that he or she should be hungry. For example, when 12:00 noon approaches, this signifies that it is lunch time and that we should be hungry. Several researchers have suggested that one difference between overweight and normal weight individuals is that overweight individuals attend more to external cues rather than internal cues as guidelines for eating. In one experiment, ten overweight and ten normal weight individuals were asked not to eat breakfast on the morning before coming to an experiment. After they arrived at the experiment, and performed several different tasks, subjects were asked to rate how hungry they felt on an 11 point scale (0 = not at all through 10 = extremely). Just prior to this, a confederate in the same room as the subject asked another confederate what time it was. For one-half the subjects, the response was 11:00 whereas for the other half, the response was 1:00. In fact, the time was 12:00. If overweight individuals attend mostly to external cues, one would expect them to report higher degrees of hunger in the 1:00 as opposed to the 11:00 time specification (since 1:00 implies they have gone longer without food). If normal weight individuals attend mostly to internal cues, then one would not expect the time manipulation to affect their ratings of hunger. The data for this experiment are presented below. Analyze these data as completely as possible using two-way between-subjects factorial analysis of variance and re-port your results in accord with the Method of Presentation section.

Overweight 11:00	Overweight 1:00	Normal weight 11:00	Normal weight 1:00
5	9	6	6
6	8	6	6
7	9	7	5
6	9	5	7
6	10	6	6

20. All of us tend to classify people into social categories (for example, a conservative businessman, a typical housewife, and so on) and frequently these categorizations influence how we behave towards individuals. Psychologists have studied factors that influence social categorization and how placing labels on people may be beneficial or harmful to them. In one experiment (based on Rubovitz and Maehr, 1973) white female undergraduates enrolled in a teacher-training course were asked to prepare a lesson for four seventh grade students. The teachers were given information about each of the students. For one-half of the teachers, one of the students was described as being "gifted" (that is, extremely intelligent) while for the other half of the teachers, the student was described as being "nongifted" (or of average intelligence). This student was also described as being either black or white, yielding a 2 × 2 factorial design (gifted versus nongifted and black versus white). The teachers were then observed during a forty minute period in which they interacted with the "target" student and three other students. The number of interactions directed toward the target student was noted, and this served as an index of how much attention the teacher gave to the student. The data are presented below. Analyze the data using a two-way between-subjects factorial analysis of variance and report your results in accord with the Method of Presentation section.

Gifted Black	Nongifted Black	Gifted White	Nongifted White
30	30	36	29
29	29	36	32
30	28	36	31
30	30	35	30
31	28	37	33

Appendix A
Squares, Square Roots,
and Reciprocals

Appendix A is a list of squares, square roots, and reciprocals of the numbers 1 to 1,000. It can be used to facilitate calculations by hand if a calculator is not available.

There are seven columns for each entry in a row. The first column, labeled N, lists the number of interest, such as 15. The second column is the square of the number of interest, or $15^2 = 225$. This column can also be used to find the square of numbers other than N. In the case of 15.0, the decimal point is after the 5. If we move the decimal point one place to the left, we obtain 1.5. The square of this number is 2.25. This was found by moving the decimal for the number in the N^2 column two places to the left. For every *one* space you move the decimal to the left or right in the N column, you should move the decimal in the N^2 column *two* spaces in the same direction. For example, the square of .15 is .0225. The square of .015 is .000225. The square of 150 is 22500.

The third column is the square root of the number of interest and the fourth column is the square root of ten times the number of interest. Thus, the square root of 15 is 3.8730 and the square root of ten times 15, or 150, is 12.2474. These columns can also be used to determine additional square roots. The principle involved is similar to the one used with respect to the squares noted above. Let us consider the $\sqrt{N}$ column first. For every *one* space you move the decimal place to the right or left in the $\sqrt{N}$ column, you should move the decimal place *two* spaces in the same direction in the N column. The square root of .15 is .38730. The square root of .0015 is .038730. The square root of 1,500 is 38.730.

The same principle applies in the $\sqrt{10N}$ column, except that the number in the N column must be viewed as $10N$. Thus, the square root of 15,000 is 122.474. The square root of 1.5 is 1.22474.

Columns five, six, and seven present the reciprocals of the corresponding numbers.

Source: J. P. Guilford, *Fundamental Statistics in Psychology and Education*, McGraw-Hill, 1965. Used with permission.

N	N²	$\sqrt{N}$	$\sqrt{10N}$	$1/N$	$1/\sqrt{N}$	$1/\sqrt{10N}$
1	1	1.0000	3.1623	1.000000	1.0000	.31623
2	4	1.4142	4.4721	.500000	.7071	.22361
3	9	1.7321	5.4772	.333333	.5774	.18257
4	16	2.0000	6.3246	.250000	.5000	.15811
5	25	2.2361	7.0711	.200000	.4472	.14142
6	36	2.4495	7.7460	.166667	.4082	.12910
7	49	2.6458	8.3666	.142857	.3780	.11952
8	64	2.8284	8.9443	.125000	.3536	.11180
9	81	3.0000	9.4868	.111111	.3333	.10541
10	1 00	3.1623	10.0000	.100000	.3162	.10000
11	1 21	3.3166	10.4881	.090909	.3015	.09535
12	1 44	3.4641	10.9545	.083333	.2887	.09129
13	1 69	3.6056	11.4018	.076923	.2774	.08771
14	1 96	3.7417	11.8322	.071429	.2673	.08452
15	2 25	3.8730	12.2474	.066667	.2582	.08165
16	2 56	4.0000	12.6491	.062500	.2500	.07906
17	2 89	4.1231	13.0384	.058824	.2425	.07670
18	3 24	4.2426	13.4164	.055556	.2357	.07454
19	3 61	4.3589	13.7840	.052632	.2294	.07255
20	4 00	4.4721	14.1421	.050000	.2236	.07071
21	4 41	4.5826	14.4914	.047619	.2182	.06901
22	4 84	4.6904	14.8324	.045455	.2132	.06742
23	5 29	4.7958	15.1658	.043478	.2085	.06594
24	5 76	4.8990	15.4919	.041667	.2041	.06455
25	6 25	5.0000	15.8114	.040000	.2000	.06325
26	6 76	5.0990	16.1245	.038462	.1961	.06202
27	7 29	5.1962	16.4317	.037037	.1925	.06086
28	7 84	5.2915	16.7332	.035714	.1890	.05976
29	8 41	5.3852	17.0294	.034483	.1857	.05872
30	9 00	5.4772	17.3205	.033333	.1826	.05774
31	9 61	5.5678	17.6068	.032258	.1796	.05680
32	10 24	5.6569	17.8885	.031250	.1768	.05590
33	10 89	5.7446	18.1659	.030303	.1741	.05505
34	11 56	5.8310	18.4391	.029412	.1715	.05423
35	12 25	5.9161	18.7083	.028571	.1690	.05345
36	12 96	6.0000	18.9737	.027778	.1667	.05270
37	13 69	6.0828	19.2354	.027027	.1644	.05199
38	14 44	6.1644	19.4936	.026316	.1622	.05130
39	15 21	6.2450	19.7484	.025641	.1601	.05064
40	16 00	6.3246	20.0000	.025000	.1581	.05000
41	16 81	6.4031	20.2485	.024390	.1562	.04939
42	17 64	6.4807	20.4939	.023810	.1543	.04880
43	18 49	6.5574	20.7364	.023256	.1525	.04822
44	19 36	6.6332	20.9762	.022727	.1508	.04767
45	20 25	6.7082	21.2132	.022222	.1491	.04714
46	21 16	6.7823	21.4476	.021739	.1474	.04663
47	22 09	6.8557	21.6795	.021277	.1459	.04613
48	23 04	6.9282	21.9089	.020833	.1443	.04564
49	24 01	7.0000	22.1359	.020408	.1429	.04518
50	25 00	7.0711	22.3607	.020000	.1414	.04472

A2   *Appendix A: Squares, Square Roots, and Reciprocals*

N	N²	√N	√10N	1/N	1/√N	1/√10N
51	26 01	7.1414	22.5832	.019608	.1400	.04428
52	27 04	7.2111	22.8035	.019231	.1387	.04385
53	28 09	7.2801	23.0217	.018868	.1374	.04344
54	29 16	7.3485	23.2379	.018519	.1361	.04303
55	30 25	7.4162	23.4521	.018182	.1348	.04264
56	31 36	7.4833	23.6643	.017857	.1336	.04226
57	32 49	7.5498	23.8747	.017544	.1325	.04189
58	33 64	7.6158	24.0832	.017241	.1313	.04152
59	34 81	7.6811	24.2899	.016949	.1302	.04117
60	36 00	7.7460	24.4949	.016667	.1291	.04082
61	37 21	7.8102	24.6982	.016393	.1280	.04049
62	38 44	7.8740	24.8998	.016129	.1270	.04016
63	39 69	7.9373	25.0998	.015873	.1260	.03984
64	40 96	8.0000	25.2982	.015625	.1250	.03953
65	42 25	8.0623	25.4951	.015385	.1240	.03922
66	43 56	8.1240	25.6905	.015152	.1231	.03892
67	44 89	8.1854	25.8844	.014925	.1222	.03863
68	46 24	8.2462	26.0768	.014706	.1213	.03835
69	47 61	8.3066	26.2679	.014493	.1204	.03807
70	49 00	8.3666	26.4575	.014286	.1195	.03780
71	50 41	8.4261	26.6458	.014085	.1187	.03753
72	51 84	8.4853	26.8328	.013889	.1179	.03727
73	53 29	8.5440	27.0185	.013699	.1170	.03701
74	54 76	8.6023	27.2029	.013514	.1162	.03676
75	56 25	8.6603	27.3861	.013333	.1155	.03651
76	57 76	8.7178	27.5681	.013158	.1147	.03627
77	59 29	8.7750	27.7489	.012987	.1140	.03604
78	60 84	8.8318	27.9285	.012821	.1132	.03581
79	62 41	8.8882	28.1069	.012658	.1125	.03558
80	64 00	8.9443	28.2843	.012500	.1118	.03536
81	65 61	9.0000	28.4605	.012346	.1111	.03514
82	67 24	9.0554	28.6356	.012195	.1104	.03492
83	68 89	9.1104	28.8097	.012048	.1098	.03471
84	70 56	9.1652	28.9828	.011905	.1091	.03450
85	72 25	9.2195	29.1548	.011765	.1085	.03430
86	73 96	9.2736	29.3258	.011628	.1078	.03410
87	75 69	9.3274	29.4958	.011494	.1072	.03390
88	77 44	9.3808	29.6648	.011364	.1066	.03371
89	79 21	9.4340	29.8329	.011236	.1060	.03352
90	81 00	9.4868	30.0000	.011111	.1054	.03333
91	82 81	9.5394	30.1662	.010989	.1048	.03315
92	84 64	9.5917	30.3315	.010870	.1043	.03297
93	86 49	9.6437	30.4959	.010753	.1037	.03279
94	88 36	9.6954	30.6594	.010638	.1031	.03262
95	90 25	9.7468	30.8221	.010526	.1026	.03244
96	92 16	9.7980	30.9839	.010417	.1021	.03227
97	94 09	9.8489	31.1448	.010309	.1015	.03211
98	96 04	9.8995	31.3050	.010204	.1010	.03194
99	98 01	9.9499	31.4643	.010101	.1005	.03178
100	1 00 00	10.0000	31.6228	.010000	.1000	.03162

N	N^2	$\sqrt{N}$	$\sqrt{10N}$	$1/N$	$1/\sqrt{N}$	$1/\sqrt{10N}$
101	1 02 01	10.0499	31.7805	.009901	.0995	.03147
102	1 04 04	10.0995	31.9374	.009804	.0990	.03131
103	1 06 09	10.1489	32.0936	.009709	.0985	.03116
104	1 08 16	10.1980	32.2490	.009615	.0981	.03101
105	1 10 25	10.2470	32.4037	.009524	.0976	.03086
106	1 12 36	10.2956	32.5576	.009434	.0971	.03071
107	1 14 49	10.3441	32.7109	.009346	.0967	.03057
108	1 16 64	10.3923	32.8634	.009259	.0962	.03043
109	1 18 81	10.4403	33.0151	.009174	.0958	.03029
110	1 21 00	10.4881	33.1662	.009091	.0953	.03015
111	1 23 21	10.5357	33.3167	.009009	.0949	.03002
112	1 25 44	10.5830	33.4664	.008929	.0945	.02988
113	1 27 69	10.6301	33.6155	.008850	.0941	.02975
114	1 29 96	10.6771	33.7639	.008772	.0937	.02962
115	1 32 25	10.7238	33.9116	.008696	.0933	.02949
116	1 34 56	10.7703	34.0588	.008621	.0928	.02936
117	1 36 89	10.8167	34.2053	.008547	.0925	.02924
118	1 39 24	10.8628	34.3511	.008475	.0921	.02911
119	1 41 61	10.9087	34.4964	.008403	.0917	.02899
120	1 44 00	10.9545	34.6410	.008333	.0913	.02887
121	1 46 41	11.0000	34.7851	.008264	.0909	.02875
122	1 48 84	11.0454	34.9285	.008197	.0905	.02863
123	1 51 29	11.0905	35.0714	.008130	.0902	.02851
124	1 53 76	11.1355	35.2136	.008065	.0898	.02840
125	1 56 25	11.1803	35.3553	.008000	.0894	.02828
126	1 58 76	11.2250	35.4965	.007937	.0891	.02817
127	1 61 29	11.2694	35.6371	.007874	.0887	.02806
128	1 63 84	11.3137	35.7771	.007813	.0884	.02795
129	1 66 41	11.3578	35.9166	.007752	.0880	.02784
130	1 69 00	11.4018	36.0555	.007692	.0877	.02774
131	1 71 61	11.4455	36.1939	.007634	.0874	.02763
132	1 74 24	11.4891	36.3318	.007576	.0870	.02752
133	1 76 89	11.5326	36.4692	.007519	.0867	.02742
134	1 79 56	11.5758	36.6060	.007463	.0864	.02732
135	1 82 25	11.6190	36.7423	.007407	.0861	.02722
136	1 84 69	11.6619	36.8782	.007353	.0857	.02712
137	1 87 69	11.7047	37.0135	.007299	.0854	.02702
138	1 90 44	11.7473	37.1484	.007246	.0851	.02692
139	1 93 21	11.7898	37.2827	.007194	.0848	.02682
140	1 96 00	11.8322	37.4166	.007143	.0845	.02673
141	1 98 81	11.8743	37.5500	.007092	.0842	.02663
142	2 01 64	11.9164	37.6829	.007042	.0839	.02654
143	2 04 49	11.9583	37.8153	.006993	.0836	.02644
144	2 07 36	12.0000	37.9473	.006944	.0833	.02635
145	2 10 25	12.0416	38.0789	.006897	.0830	.02626
146	2 13 16	12.0830	38.2099	.006849	.0828	.02617
147	2 16 09	12.1244	38.3406	.006803	.0825	.02608
148	2 19 04	12.1655	38.4708	.006757	.0822	.02599
149	2 22 01	12.2066	38.6005	.006711	.0819	.02591
150	2 25 00	12.2474	38.7298	.006667	.0816	.02582

N	N²	√N	√10N	1/N	1/√N	1/√10N
151	2 28 01	12.2882	38.8587	.006623	.0814	.02573
152	2 31 04	12.3288	38.9872	.006579	.0811	.02565
153	2 34 09	12.3693	39.1152	.006536	.0808	.02557
154	2 37 16	12.4097	39.2428	.006494	.0806	.02548
155	2 40 25	12.4499	39.3700	.006452	.0803	.02540
156	2 43 36	12.4900	39.4968	.006410	.0801	.02532
157	2 46 49	12.5300	39.6232	.006369	.0798	.02524
158	2 49 64	12.5698	39.7492	.006329	.0796	.02516
159	2 52 81	12.6095	39.8748	.006289	.0793	.02508
160	2 56 00	12.6491	40.0000	.006250	.0791	.02500
161	2 59 21	12.6886	40.1248	.006211	.0788	.02492
162	2 62 44	12.7279	40.2492	.006173	.0786	.02485
163	2 65 69	12.7671	40.3733	.006135	.0783	.02477
164	2 68 96	12.8062	40.4969	.006098	.0781	.02469
165	2 72 25	12.8452	40.6202	.006061	.0778	.02462
166	2 75 56	12.8841	40.7431	.006024	.0776	.02454
167	2 78 89	12.9228	40.8656	.005988	.0774	.02447
168	2 82 24	12.9615	40.9878	.005952	.0772	.02440
169	2 85 61	13.0000	41.1096	.005917	.0769	.02433
170	2 89 00	13.0384	41.2311	.005882	.0767	.02425
171	2 92 41	13.0767	41.3521	.005848	.0765	.02418
172	2 95 84	13.1149	41.4729	.005814	.0762	.02411
173	2 99 29	13.1529	41.5933	.005780	.0760	.02404
174	3 02 76	13.1909	41.7133	.005747	.0758	.02397
175	3 06 25	13.2288	41.8330	.005714	.0756	.02390
176	3 09 76	13.2665	41.9524	.005682	.0754	.02384
177	3 13 29	13.3041	42.0714	.005650	.0752	.02377
178	3 16 84	13.3417	42.1900	.005618	.0750	.02370
179	3 20 41	13.3791	42.3084	.005587	.0747	.02364
180	3 24 00	13.4164	42.4264	.005556	.0745	.02357
181	3 27 61	13.4536	42.5441	.005525	.0743	.02351
182	3 31 24	13.4907	42.6615	.005495	.0741	.02344
183	3 34 89	13.5277	42.7785	.005464	.0739	.02338
184	3 38 56	13.5647	42.8952	.005435	.0737	.02331
185	3 42 25	13.6015	43.0116	.005405	.0735	.02325
186	3 45 96	13.6382	43.1277	.005376	.0733	.02319
187	3 49 69	13.6748	43.2435	.005348	.0731	.02312
188	3 53 44	13.7113	43.3590	.005319	.0729	.02306
189	3 57 21	13.7477	43.4741	.005291	.0727	.02300
190	3 61 00	13.7840	43.5890	.005263	.0725	.02294
191	3 64 81	13.8203	43.7035	.005236	.0724	.02288
192	3 68 64	13.8564	43.8178	.005208	.0722	.02282
193	3 72 49	13.8924	43.9318	.005181	.0720	.02276
194	3 76 36	13.9284	44.0454	.005155	.0718	.02270
195	3 80 25	13.9642	44.1588	.005128	.0716	.02265
196	3 84 16	14.0000	44.2719	.005102	.0714	.02259
197	3 88 09	14.0357	44.3847	.005076	.0712	.02253
198	3 92 04	14.0712	44.4972	.005051	.0711	.02247
199	3 96 01	14.1067	44.6094	.005025	.0709	.02242
200	4 00 00	14.1421	44.7214	.005000	.0707	.02236

N	N^2	$\sqrt{N}$	$\sqrt{10N}$	$1/N$	$1/\sqrt{N}$	$1/\sqrt{10N}$
201	4 04 01	14.1774	44.8330	.004975	.0705	.02230
202	4 08 04	14.2127	44.9444	.004950	.0704	.02225
203	4 12 09	14.2478	45.0555	.004926	.0702	.02219
204	4 16 16	14.2829	45.1664	.004902	.0700	.02214
205	4 20 25	14.3178	45.2769	.004878	.0698	.02209
206	4 24 36	14.3527	45.3872	.004854	.0697	.02203
207	4 28 49	14.3875	45.4973	.004831	.0695	.02198
208	4 32 64	14.4222	45.6070	.004801	.0693	.02193
209	4 36 81	14.4568	45.7165	.004785	.0692	.02187
210	4 41 00	14.4914	45.8258	.004762	.0690	.02182
211	4 45 21	14.5258	45.9347	.004739	.0688	.02177
212	4 49 44	14.5602	46.0435	.004717	.0687	.02172
213	4 53 69	14.5945	46.1519	.004695	.0685	.02167
214	4 57 96	14.6287	46.2601	.004673	.0684	.02162
215	4 62 25	14.6629	46.3681	.004651	.0682	.02157
216	4 66 56	14.6969	46.4758	.004630	.0680	.02152
217	4 70 89	14.7309	46.5833	.004608	.0679	.02147
218	4 75 24	14.7648	46.6905	.004587	.0677	.02142
219	4 79 61	14.7986	46.7974	.004566	.0676	.02137
220	4 84 00	14.8324	46.9042	.004545	.0674	.02132
221	4 88 41	14.8661	47.0106	.004525	.0673	.02127
222	4 92 84	14.8997	47.1169	.004505	.0671	.02122
223	4 97 29	14.9332	47.2229	.004484	.0670	.02118
224	5 01 76	14.9666	47.3286	.004464	.0668	.02113
225	5 06 25	15.0000	47.4342	.004444	.0667	.02108
226	5 10 76	15.0333	47.5395	.004425	.0665	.02104
227	5 15 29	15.0665	47.6445	.004405	.0664	.02099
228	5 19 84	15.0997	47.7493	.004386	.0662	.02094
229	5 24 41	15.1327	47.8539	.004367	.0661	.02090
230	5 29 00	15.1658	47.9583	.004348	.0659	.02085
231	5 33 61	15.1987	48.0625	.004329	.0658	.02081
232	5 38 24	15.2315	48.1664	.004310	.0657	.02076
233	5 42 89	15.2643	48.2701	.004292	.0655	.02072
234	5 47 56	15.2971	48.3735	.004274	.0654	.02067
235	5 52 25	15.3297	48.4768	.004255	.0652	.02063
236	5 56 96	15.3623	48.5798	.004237	.0651	.02058
237	5 61 69	15.3948	48.6826	.004219	.0650	.02054
238	5 66 44	15.4272	48.7852	.004202	.0648	.02050
239	5 71 21	15.4596	48.8876	.004184	.0647	.02046
240	5 76 00	15.4919	48.9898	.004167	.0645	.02041
241	5 80 81	15.5242	49.0918	.004149	.0644	.02037
242	5 85 64	15.5563	49.1935	.004132	.0643	.02033
243	5 90 49	15.5885	49.2950	.004115	.0642	.02029
244	5 95 36	15.6205	49.3964	.004098	.0640	.02024
245	6 00 25	15.6525	49.4975	.004082	.0639	.02020
246	6 05 16	15.6844	49.5984	.004065	.0638	.02016
247	6 10 09	15.7162	49.6991	.004049	.0636	.02012
248	6 15 04	15.7480	49.7996	.004032	.0635	.02008
249	6 20 01	15.7797	49.8999	.004016	.0634	.02004
250	6 25 00	15.8114	50.0000	.004000	.0632	.02000

N	N²	$\sqrt{N}$	$\sqrt{10N}$	$1/N$	$1/\sqrt{N}$	$1/\sqrt{10N}$
251	6 30 01	15.8430	50.0999	.003984	.0631	.01996
252	6 35 04	15.8745	50.1996	.003968	.0630	.01992
253	6 40 09	15.9060	50.2991	.003953	.0629	.01988
254	6 45 16	15.9374	50.3984	.003937	.0627	.01984
255	6 50 25	15.9687	50.4975	.003922	.0626	.01980
256	6 55 36	16.0000	50.5964	.003906	.0625	.01976
257	6 60 49	16.0312	50.6952	.003891	.0624	.01973
258	6 65 64	16.0624	50.7937	.003876	.0623	.01969
259	6 70 81	16.0935	50.8920	.003861	.0621	.01965
260	6 76 00	16.1245	50.9902	.003846	.0620	.01961
261	6 81 21	16.1555	51.0882	.003831	.0619	.01957
262	6 86 44	16.1864	51.1859	.003817	.0618	.01954
263	6 91 69	16.2173	51.2835	.003802	.0617	.01950
264	6 96 96	16.2481	51.3809	.003788	.0615	.01946
265	7 02 25	16.2788	51.4782	.003774	.0614	.01943
266	7 07 56	16.3095	51.5752	.003759	.0613	.01939
267	7 12 89	16.3401	51.6720	.003745	.0612	.01935
268	7 18 24	16.3707	51.7687	.003731	.0611	.01932
269	7 23 61	16.4012	51.8652	.003717	.0610	.01928
270	7 29 00	16.4317	51.9615	.003704	.0609	.01925
271	7 34 41	16.4621	52.0577	.003690	.0607	.01921
272	7 39 84	16.4924	52.1536	.003676	.0606	.01917
273	7 45 29	16.5227	52.2494	.003663	.0605	.01914
274	7 50 76	16.5529	52.3450	.003650	.0604	.01910
275	7 56 25	16.5831	52.4404	.003636	.0603	.01907
276	7 61 76	16.6132	52.5357	.003623	.0602	.01903
277	7 67 29	16.6433	52.6308	.003610	.0601	.01900
278	7 72 84	16.6733	52.7257	.003597	.0600	.01897
279	7 78 41	16.7033	52.8205	.003584	.0599	.01893
280	7 84 00	16.7332	52.9150	.003571	.0598	.01890
281	7 89 61	16.7631	53.0094	.003559	.0597	.01886
282	7 95 24	16.7929	53.1037	.003546	.0595	.01883
283	8 00 89	16.8226	53.1977	.003534	.0594	.01880
284	8 06 56	16.8523	53.2917	.003521	.0593	.01876
285	8 12 25	16.8819	53.3854	.003509	.0592	.01873
286	8 17 96	16.9115	53.4790	.003497	.0591	.01870
287	8 23 69	16.9411	53.5724	.003484	.0590	.01867
288	8 29 44	16.9706	53.6656	.003472	.0589	.01863
289	8 35 21	17.0000	53.7587	.003460	.0588	.01860
290	8 41 00	17.0294	53.8516	.003448	.0587	.01857
291	8 46 81	17.0587	53.9444	.003436	.0586	.01854
292	8 52 64	17.0880	54.0370	.003425	.0585	.01851
293	8 58 49	17.1172	54.1295	.003413	.0584	.01847
294	8 64 36	17.1464	54.2218	.003401	.0583	.01844
295	8 70 25	17.1756	54.3139	.003390	.0582	.01841
296	8 76 16	17.2047	54.4059	.003378	.0581	.01838
297	8 82 09	17.2337	54.4977	.003367	.0580	.01835
298	8 88 04	17.2627	54.5894	.003356	.0579	.01832
299	8 94 01	17.2916	54.6809	.003344	.0578	.01829
300	9 00 00	17.3205	54.7723	.003333	.0577	.01826

N	N^2	$\sqrt{N}$	$\sqrt{10N}$	$1/N$	$1/\sqrt{N}$	$1/\sqrt{10N}$
301	9 06 01	17.3494	54.8635	.003322	.0576	.01823
302	9 12 04	17.3781	54.9545	.003311	.0575	.01820
303	9 18 09	17.4069	55.0454	.003300	.0574	.01817
304	9 24 16	17.4356	55.1362	.003289	.0574	.01814
305	9 30 25	17.4642	55.2268	.003279	.0573	.01811
306	9 36 36	17.4929	55.3173	.003268	.0572	.01808
307	9 42 49	17.5214	55.4076	.003257	.0571	.01805
308	9 48 64	17.5499	55.4977	.003247	.0570	.01802
309	9 54 81	17.5784	55.5878	.003236	.0569	.01799
310	9 61 00	17.6068	55.6776	.003226	.0568	.01796
311	9 67 21	17.6352	55.7674	.003215	.0567	.01793
312	9 73 44	17.6635	55.8570	.003205	.0566	.01790
313	9 79 69	17.6918	55.9464	.003195	.0565	.01787
314	9 85 96	17.7200	56.0357	.003185	.0564	.01785
315	9 92 25	17.7482	56.1249	.003175	.0563	.01782
316	9 98 56	17.7764	56.2139	.003165	.0563	.01779
317	10 04 89	17.8045	56.3028	.003155	.0562	.01776
318	10 11 24	17.8326	56.3915	.003145	.0561	.01773
319	10 17 61	17.8606	56.4801	.003135	.0560	.01771
320	10 24 00	17.8885	56.5685	.003125	.0559	.01768
321	10 30 41	17.9165	56.6569	.003115	.0558	.01765
322	10 36 84	17.9444	56.7450	.003106	.0557	.01762
323	10 43 29	17.9722	56.8331	.003096	.0556	.01760
324	10 49 76	18.0000	56.9210	.003086	.0556	.01757
325	10 56 25	18.0278	57.0088	.003077	.0555	.01754
326	10 62 76	18.0555	57.0964	.003067	.0554	.01751
327	10 69 29	18.0831	57.1839	.003058	.0553	.01749
328	10 75 84	18.1108	57.2713	.003049	.0552	.01746
329	10 82 41	18.1384	57.3585	.003040	.0551	.01743
330	10 89 00	18.1659	57.4456	.003030	.0550	.01741
331	10 95 61	18.1934	57.5326	.003021	.0550	.01738
332	11 02 24	18.2209	57.6194	.003012	.0549	.01736
333	11 08 89	18.2483	57.7062	.003003	.0548	.01733
334	11 15 56	18.2757	57.7927	.002994	.0547	.01730
335	11 22 25	18.3030	57.8792	.002985	.0546	.01728
336	11 28 96	18.3303	57.9655	.002976	.0546	.01725
337	11 35 69	18.3576	58.0517	.002967	.0545	.01723
338	11 42 44	18.3848	58.1378	.002959	.0544	.01720
339	11 49 21	18.4120	58.2237	.002950	.0543	.01718
340	11 56 00	18.4391	58.3095	.002941	.0542	.01715
341	11 62 81	18.4662	58.3952	.002933	.0542	.01712
342	11 69 64	18.4932	58.4808	.002924	.0541	.01710
343	11 76 49	18.5203	58.5662	.002915	.0540	.01707
344	11 83 36	18.5472	58.6515	.002907	.0539	.01705
345	11 90 25	18.5742	58.7367	.002899	.0538	.01703
346	11 97 16	18.6011	58.8218	.002890	.0538	.01700
347	12 04 09	18.6279	58.9067	.002882	.0537	.01698
348	12 11 04	18.6548	58.9915	.002874	.0536	.01695
349	12 18 01	18.6815	59.0762	.002865	.0535	.01693
350	12 25 00	18.7083	59.1608	.002857	.0535	.01690

N	N²	$\sqrt{N}$	$\sqrt{10N}$	1/N	$1/\sqrt{N}$	$1/\sqrt{10N}$
351	12 32 01	18.7350	59.2453	.002849	.0534	.01688
352	12 39 04	18.7617	59.3296	.002841	.0533	.01685
353	12 46 09	18.7883	59.4138	.002833	.0532	.01683
354	12 53 16	18.8149	59.4979	.002825	.0531	.01681
355	12 60 25	18.8414	59.5819	.002817	.0531	.01678
356	12 67 36	18.8680	59.6657	.002809	.0530	.01676
357	12 74 49	18.8944	59.7495	.002801	.0529	.01674
358	12 81 64	18.9209	59.8331	.002793	.0529	.01671
359	12 88 81	18.9473	59.9166	.002786	.0528	.01669
360	12 96 00	18.9737	60.0000	.002778	.0527	.01667
361	13 03 21	19.0000	60.0833	.002770	.0526	.01664
362	13 10 44	19.0263	60.1664	.002762	.0526	.01662
363	13 17 69	19.0526	60.2495	.002755	.0525	.01660
364	13 24 96	19.0788	60.3324	.002747	.0524	.01657
365	13 32 25	19.1050	60.4152	.002740	.0523	.01655
366	13 39 56	19.1311	60.4979	.002732	.0523	.01653
367	13 46 89	19.1572	60.5805	.002725	.0522	.01651
368	13 54 24	19.1833	60.6630	.002717	.0521	.01648
369	13 61 61	19.2094	60.7454	.002710	.0521	.01646
370	13 69 00	19.2354	60.8276	.002703	.0520	.01644
371	13 76 41	19.2614	60.9098	.002695	.0519	.01642
372	13 83 84	19.2873	60.9918	.002688	.0518	.01640
373	13 91 29	19.3132	61.0737	.002681	.0518	.01637
374	13 98 76	19.3391	61.1555	.002674	.0517	.01635
375	14 06 25	19.3649	61.2372	.002667	.0516	.01633
376	14 13 76	19.3907	61.3188	.002660	.0516	.01631
377	14 21 29	19.4165	61.4003	.002653	.0515	.01629
378	14 28 84	19.4422	61.4817	.002646	.0514	.01627
379	14 36 41	19.4679	61.5630	.002639	.0514	.01624
380	14 44 00	19.4936	61.6441	.002632	.0513	.01622
381	14 51 61	19.5192	61.7252	.002625	.0512	.01620
382	14 59 24	19.5448	61.8061	.002618	.0512	.01618
383	14 66 89	19.5704	61.8870	.002611	.0511	.01616
384	14 74 56	19.5959	61.9677	.002604	.0510	.01614
385	14 82 25	19.6214	62.0484	.002597	.0510	.01612
386	14 89 96	19.6469	62.1289	.002591	.0509	.01610
387	14 97 69	19.6723	62.2093	.002584	.0508	.01607
388	15 05 44	19.6977	62.2896	.002577	.0508	.01605
389	15 13 21	19.7231	62.3699	.002571	.0507	.01603
390	15 21 00	19.7484	62.4500	.002564	.0506	.01601
391	15 28 81	19.7737	62.5300	.002558	.0506	.01599
392	15 36 64	19.7990	62.6099	.002551	.0505	.01597
393	15 44 49	19.8242	62.6897	.002545	.0504	.01595
394	15 52 36	19.8494	62.7694	.002538	.0504	.01593
395	15 60 25	19.8746	62.8490	.002532	.0503	.01591
396	15 68 16	19.8997	62.9285	.002525	.0503	.01589
397	15 76 09	19.9249	63.0079	.002519	.0502	.01587
398	15 84 04	19.9499	63.0872	.002513	.0501	.01585
399	15 92 01	19.9750	63.1664	.002506	.0501	.01583
400	16 00 00	20.0000	63.2456	.002500	.0500	.01581

N	N^2	$\sqrt{N}$	$\sqrt{10N}$	$1/N$	$1/\sqrt{N}$	$1/\sqrt{10N}$
401	16 08 01	20.0250	63.3246	.002494	.0499	.01579
402	16 16 04	20.0499	63.4035	.002488	.0499	.01577
403	16 24 09	20.0749	63.4823	.002481	.0498	.01575
404	16 32 16	20.0998	63.5610	.002475	.0498	.01573
405	16 40 25	20.1246	63.6396	.002469	.0497	.01571
406	16 48 36	20.1494	63.7181	.002463	.0496	.01569
407	16 56 49	20.1742	63.7966	.002457	.0496	.01567
408	16 64 64	20.1990	63.8749	.002451	.0495	.01566
409	16 72 81	20.2237	63.9531	.002445	.0494	.01564
410	16 81 00	20.2485	64.0312	.002439	.0494	.01562
411	16 89 21	20.2731	64.1093	.002433	.0493	.01560
412	16 97 44	20.2978	64.1872	.002427	.0493	.01558
413	17 05 69	20.3224	64.2651	.002421	.0492	.01556
414	17 13 96	20.3470	64.3428	.002415	.0491	.01554
415	17 22 25	20.3715	64.4205	.002410	.0491	.01552
416	17 30 56	20.3961	64.4981	.002404	.0490	.01550
417	17 38 89	20.4206	64.5755	.002398	.0490	.01549
418	17 47 24	20.4450	64.6529	.002392	.0489	.01547
419	17 55 61	20.4695	64.7302	.002387	.0489	.01545
420	17 64 00	20.4939	64.8074	.002381	.0488	.01543
421	17 72 41	20.5183	64.8845	.002375	.0487	.01541
422	17 80 84	20.5426	64.9615	.002370	.0487	.01539
423	17 89 29	20.5670	65.0385	.002364	.0486	.01538
424	17 97 76	20.5913	65.1153	.002358	.0486	.01536
425	18 06 25	20.6155	65.1920	.002353	.0485	.01534
426	18 14 76	20.6398	65.2687	.002347	.0485	.01532
427	18 23 29	20.6640	65.3452	.002342	.0484	.01530
428	18 31 84	20.6882	65.4217	.002336	.0483	.01529
429	18 40 41	20.7123	65.4981	.002331	.0483	.01527
430	18 49 00	20.7364	65.5744	.002326	.0482	.01525
431	18 57 61	20.7605	65.6506	.002320	.0482	.01523
432	18 66 24	20.7846	65.7267	.002315	.0481	.01521
433	18 74 89	20.8087	65.8027	.002309	.0481	.01520
434	18 83 56	20.8327	65.8787	.002304	.0480	.01518
435	18 92 25	20.8567	65.9545	.002299	.0479	.01516
436	19 00 06	20.8806	66.0303	.002294	.0479	.01514
437	19 09 69	20.9045	66.1060	.002288	.0478	.01513
438	19 18 44	20.9284	66.1816	.002283	.0478	.01511
439	19 27 21	20.9523	66.2571	.002278	.0477	.01509
440	19 36 00	20.9762	66.3325	.002273	.0477	.01508
441	19 44 81	21.0000	66.4078	.002268	.0476	.01506
442	19 53 64	21.0238	66.4831	.002262	.0476	.01504
443	19 62 49	21.0476	66.5582	.002257	.0475	.01502
444	19 71 36	21.0713	66.6333	.002252	.0475	.01501
445	19 80 25	21.0950	66.7083	.002247	.0474	.01499
446	19 89 16	21.1187	66.7832	.002242	.0474	.01497
447	19 98 09	21.1424	66.8581	.002237	.0473	.01496
448	20 07 04	21.1660	66.9328	.002232	.0472	.01494
449	20 16 01	21.1896	67.0075	.002227	.0472	.01492
450	20 25 00	21.2132	67.0820	.002222	.0471	.01491

N	N^2	$\sqrt{N}$	$\sqrt{10N}$	$1/N$	$1/\sqrt{N}$	$1/\sqrt{10N}$
451	20 34 01	21.2368	67.1565	.002217	.0471	.01489
452	20 43 04	21.2603	67.2309	.002212	.0470	.01487
453	20 52 09	21.2838	67.3053	.002208	.0470	.01486
454	20 61 16	21.3073	67.3795	.002203	.0469	.01484
455	20 70 25	21.3307	67.4537	.002198	.0469	.01482
456	20 79 36	21.3542	67.5278	.002193	.0468	.01481
457	20 88 49	21.3776	67.6018	.002188	.0468	.01479
458	20 97 64	21.4009	67.6757	.002183	.0467	.01478
459	21 06 81	21.4243	67.7495	.002179	.0467	.01476
460	21 16 00	21.4476	67.8233	.002174	.0466	.01474
461	21 25 21	21.4709	67.8970	.002169	.0466	.01473
462	21 34 44	21.4942	67.9706	.002165	.0465	.01471
463	21 43 69	21.5174	68.0441	.002160	.0465	.01470
464	21 52 96	21.5407	68.1175	.002155	.0464	.01468
465	21 62 25	21.5639	68.1909	.002151	.0464	.01466
466	21 71 56	21.5870	68.2642	.002146	.0463	.01465
467	21 80 89	21.6102	68.3374	.002141	.0463	.01463
468	21 90 24	21.6333	68.4105	.002137	.0462	.01462
469	21 99 61	21.6564	68.4836	.002132	.0462	.01460
470	22 09 00	21.6795	68.5565	.002128	.0461	.01459
471	22 18 41	21.7025	68.6294	.002123	.0461	.01457
472	22 27 84	21.7256	68.7023	.002119	.0460	.01456
473	22 37 29	21.7486	68.7750	.002114	.0460	.01454
474	22 46 76	21.7715	68.8477	.002110	.0459	.01452
475	22 56 25	21.7945	68.9202	.002105	.0459	.01451
476	22 65 76	21.8174	68.9928	.002101	.0458	.01449
477	22 75 29	21.8403	69.0652	.002096	.0458	.01448
478	22 84 84	21.8632	69.1375	.002092	.0457	.01446
479	22 94 41	21.8861	69.2098	.002088	.0457	.01445
480	23 04 00	21.9089	69.2820	.002083	.0456	.01443
481	23 13 61	21.9317	69.3542	.002079	.0456	.01442
482	23 23 24	21.9545	69.4262	.002075	.0455	.01440
483	23 32 89	21.9773	69.4982	.002070	.0455	.01439
484	23 42 56	22.0000	69.5701	.002066	.0455	.01437
485	23 52 25	22.0227	69.6419	.002062	.0454	.01436
486	23 61 96	22.0454	69.7137	.002058	.0454	.01434
487	23 71 69	22.0681	69.7854	.002053	.0453	.01433
488	23 81 44	22.0907	69.8570	.002049	.0453	.01431
489	23 91 21	22.1133	69.9285	.002045	.0452	.01430
490	24 01 00	22.1359	70.0000	.002041	.0452	.01429
491	24 10 81	22.1585	70.0714	.002037	.0451	.01427
492	24 20 64	22.1811	70.1427	.002033	.0451	.01426
493	24 30 49	22.2036	70.2140	.002028	.0450	.01424
494	24 40 36	22.2261	70.2851	.002024	.0450	.01423
495	24 50 25	22.2486	70.3562	.002020	.0449	.01421
496	24 60 16	22.2711	70.4273	.002016	.0449	.01420
497	24 70 09	22.2935	70.4982	.002012	.0449	.01418
498	24 80 04	22.3159	70.5691	.002008	.0448	.01417
499	24 90 01	22.3383	70.6399	.002004	.0448	.01416
500	25 00 00	22.3607	70.7107	.002000	.0447	.01414

N	N²	√N	√10N	1/N	1/√N	1/√10N
501	25 10 01	22.3830	70.7814	.001996	.0447	.01413
502	25 20 04	22.4054	70.8520	.001992	.0446	.01411
503	25 30 09	22.4277	70.9225	.001988	.0446	.01410
504	25 40 16	22.4499	70.9930	.001984	.0445	.01409
505	25 50 25	22.4722	71.0634	.001980	.0445	.01407
506	25 60 36	22.4944	71.1337	.001976	.0445	.01406
507	25 70 49	22.5167	71.2039	.001972	.0444	.01404
508	25 80 64	22.5389	71.2741	.001969	.0444	.01403
509	25 90 81	22.5610	71.3442	.001965	.0443	.01402
510	26 01 00	22.5832	71.4143	.001961	.0443	.01400
511	26 11 21	22.6053	71.4843	.001957	.0442	.01399
512	26 21 44	22.6274	71.5542	.001953	.0442	.01398
513	26 31 69	22.6495	71.6240	.001949	.0442	.01396
514	26 41 96	22.6716	71.6938	.001946	.0441	.01395
515	26 52 25	22.6936	71.7635	.001942	.0441	.01393
516	26 62 56	22.7156	71.8331	.001938	.0440	.01392
517	26 72 89	22.7376	71.9027	.001934	.0440	.01391
518	26 83 24	22.7596	71.9722	.001931	.0439	.01389
519	26 93 61	22.7816	72.0417	.001927	.0439	.01388
520	27 04 00	22.8035	72.1110	.001923	.0439	.01387
521	27 14 41	22.8254	72.1803	.001919	.0438	.01385
522	27 24 84	22.8473	72.2496	.001916	.0438	.01384
523	27 35 29	22.8692	72.3187	.001912	.0437	.01383
524	27 45 76	22.8910	72.3878	.001908	.0437	.01381
525	27 56 25	22.9129	72.4569	.001905	.0436	.01380
526	27 66 76	22.9347	72.5259	.001901	.0436	.01379
527	27 77 29	22.9565	72.5948	.001898	.0436	.01378
528	27 87 84	22.9783	72.6636	.001894	.0435	.01376
529	27 98 41	23.0000	72.7324	.001890	.0435	.01375
530	28 09 00	23.0217	72.8011	.001887	.0434	.01374
531	28 19 61	23.0434	72.8697	.001883	.0434	.01372
532	28 30 24	23.0651	72.9383	.001880	.0434	.01371
533	28 40 89	23.0868	73.0068	.001876	.0433	.01370
534	28 51 56	23.1084	73.0753	.001873	.0433	.01368
535	28 62 25	23.1301	73.1437	.001869	.0432	.01367
536	28 72 96	23.1517	73.2120	.001866	.0432	.01366
537	28 83 69	23.1733	73.2803	.001862	.0432	.01365
538	28 94 44	23.1948	73.3485	.001859	.0431	.01363
539	29 05 21	23.2164	73.4166	.001855	.0431	.01362
540	29 16 00	23.2379	73.4847	.001852	.0430	.01361
541	29 26 81	23.2594	73.5527	.001848	.0430	.01360
542	29 37 64	23.2809	73.6206	.001845	.0430	.01358
543	29 48 49	23.3024	73.6885	.001842	.0429	.01357
544	29 59 36	23.3238	73.7564	.001838	.0429	.01356
545	29 70 25	23.3452	73.8241	.001835	.0428	.01355
546	29 81 16	23.3666	73.8918	.001832	.0428	.01353
547	29 92 09	23.3880	73.9594	.001828	.0428	.01352
548	30 03 04	23.4094	74.0270	.001825	.0427	.01351
549	30 14 01	23.4307	74.0945	.001821	.0427	.01350
550	30 25 00	23.4521	74.1620	.001818	.0426	.01348

N	N²	√N	√10N	1/N	1/√N	1/√10N
551	30 36 01	23.4734	74.2294	.001815	.0426	.01347
552	30 47 04	23.4947	74.2967	.001812	.0426	.01346
553	30 58 09	23.5160	74.3640	.001808	.0425	.01345
554	30 69 16	23.5372	74.4312	.001805	.0425	.01344
555	30 80 25	23.5584	74.4983	.001802	.0424	.01342
556	30 91 36	23.5797	74.5654	.001799	.0424	.01341
557	31 02 49	23.6008	74.6324	.001795	.0424	.01340
558	31 13 64	23.6220	74.6994	.001792	.0423	.01339
559	31 24 81	23.6432	74.7663	.001789	.0423	.01338
560	31 36 00	23.6643	74.8331	.001786	.0423	.01336
561	31 47 21	23.6854	74.8999	.001783	.0422	.01335
562	31 58 44	23.7065	74.9667	.001779	.0422	.01334
563	31 69 69	23.7276	75.0333	.001776	.0421	.01333
564	31 80 96	23.7487	75.0999	.001773	.0421	.01332
565	31 92 25	23.7697	75.1665	.001770	.0421	.01330
566	32 03 56	23.7908	75.2330	.001767	.0420	.01329
567	32 14 89	23.8118	75.2994	.001764	.0420	.01328
568	32 26 24	23.8328	75.3658	.001761	.0420	.01327
569	32 37 61	23.8537	75.4321	.001757	.0419	.01326
570	32 49 00	23.8747	75.4983	.001754	.0419	.01325
571	32 60 41	23.8956	75.5645	.001751	.0418	.01323
572	32 71 84	23.9165	75.6307	.001748	.0418	.01322
573	32 83 29	23.9374	75.6968	.001745	.0418	.01321
574	32 94 76	23.9583	75.7628	.001742	.0417	.01320
575	30 06 25	23.9792	75.8288	.001739	.0417	.01319
576	33 17 76	24.0000	75.8947	.001736	.0417	.01318
577	33 29 29	24.0208	75.9605	.001733	.0416	.01316
578	33 40 84	24.0416	76.0263	.001730	.0416	.01315
579	33 52 41	24.0624	76.0920	.001727	.0416	.01314
580	33 64 00	24.0832	76.1577	.001724	.0415	.01313
581	33 75 61	24.1039	76.2234	.001721	.0415	.01312
582	33 87 24	24.1247	76.2889	.001718	.0415	.01311
583	33 98 89	24.1454	76.3544	.001715	.0414	.01310
584	34 10 56	24.1661	76.4199	.001712	.0414	.01309
585	34 22 25	24.1868	76.4853	.001709	.0413	.01307
586	34 33 96	24.2074	76.5506	.001706	.0413	.01306
587	34 45 69	24.2281	76.6159	.001704	.0413	.01305
588	34 57 44	24.2487	76.6812	.001701	.0412	.01304
589	34 69 21	24.2693	76.7463	.001698	.0412	.01303
590	34 81 00	24.2899	76.8115	.001695	.0412	.01302
591	34 92 81	24.3105	76.8765	.001692	.0411	.01301
592	35 04 64	24.3311	76.9415	.001689	.0411	.01300
593	35 16 49	24.3516	77.0065	.001686	.0411	.01299
594	35 28 36	24.3721	77.0714	.001684	.0410	.01297
595	35 40 25	24.3926	77.1362	.001681	.0410	.01296
596	35 52 16	24.4131	77.2010	.001678	.0410	.01295
597	35 64 09	24.4336	77.2658	.001675	.0409	.01294
598	35 76 04	24.4540	77.3305	.001672	.0409	.01293
599	35 88 01	24.4745	77.3951	.001669	.0409	.01292
600	36 00 00	24.4949	77.4597	.001667	.0408	.01291

Appendix A: Squares, Square Roots, and Reciprocals **A13**

N	N²	$\sqrt{N}$	$\sqrt{10N}$	1/N	$1/\sqrt{N}$	$1/\sqrt{10N}$
601	36 12 01	24.5153	77.5242	.001664	.0408	.01290
602	36 24 04	24.5357	77.5887	.001661	.0408	.01289
603	36 36 09	24.5561	77.6531	.001658	.0407	.01288
604	36 48 16	24.5764	77.7174	.001656	.0407	.01287
605	36 60 25	24.5967	77.7817	.001653	.0407	.01286
606	36 72 36	24.6171	77.8460	.001650	.0406	.01285
607	36 84 49	24.6374	77.9102	.001647	.0406	.01284
608	36 96 64	24.6577	77.9744	.001645	.0406	.01282
609	37 08 81	24.6779	78.0385	.001642	.0405	.01281
610	37 21 00	24.6982	78.1025	.001639	.0405	.01280
611	37 33 21	24.7184	78.1665	.001637	.0405	.01279
612	37 45 44	24.7386	78.2304	.001634	.0404	.01278
613	37 57 69	24.7588	78.2943	.001631	.0404	.01277
614	37 69 96	24.7790	78.3582	.001629	.0404	.01276
615	37 82 25	24.7992	78.4219	.001626	.0403	.01275
616	37 94 56	24.8193	78.4857	.001623	.0403	.01274
617	38 06 89	24.8395	78.5493	.001621	.0403	.01273
618	38 19 24	24.8596	78.6130	.001618	.0402	.01272
619	38 31 61	24.8797	78.6766	.001616	.0402	.01271
620	38 44 00	24.8998	78.7401	.001613	.0402	.01270
621	38 56 41	24.9199	78.8036	.001610	.0401	.01269
622	38 68 84	24.9399	78.8670	.001608	.0401	.01268
623	38 81 29	24.9600	78.9303	.001605	.0401	.01267
624	38 93 76	24.9800	78.9937	.001603	.0400	.01266
625	39 06 25	25.0000	79.0569	.001600	.0400	.01265
626	39 18 76	25.0200	79.1202	.001597	.0400	.01264
627	39 31 29	25.0400	79.1833	.001595	.0399	.01263
628	39 43 84	25.0599	79.2465	.001592	.0399	.01262
629	39 56 41	25.0799	79.3095	.001590	.0399	.01261
630	39 69 00	25.0998	79.3725	.001587	.0398	.01260
631	39 81 61	25.1197	79.4355	.001585	.0398	.01259
632	39 94 24	25.1396	79.4984	.001582	.0398	.01258
633	40 06 89	25.1595	79.5613	.001580	.0397	.01257
634	40 19 56	25.1794	79.6241	.001577	.0397	.01256
635	40 32 25	25.1992	79.6869	.001575	.0397	.01255
636	40 44 96	25.2190	79.7496	.001572	.0397	.01254
637	40 57 69	25.2389	79.8123	.001570	.0396	.01253
638	40 70 44	25.2587	79.8749	.001567	.0396	.01252
639	40 83 21	25.2784	79.9375	.001565	.0396	.01251
640	40 96 00	25.2982	80.0000	.001562	.0395	.01250
641	41 08 81	25.3180	80.0625	.001560	.0395	.01249
642	41 21 64	25.3377	80.1249	.001558	.0395	.01248
643	41 34 49	25.3574	80.1873	.001555	.0394	.01247
644	41 47 36	25.3772	80.2496	.001553	.0394	.01246
645	41 60 25	25.3969	80.3119	.001550	.0394	.01245
646	41 73 16	25.4165	80.3714	.001548	.0393	.01244
647	41 86 09	25.4362	80.4363	.001546	.0393	.01243
648	41 99 04	25.4558	80.4984	.001543	.0393	.01242
649	42 12 01	25.4775	80.5605	.001541	.0393	.01241
650	42 25 00	25.4951	80.6226	.001538	.0392	.01240

N	N²	√N	√10N	1/N	1/√N	1/√10N
651	42 38 01	25.5147	80.6846	.001536	.0392	.01239
652	42 51 04	25.5343	80.7465	.001534	.0392	.01238
653	42 64 09	25.5539	80.8084	.001531	.0391	.01237
654	42 77 16	25.5734	80.8703	.001529	.0391	.01237
655	42 90 25	25.5930	80.9321	.001527	.0391	.01236
656	43 03 36	25.6125	80.9938	.001524	.0390	.01235
657	43 16 49	25.6320	81.0555	.001522	.0390	.01234
658	43 29 64	25.6515	81.1172	.001520	.0390	.01233
659	43 42 81	25.6710	81.1788	.001517	.0390	.01232
660	43 56 00	25.6905	81.2404	.001515	.0389	.01231
661	43 69 21	25.7099	81.3019	.001513	.0389	.01230
662	43 82 44	25.7294	81.3634	.001511	.0389	.01229
663	43 95 69	25.7488	81.4248	.001508	.0388	.01228
664	44 08 96	25.7682	81.4862	.001506	.0388	.01227
665	44 22 25	25.7876	81.5475	.001504	.0388	.01226
666	44 35 56	25.8070	81.6088	.001502	.0387	.01225
667	44 48 89	25.8263	81.6701	.001499	.0387	.01224
668	44 62 24	25.8457	81.7313	.001497	.0387	.01224
669	44 75 61	25.8650	81.7924	.001495	.0387	.01223
670	44 89 00	25.8844	81.8535	.001493	.0386	.01222
671	45 02 41	25.9037	81.9146	.001490	.0386	.01221
672	45 15 84	25.9230	81.9756	.001488	.0386	.01220
673	45 29 29	25.9422	82.0366	.001486	.0385	.01219
674	45 42 76	25.9615	82.0975	.001484	.0385	.01218
675	45 56 25	25.9808	82.1584	.001481	.0385	.01217
676	45 69 76	26.0000	82.2192	.001479	.0385	.01216
677	45 83 29	26.0192	82.2800	.001477	.0384	.01215
678	45 96 84	26.0384	82.3408	.001475	.0384	.01214
679	46 10 41	26.0576	82.4015	.001473	.0384	.01214
680	46 24 00	26.0768	82.4621	.001471	.0383	.01213
681	46 37 61	26.0960	82.5227	.001468	.0383	.01212
682	46 51 24	26.1151	82.5833	.001466	.0383	.01211
683	46 64 89	26.1343	82.6438	.001464	.0383	.01210
684	46 78 56	26.1534	82.7043	.001462	.0382	.01209
685	46 92 25	26.1725	82.7647	.001460	.0382	.01208
686	47 05 96	26.1916	82.8251	.001458	.0382	.01207
687	47 19 69	26.2107	82.8855	.001456	.0382	.01206
688	47 33 44	26.2298	82.9458	.001453	.0381	.01206
689	47 47 21	26.2488	83.0060	.001451	.0381	.01205
690	47 61 00	26.2679	83.0662	.001449	.0381	.01204
691	47 74 81	26.2869	83.1264	.001447	.0380	.01203
692	47 88 64	26.3059	83.1865	.001445	.0380	.01202
693	48 02 49	26.3249	83.2466	.001443	.0380	.01201
694	48 16 36	26.3439	83.3067	.001441	.0380	.01200
695	48 30 25	26.3629	83.3667	.001439	.0379	.01200
696	48 44 16	26.3818	83.4266	.001437	.0379	.01199
697	48 58 09	26.4008	83.4865	.001435	.0379	.01198
698	48 72 04	26.4197	83.5464	.001433	.0379	.01197
699	48 86 01	26.4386	83.6062	.001431	.0378	.01196
700	49 00 00	26.4575	83.6660	.001429	.0378	.01195

N	N²	√N	√10N	1/N	1/√N	1/√10N
701	49 14 01	26.4764	83.7257	.001427	.0378	.01194
702	49 28 04	26.4953	83.7854	.001425	.0377	.01194
703	49 42 09	26.5141	83.8451	.001422	.0377	.01193
704	49 56 16	26.5330	83.9047	.001420	.0377	.01192
705	49 70 25	26.5518	83.9643	.001418	.0377	.01191
706	49 84 36	26.5707	84.0238	.001416	.0376	.01190
707	49 98 49	26.5895	84.0833	.001414	.0376	.01189
708	50 12 64	26.6083	84.1427	.001412	.0376	.01188
709	50 26 81	26.6271	84.2021	.001410	.0376	.01188
710	50 41 00	26.6458	84.2615	.001408	.0375	.01187
711	50 55 21	26.6646	84.3208	.001406	.0375	.01186
712	50 69 44	26.6833	84.3801	.001404	.0375	.01185
713	50 83 69	26.7021	84.4393	.001403	.0375	.01184
714	50 97 96	26.7208	84.4985	.001401	.0374	.01183
715	51 12 25	26.7395	84.5577	.001399	.0374	.01183
716	51 26 56	26.7582	84.6168	.001397	.0374	.01182
717	51 40 89	26.7769	84.6759	.001395	.0373	.01181
718	51 55 24	26.7955	84.7349	.001393	.0373	.01180
719	51 69 61	26.8142	84.7939	.001391	.0373	.01179
720	51 84 00	26.8328	84.8528	.001389	.0373	.01179
721	51 98 41	26.8514	84.9117	.001387	.0372	.01178
722	52 12 84	26.8701	84.9706	.001385	.0372	.01177
723	52 27 29	26.8887	85.0294	.001383	.0372	.01176
724	52 41 76	26.9072	85.0882	.001381	.0372	.01175
725	52 56 25	26.9258	85.1469	.001379	.0371	.01174
726	52 70 76	26.9444	85.2056	.001377	.0371	.01174
727	52 85 29	26.9629	85.2643	.001376	.0371	.01173
728	52 99 84	26.9815	85.3229	.001374	.0371	.01172
729	53 14 41	27.0000	85.3815	.001372	.0370	.01171
730	53 29 00	27.0185	85.4400	.001370	.0370	.01170
731	53 43 61	27.0370	85.4985	.001368	.0370	.01170
732	53 58 24	27.0555	85.5570	.001366	.0370	.01169
733	53 72 89	27.0740	85.6154	.001364	.0369	.01168
734	53 87 56	27.0924	85.6738	.001362	.0369	.01167
735	54 02 25	27.1109	85.7321	.001361	.0369	.01166
736	54 16 96	27.1293	85.7904	.001359	.0369	.01166
737	54 31 69	27.1477	85.8487	.001357	.0368	.01165
738	54 46 44	27.1662	85.9069	.001355	.0368	.01164
739	54 61 21	27.1846	85.9651	.001353	.0368	.01163
740	54 76 00	27.2029	86.0233	.001351	.0368	.01162
741	54 90 81	27.2213	86.0814	.001350	.0367	.01162
742	55 05 64	27.2397	86.1394	.001348	.0367	.01161
743	55 20 49	27.2580	86.1974	.001346	.0367	.01160
744	55 35 36	27.2764	86.2554	.001344	.0367	.01159
745	55 50 25	27.2947	86.3134	.001342	.0366	.01159
746	55 65 16	27.3130	86.3713	.001340	.0366	.01158
747	55 80 09	27.3313	86.4292	.001339	.0366	.01157
748	55 95 04	27.3496	86.4870	.001337	.0366	.01156
749	56 10 01	27.3679	86.5448	.001335	.0365	.01155
750	56 25 00	27.3861	86.6025	.001333	.0365	.01155

N	N²	√N	√10N	1/N	1/√N	1/√10N
751	56 40 01	27.4044	86.6603	.001332	.0365	.01154
752	56 55 04	27.4226	86.7179	.001330	.0365	.01153
753	56 70 09	⁻7.4408	86.7756	.001328	.0364	.01152
754	56 85 16	27.4591	86.8332	.001326	.0364	.01152
755	57 00 25	27.4773	86.8907	.001325	.0364	.01151
756	57 15 36	27.4955	86.9483	.001323	.0364	.01150
757	57 30 49	27.5136	87.0057	.001321	.0363	.01149
758	57 45 64	27.5318	87.0632	.001319	.0363	.01149
759	57 60 81	27.5500	87.1206	.001318	.0363	.01148
760	57 76 00	27.5681	87.1780	.001316	.0363	.01147
761	57 91 21	27.5862	87.2353	.001314	.0362	.01146
762	58 06 44	27.6043	87.2926	.001312	.0362	.01146
763	58 21 69	27.6225	87.3499	.001311	.0362	.01145
764	58 36 96	27.6405	87.4071	.001309	.0362	.01144
765	58 52 25	27.6586	87.4643	.001307	.0362	.01143
766	58 67 56	27.6767	87.5214	.001305	.0361	.01143
767	58 82 89	27.6948	87.5785	.001304	.0361	.01142
768	58 98 24	27.7128	87.6356	.001302	.0361	.01141
769	59 13 61	27.7308	87.6926	.001300	.0361	.01140
770	59 29 00	27.7489	87.7496	.001299	.0360	.01140
771	59 44 41	27.7669	87.8066	.001297	.0360	.01139
772	59 59 84	27.7849	87.8635	.001295	.0360	.01138
773	59 75 29	27.8029	87.9204	.001294	.0360	.01137
774	59 90 76	27.8209	87.9773	.001292	.0359	.01137
775	60 06 25	27.8388	88.0341	.001290	.0359	.01136
776	60 21 76	27.8568	88.0909	.001289	.0359	.01135
777	60 37 29	27.8747	88.1476	.001287	.0359	.01134
778	60 52 84	27.8927	88.2043	.001285	.0359	.01134
779	60 68 41	27.9106	88.2610	.001284	.0358	.01133
780	60 84 00	27.9285	88.3176	.001282	.0358	.01132
781	60 99 61	27.9464	88.3742	.001280	.0358	.01132
782	61 15 24	27.9643	88.4308	.001279	.0358	.01131
783	61 30 89	27.9821	88.4873	.001277	.0357	.01130
784	61 46 56	28.0000	88.5438	.001276	.0357	.01129
785	61 62 25	28.0179	88.6002	.001274	.0357	.01129
786	61 77 96	28.0357	88.6566	.001272	.0357	.01128
787	61 93 69	28.0535	88.7130	.001271	.0356	.01127
788	62 09 44	28.0713	88.7694	.001269	.0356	.01127
789	62 25 21	28.0891	88.8257	.001267	.0356	.01126
790	62 41 00	28.1069	88.8819	.001266	.0356	.01125
791	62 56 81	28.1247	88.9382	.001264	.0356	.01124
792	62 72 64	28.1425	88.9944	.001263	.0355	.01124
793	62 88 49	28.1603	89.0505	.001261	.0355	.01123
794	63 04 36	28.1780	89.1067	.001259	.0355	.01122
795	63 20 25	28.1957	89.1628	.001258	.0355	.01122
796	63 36 16	28.2135	89.2188	.001256	.0354	.01121
797	63 52 09	28.2312	89.2749	.001255	.0354	.01120
798	63 68 04	28.2489	89.3308	.001253	.0354	.01119
799	63 84 01	28.2666	89.3868	.001252	.0354	.01119
800	64 00 00	28.2843	89.4427	.001250	.0354	.01118

N	N²	$\sqrt{N}$	$\sqrt{10N}$	$1/N$	$1/\sqrt{N}$	$1/\sqrt{10N}$
801	64 16 01	28.3019	89.4986	.001248	.0353	.01117
802	64 32 04	28.3196	89.5545	.001247	.0353	.01117
803	64 48 09	28.3373	89.6103	.001245	.0353	.01116
804	64 64 16	28.3549	89.6660	.001244	.0353	.01115
805	64 80 25	28.3725	89.7218	.001242	.0352	.01115
806	64 96 36	28.3901	89.7775	.001241	.0352	.01114
807	65 12 49	28.4077	89.8332	.001239	.0352	.01113
808	65 28 64	28.4253	89.8888	.001238	.0352	.01112
809	65 44 81	28.4429	89.9444	.001236	.0352	.01112
810	65 61 00	28.4605	90.0000	.001235	.0351	.01111
811	65 77 21	28.4781	90.0555	.001233	.0351	.01110
812	65 93 44	28.4956	90.1110	.001232	.0351	.01110
813	66 09 69	28.5132	90.1665	.001230	.0351	.01109
814	66 25 96	28.5307	90.2219	.001229	.0350	.01108
815	66 42 25	28.5482	90.2774	.001227	.0350	.01108
816	66 58 56	28.5657	90.3327	.001225	.0350	.01107
817	66 74 89	28.5832	90.3881	.001224	.0350	.01106
818	66 91 24	28.6007	90.4434	.001222	.0350	.01106
819	67 07 61	28.6182	90.4986	.001221	.0349	.01105
820	67 24 00	28.6356	90.5539	.001220	.0349	.01104
821	67 40 41	28.6531	90.6091	.001218	.0349	.01104
822	67 56 84	28.6705	90.6642	.001217	.0349	.01103
823	67 73 29	28.6880	90.7193	.001215	.0349	.01102
824	67 89 76	28.7054	90.7744	.001214	.0348	.01102
825	68 06 25	28.7228	90.8295	.001212	.0348	.01101
826	68 22 76	28.7402	90.8845	.001211	.0348	.01100
827	68 39 29	28.7576	90.9395	.001209	.0348	.01100
828	68 55 84	28.7750	90.9945	.001208	.0348	.01099
829	68 72 41	28.7924	91.0494	.001206	.0347	.01098
830	68 89 00	28.8097	91.1043	.001205	.0347	.01098
831	69 05 61	28.8271	91.1592	.001203	.0347	.01097
832	69 22 24	28.8444	91.2140	.001202	.0347	.01096
833	69 38 89	28.8617	91.2688	.001200	.0346	.01096
834	69 55 56	28.8791	91.3236	.001199	.0346	.01095
835	69 72 25	28.8964	91.3783	.001198	.0346	.01094
836	69 88 96	28.9137	91.4330	.001196	.0346	.01094
837	70 05 69	28.9310	91.4877	.001195	.0346	.01093
838	70 22 44	28.9482	91.5423	.001193	.0345	.01092
839	70 39 21	28.9655	91.5969	.001192	.0345	.01092
840	70 56 00	28.9828	91.6515	.001190	.0345	.01091
841	70 72 81	29.0000	91.7061	.001189	.0345	.01090
842	70 89 64	29.0172	91.7606	.001188	.0345	.01090
843	71 06 49	29.0345	91.8150	.001186	.0344	.01089
844	71 23 36	29.0517	91.8695	.001185	.0344	.01089
845	71 40 25	29.0689	91.9239	.001183	.0344	.01088
846	71 57 16	29.0861	91.9783	.001182	.0344	.01087
847	71 74 09	29.1033	92.0326	.001181	.0344	.01087
848	71 91 04	29.1204	92.0869	.001179	.0343	.01086
849	72 08 01	29.1376	92.1412	.001178	.0343	.01085
850	72 25 00	29.1548	92.1954	.001176	.0343	.01085

A18 *Appendix A: Squares, Square Roots, and Reciprocals*

N	N^2	$\sqrt{N}$	$\sqrt{10N}$	$1/N$	$1/\sqrt{N}$	$1/\sqrt{10N}$
851	72 42 01	29.1719	92.2497	.001175	.0343	.01084
852	72 59 04	29.1890	92.3038	.001174	.0343	.01083
853	72 76 09	29.2062	92.3580	.001172	.0342	.01083
854	72 93 16	29.2233	92.4121	.001171	.0342	.01082
855	73 10 25	29.2404	92.4662	.001170	.0342	.01081
856	73 27 36	29.2575	92.5203	.001168	.0342	.01081
857	73 44 49	29.2746	92.5743	.001167	.0342	.01080
858	73 61 64	29.2916	92.6283	.001166	.0341	.01080
859	73 78 81	29.3087	92.6823	.001164	.0341	.01079
860	73 96 00	29.3258	92.7362	.001163	.0341	.01078
861	74 13 21	29.3428	92.7901	.001161	.0341	.01078
862	74 30 44	29.3598	92.8440	.001160	.0341	.01077
863	74 47 69	29.3769	92.8978	.001159	.0340	.01076
864	74 64 96	29.3939	92.9516	.001157	.0340	.01076
865	74 82 25	29.4109	93.0054	.001156	.0340	.01075
866	74 99 56	29.4279	93.0591	.001155	.0340	.01075
867	75 16 89	29.4449	93.1128	.001153	.0340	.01074
868	75 34 24	29.4618	93.1665	.001152	.0339	.01073
869	75 51 61	29.4788	93.2202	.001151	.0339	.01073
870	75 69 00	29.4958	93.2738	.001149	.0339	.01072
871	75 86 41	29.5127	93.3274	.001148	.0339	.01071
872	76 03 84	29.5296	93.3809	.001147	.0339	.01071
873	76 21 29	29.5466	93.4345	.001145	.0338	.01070
874	76 38 76	29.5635	93.4880	.001144	.0338	.01070
875	76 56 25	29.5804	93.5414	.001143	.0338	.01069
876	76 73 76	29.5973	93.5949	.001142	.0338	.01068
877	76 91 29	29.6142	93.6483	.001140	.0338	.01068
878	77 08 84	29.6311	93.7017	.001139	.0337	.01067
879	77 26 41	29.6479	93.7550	.001138	.0337	.01067
880	77 44 00	29.6648	93.8083	.001136	.0337	.01066
881	77 61 61	29.6816	93.8616	.001135	.0337	.01065
882	77 79 24	29.6985	93.9149	.001134	.0337	.01065
883	77 96 89	29.7153	93.9681	.001133	.0337	.01064
884	78 14 56	29.7321	94.0213	.001131	.0336	.01064
885	78 32 25	29.7489	94.0744	.001130	.0336	.01063
886	78 49 96	29.7658	94.1276	.001129	.0336	.01062
887	78 67 69	29.7825	94.1807	.001127	.0336	.01062
888	78 85 44	29.7993	94.2338	.001126	.0336	.01061
889	79 03 21	29.8161	94.2868	.001125	.0335	.01061
890	79 21 00	29.8329	94.3398	.001124	.0335	.01060
891	79 38 81	29.8496	94.3928	.001122	.0335	.01059
892	79 56 64	29.8664	94.4458	.001121	.0335	.01059
893	79 74 49	29.8831	94.4987	.001120	.0335	.01058
894	79 92 36	29.8998	94.5516	.001119	.0334	.01058
895	80 10 25	29.9166	94.6044	.001117	.0334	.01057
896	80 28 16	29.9333	94.6573	.001116	.0334	.01056
897	80 46 09	29.9500	94.7101	.001115	.0334	.01056
898	80 64 04	29.9666	94.7629	.001114	.0334	.01055
899	80 82 01	29.9833	94.8156	.001112	.0334	.01055
900	81 00 00	30.0000	94.8683	.001111	.0333	.01054

N	N^2	$\sqrt{N}$	$\sqrt{10N}$	$1/N$	$1/\sqrt{N}$	$1/\sqrt{10N}$
901	81 18 01	30.0167	94.9210	.001110	.0333	.01054
902	81 36 04	30.0333	94.9737	.001109	.0333	.01053
903	81 54 09	30.0500	95.0263	.001107	.0333	.01052
904	81 72 16	30.0666	95.0789	.001106	.0333	.01052
905	81 90 25	30.0832	95.1315	.001105	.0332	.01051
906	82 08 36	30.0998	95.1840	.001104	.0332	.01051
907	82 26 49	30.1164	95.2365	.001103	.0332	.01050
908	82 44 64	30.1330	95.2890	.001101	.0332	.01049
909	82 62 81	30.1496	95.3415	.001100	.0332	.01049
910	82 81 00	30.1662	95.3939	.001099	.0331	.01048
911	82 99 21	30.1828	95.4463	.001098	.0331	.01048
912	83 17 44	30.1993	95.4987	.001096	.0331	.01047
913	83 35 69	30.2159	95.5510	.001095	.0331	.01047
914	83 53 96	30.2324	95.6033	.001094	.0331	.01046
915	83 72 25	30.2490	95.6556	.001093	.0331	.01045
916	83 90 56	30.2655	95.7079	.001092	.0330	.01045
917	84 08 89	30.2820	95.7601	.001091	.0330	.01044
918	84 27 24	30.2985	95.8123	.001089	.0330	.01044
919	84 45 61	30.3150	95.8645	.001088	.0330	.01043
920	84 64 00	30.3315	95.9166	.001087	.0330	.01043
921	84 82 41	30.3480	95.9687	.001086	.0330	.01042
922	85 00 84	30.3645	96.0208	.001085	.0329	.01041
923	85 19 29	30.3809	96.0729	.001083	.0329	.01041
924	85 37 76	30.3974	96.1249	.001082	.0329	.01040
925	85 56 25	30.4138	96.1769	.001081	.0329	.01040
926	85 74 76	30.4302	96.2289	.001080	.0329	.01039
927	85 93 29	30.4467	96.2808	.001079	.0328	.01039
928	86 11 84	30.4631	96.3328	.001078	.0328	.01038
929	86 30 41	30.4795	96.3846	.001076	.0328	.01038
930	86 49 00	30.4959	96.4365	.001075	.0328	.01037
931	86 67 61	30.5123	96.4883	.001074	.0328	.01036
932	86 86 24	30.5287	96.5401	.001073	.0328	.01036
933	87 04 89	30.5450	96.5919	.001072	.0327	.01035
934	87 23 56	30.5614	96.6437	.001071	.0327	.01055
935	87 42 25	30.5778	96.6954	.001070	.0327	.01034
936	87 60 96	30.5941	96.7471	.001068	.0327	.01034
937	87 79 69	30.6105	96.7988	.001067	.0327	.01033
938	87 98 44	30.6268	96.8504	.001066	.0327	.01033
939	88 17 21	30.6431	96.9020	.001065	.0326	.01032
940	88 36 00	30.6594	96.9536	.001064	.0326	.01031
941	88 54 81	30.6757	97.0052	.001063	.0326	.01031
942	88 73 64	30.6920	97.0567	.001062	.0326	.01030
943	88 92 49	30.7083	97.1082	.001060	.0326	.01030
944	89 11 36	30.7246	97.1597	.001059	.0325	.01029
945	89 30 25	30.7409	97.2111	.001058	.0325	.01029
946	89 49 16	30.7571	97.2625	.001057	.0325	.01028
947	89 68 09	30.7734	97.3139	.001056	.0325	.01028
948	89 87 04	30.7896	97.3653	.001055	.0325	.01027
949	90 06 01	30.8058	97.4166	.001054	.0325	.01027
950	90 25 00	30.8221	97.4679	.001053	.0324	.01026

N	N²	√N	√10N	1/N	1/√N	1/√10N
951	90 44 01	30.8383	97.5192	.001052	.0324	.01025
952	90 63 04	30.8545	97.5705	.001050	.0324	.01025
953	90 82 09	30.8707	97.6217	.001049	.0324	.01024
954	91 01 16	30.8869	97.6729	.001048	.0324	.01024
955	91 20 25	30.9031	97.7241	.001047	.0324	.01023
956	91 39 36	30.9192	97.7753	.001046	.0323	.01023
957	91 58 49	30.9354	97.8264	.001045	.0323	.01022
958	91 77 64	30.9516	97.8775	.001044	.0323	.01022
959	91 96 81	30.9677	97.9285	.001043	.0323	.01021
960	92 16 00	30.9839	97.9796	.001042	.0323	.01021
961	92 35 21	31.0000	98.0306	.001041	.0323	.01020
962	92 54 44	31.0161	98.0816	.001040	.0322	.01020
963	92 73 69	31.0322	98.1326	.001038	.0322	.01019
964	92 92 96	31.0483	98.1835	.001037	.0322	.01019
965	93 12 25	31.0644	98.2344	.001036	.0322	.01018
966	93 31 56	31.0805	98.2853	.001035	.0322	.01017
967	93 50 89	31.0966	98.3362	.001034	.0322	.01017
968	93 70 24	31.1127	98.3870	.001033	.0321	.01016
969	93 89 61	31.1288	98.4378	.001032	.0321	.01016
970	94 09 00	31.1448	98.4886	.001031	.0321	.01015
971	94 28 41	31.1609	98.5393	.001030	.0321	.01015
972	94 47 84	31.1769	98.5901	.001029	.0321	.01014
973	94 67 29	31.1929	98.6408	.001028	.0321	.01014
974	94 86 76	31.2090	98.6914	.001027	.0320	.01013
975	95 06 25	31.2250	98.7421	.001026	.0320	.01013
976	95 25 76	31.2410	98.7927	.001025	.0320	.01012
977	95 45 29	31.2570	98.8433	.001024	.0320	.01012
978	95 64 84	31.2730	98.8939	.001022	.0320	.01011
979	95 84 41	31.2890	98.9444	.001021	.0320	.01011
980	96 04 00	31.3050	98.9949	.001020	.0319	.01010
981	96 23 61	31.3209	99.0454	.001019	.0319	.01010
982	96 43 24	31.3369	99.0959	.001018	.0319	.01009
983	96 62 89	31.3528	99.1464	.001017	.0319	.01009
984	96 82 56	31.3688	99.1968	.001016	.0319	.01008
985	97 02 25	31.3847	99.2472	.001015	.0319	.01008
986	97 21 96	31.4006	99.2975	.001014	.0318	.01007
987	97 41 69	31.4166	99.3479	.001013	.0318	.01007
988	97 61 44	31.4325	99.3982	.001012	.0318	.01006
989	97 81 21	31.4484	99.4485	.001011	.0318	.01006
990	98 01 00	31.4643	99.4987	.001010	.0318	.01005
991	98 20 81	31.4802	99.5490	.001009	.0318	.01005
992	98 40 64	31.4960	99.5992	.001008	.0318	.01004
993	98 60 49	31.5119	99.6494	.001007	.0317	.01004
994	98 80 36	31.5278	99.6995	.001006	.0317	.01003
995	99 00 25	31.5436	99.7497	.001005	.0317	.01003
996	99 20 16	31.5595	99.7998	.001004	.0317	.01002
997	99 40 09	31.5753	99.8499	.001003	.0317	.01002
998	99 60 04	31.5911	99.8999	.001002	.0317	.01001
999	99 80 01	31.6070	99.9500	.001001	.0316	.01001
1,000	1 00 00 00	31.6228	100.0000	.001000	.0316	.01000

Appendix B
Table of Random Numbers

Appendix B is a table of random numbers. It has many different uses. For example, suppose one had a list of 350 people in a population and wanted to select a random sample of 50 individuals. The table can help accomplish this as follows: Arbitrarily number the 350 individuals from 1 to 350. Then, enter the random number table, at any place. For convenience, start in the upper left-hand corner where the digits 19612 appear. Since we are concerned, at most, with a three-digit number, we will examine only the left-most three digits. This yields the number 196. Therefore, the individual who was numbered 196 is included in the sample. Next, move down one row. (We could have moved across one column if we wanted to.) The three-digit number is 391. Since there is no individual with this number, ignore it and continue down the column. The next "valid" number is 035, meaning the individual numbered 35 would be included in the sample. Continue this process until 50 individuals have been selected.

Source: Reprinted from pages 5 and 3 of *A Million Random Digits with 100,000 Normal Deviates,* by The Rand Corporation. New York: The Free Press, 1955. Copyright 1955 by The Rand Corporation. Used by permission.

```
19612  78430    11661  94770    77603  65669    86868  12665    30012  75989
39141  77400    28000  64238    73258  71794    31340  26256    66453  37016
64756  80457    08747  12836    03469  50678    03274  43423    66677  82556
92901  51878    56441  22998    29718  38447    06453  25311    07565  53771
03551  90070    09483  94050    45938  18135    36908  43321    11073  51803

98884  66209    06830  53656    14663  56346    71430  04909    19818  05707
27369  86882    53473  07541    53633  70863    03748  12822    19360  49088
59066  75974    63335  20483    43514  37481    58278  26967    49325  43951
91647  93783    64169  49022    98588  09495    49829  59068    38831  04838
83605  92419    39542  07772    71568  75673    35185  89759    44901  74291

24895  88530    70774  35439    46758  70472    70207  92675    91623  61275
35720  26556    95596  20094    73750  85788    34264  01703    46833  65248
14141  53410    38649  06343    57256  61342    72709  75318    90379  37562
27416  75670    92176  72535    93119  56077    06886  18244    92344  31374
82071  07429    81007  47749    40744  56974    23336  88821    53841  10536

21445  82793    24831  93241    14199  76268    70883  68002    03829  17443
72513  76400    52225  92348    62308  98481    29744  33165    33141  61020
71479  45027    76160  57411    13780  13632    52308  77762    88874  33697
83210  51466    09088  50395    26743  05306    21706  70001    99439  80767
68749  95148    94897  78636    96750  09024    94538  91143    96693  61886

05184  75763    47075  88158    05313  53439    14908  08830    60096  21551
13651  62546    96892  25240    47511  58483    87342  78818    07855  39269
00566  21220    00292  24069    25072  29519    52548  54091    21282  21296
50958  17695    58072  68990    60329  95955    71586  63417    35947  67807
57621  64547    46850  37981    38527  09037    64756  03324    04986  83666

09282  25844    79139  78435    35428  43561    69799  63314    12991  93516
23394  94206    93432  37836    94919  26846    02555  74410    94915  48199
05280  37470    93622  04345    15092  19510    18094  16613    78234  50001
95491  97976    38306  32192    82639  54624    72434  92606    23191  74693
78521  00104    18248  75583    90326  50785    54034  66251    35774  14692

96345  44579    85932  44053    75704  20840    86583  83944    52456  73766
77963  31151    32364  91691    47357  40338    23435  24065    08458  95366
07520  11294    23238  01748    41690  67328    54814  37777    10057  42332
38423  02309    70703  85736    46148  14258    29236  12152    05088  65825
02463  65533    21199  60555    33928  01817    07396  89215    30722  22102

15880  92261    17292  88190    61781  48898    92525  21283    88581  60098
71926  00819    59144  00224    30570  90194    18329  06999    26857  19238
64425  28108    16554  16016    00042  83229    10333  36168    65617  94834
79782  23924    49440  30432    81077  31543    95216  64865    13658  51081
35337  74538    44553  64672    90960  41849    93865  44608    93176  34851

05249  29329    19715  94082    14738  86667    43708  66354    93692  25527
56463  99380    38793  85774    19056  13939    46062  27647    66146  63210
96296  33121    54196  34108    75814  85986    71171  15102    28992  63165
98380  36269    60014  07201    62448  46385    42175  88350    46182  49126
52567  64350    16315  53969    80395  81114    54358  64578    47269  15747

78498  90830    25955  99236    43286  91064    99969  95144    64424  77377
49553  24241    08150  89535    08703  91041    77323  81079    45127  93686
32151  07075    83155  10252    73100  88618    23891  87418    45417  20268
11314  50363    26860  27799    49416  83534    19187  08059    76677  02110
12364  71210    87052  50241    90785  97889    81399  58130    64439  05614
```

03991	10461	93716	16894	66083	24653	84609	58232	88618	19161
38555	95554	32886	59780	08355	60860	29735	47762	71299	23853
17546	73704	92052	46215	55121	29281	59076	07936	27954	58909
32643	52861	95819	06831	00911	98936	76355	93779	80863	00514
69572	68777	39510	35905	14060	40619	29549	69616	33564	60780
24122	66591	27699	06494	14845	46672	61958	77100	90899	75754
61196	30231	92962	61773	41839	55382	17267	70943	78038	70267
30532	21704	10274	12202	39685	23309	10061	68829	55986	66485
03788	97599	75867	20717	74416	53166	35208	33374	87539	08823
48228	63379	85783	47619	53152	67433	35663	52972	16818	60311
60365	94653	35075	33949	42614	29297	01918	28316	98953	73231
83799	42402	56623	34442	34994	41374	70071	14736	09958	18065
32960	07405	36409	83232	99385	41600	11133	07586	15917	06253
19322	53845	57620	52606	66497	68646	78138	66559	19640	99413
11220	94747	07399	37408	48509	23929	27482	45476	85244	35159
31751	57260	68980	05339	15470	48355	88651	22596	03152	19121
88492	99382	14454	04504	20094	98977	74843	93413	22109	78508
30934	47744	07481	83828	73788	06533	28597	20405	94205	20380
22888	48893	27499	98748	60530	45128	74022	84617	82037	10268
78212	16993	35902	91386	44372	15486	65741	14014	87481	37220
41849	84547	46850	52326	34677	58300	74910	64345	19325	81549
46352	33049	69248	93460	45305	07521	61318	31855	14413	70951
11087	96294	14013	31792	59747	67277	76503	34513	39663	77544
52701	08337	56303	87315	16520	69676	11654	99893	02181	68161
57275	36898	81304	48585	68652	27376	92852	55866	88448	03584
20857	73156	70284	24326	79375	95220	01159	63267	10622	48391
15633	84924	90415	93614	33521	26665	55823	47641	86225	31704
92694	48297	39904	02115	59589	49067	66821	41575	49767	04037
77613	19019	88152	00080	20554	91409	96277	48257	50816	97616
38688	32486	45134	63545	59404	72059	43947	51680	43852	59693
25163	01889	70014	15021	41290	67312	71857	15957	68971	11403
65251	07629	37239	33295	05870	01119	92784	26340	18477	65622
36815	43625	18637	37509	82444	99005	04921	73701	14707	93997
64397	11692	05327	82162	20247	81759	45197	25332	83745	22567
04515	25624	95096	67946	48460	85558	15191	18782	16930	33361
83761	60873	43253	84145	60833	25983	01291	41349	20368	07126
14387	06345	80854	09279	43529	06318	38384	74761	41196	37480
51321	92246	80088	77074	88722	56736	66164	49431	66919	31678
72472	00008	80890	18002	94813	31900	54155	83436	35352	54131
05466	55306	93128	18464	74457	90561	72848	11834	79982	68416
39528	72484	82474	25593	48545	35247	18619	13674	18611	19241
81616	18711	53342	44276	75122	11724	74627	73707	58319	15997
07586	16120	82641	22820	92904	13141	32392	19763	61199	67940
90767	04235	13574	17200	69902	63742	78464	22501	18627	90872
40188	28193	29593	88627	94972	11598	62095	36787	00441	58997
34414	82157	86887	55087	19152	00023	12302	80783	32624	68691
63439	75363	44989	16822	36024	00867	76378	41605	65961	73488
67049	09070	93399	45547	94458	74284	05041	49807	20288	34060
79495	04146	52162	90286	54158	34243	46978	35482	59362	95938
91704	30552	04737	21031	75051	93029	47665	64382	99782	93478

Appendix C
Areas Under the Standard
Normal Distribution

The following table reports probabilities of obtaining selected ranges of z scores. Column 1 lists the z score of interest. Column 2 presents the probability of observing z scores greater than or equal to $-z$ and less than or equal to z. Column 3 presents the probability of obtaining z scores greater than or equal to z. This column can also be used to determine the probability of obtaining z scores less than or equal to $-z$. Column 4 presents the probability of obtaining a z score greater than or equal to z and less than or equal to $-z$. Column 5 presents the probability of obtaining a z score between 0 and the z score of interest.

As an example, consider the z score 1.96. In column 2, we see that the probability of obtaining a score between -1.96 and $+1.96$ is .9500. In column 3, we see that the probability of obtaining a z greater than or equal to 1.96 is .0250. Since the normal distribution is symmetrical, this column also indicates the probability of obtaining a z less than or equal to -1.96. It also is .0250. In column 4, we see that the probability of obtaining a z score that is greater than or equal to 1.96 *or* which is less than or equal to -1.96 is .0500. In column 5, we see that the probability of obtaining a z score between 0 and 1.96 is .4750. Again, because the normal distribution is symmetrical, this also represents the probability of obtaining a z score between 0 and -1.96.

Source: Adapted from R. B. McCall, *Fundamental Statistics for Psychology.* New York: Harcourt Brace Jovanovich, 1970. Used with permission.

z	$\geq -z$ and $\leq z$	$\geq z$ (also use to find $\leq -z$)	$\geq z$ and $\leq -z$	≥ 0 and $\leq z$ (also use to find $\geq -z$ and ≤ 0)
0.00	.0000	.5000	1.0000	.0000
0.01	.0080	.4960	.9920	.0040
0.02	.0160	.4920	.9840	.0080
0.03	.0240	.4880	.9768	.0170
0.04	.0320	.4840	.9680	.0160
0.05	.0398	.4801	.9602	.0199
0.06	.0478	.4761	.9522	.0239
0.07	.0558	.4721	.9442	.0279
0.08	.0638	.4681	.9362	.0319
0.09	.0718	.4641	.9282	.0359
0.10	.0796	.4602	.9204	.0398
0.11	.0876	.4562	.9124	.0438
0.12	.0956	.4522	.9044	.0478
0.13	.1034	.4483	.8966	.0517
0.14	.1114	.4443	.8886	.0557
0.15	.1192	.4404	.8808	.0596
0.16	.1272	.4364	.8728	.0636
0.17	.1350	.4325	.8650	.0675
0.18	.1428	.4286	.8572	.0714
0.19	.1506	.4247	.8494	.0753
0.20	.1586	.4207	.8414	.0793
0.21	.1664	.4168	.8336	.0832
0.22	.1742	.4129	.8258	.0871
0.23	.1820	.4090	.8180	.0910
0.24	.1896	.4052	.8104	.0948
0.25	.1974	.4013	.8026	.0987
0.26	.2052	.3974	.7948	.1026
0.27	.2128	.3936	.7872	.1064
0.28	.2206	.3897	.7794	.1103
0.29	.2282	.3859	.7718	.1141
0.30	.2358	.3821	.7642	.1179
0.31	.2434	.3783	.7566	.1217
0.32	.2510	.3745	.7490	.1255
0.33	.2586	.3707	.7414	.1293
0.34	.2662	.3669	.7338	.1331
0.35	.2736	.3632	.7264	.1368
0.36	.2812	.3594	.7188	.1406
0.37	.2886	.3557	.7114	.1443
0.38	.2960	.3520	.7040	.1480
0.39	.3034	.3483	.6966	.1517
0.40	.3108	.3446	.6892	.1554
0.41	.3182	.3409	.6818	.1591
0.42	.3256	.3372	.6744	.1628
0.43	.3328	.3336	.6672	.1664
0.44	.3400	.3300	.6600	.1700

z	≥ −z and ≤ z	≥ z (also use to find ≤ −z)	≥ z and ≤ −z	≥ 0 and ≤ z (also use to find ≥ −z and ≤ 0)
0.45	.3472	.3264	.6528	.1736
0.46	.3544	.3228	.6456	.1772
0.47	.3616	.3192	.6384	.1808
0.48	.3688	.3156	.6312	.1844
0.49	.3758	.3121	.6242	.1879
0.50	.3830	.3085	.6170	.1915
0.51	.3900	.3050	.6100	.1950
0.52	.3970	.3015	.6030	.1985
0.53	.4038	.2981	.5962	.2019
0.54	.4108	.2946	.5892	.2054
0.55	.4176	.2912	.5824	.2088
0.56	.4246	.2877	.5754	.2123
0.57	.4314	.2843	.5686	.2157
0.58	.4380	.2810	.5670	.2190
0.59	.4448	.2776	.5552	.2224
0.60	.4514	.2743	.5486	.2257
0.61	.4582	.2709	.5418	.2291
0.62	.4648	.2676	.5352	.2324
0.63	.4714	.2643	.5286	.2357
0.64	.4778	.2611	.5222	.2389
0.65	.4844	.2578	.5156	.2422
0.66	.4908	.2546	.5092	.2454
0.67	.4972	.2514	.5028	.2486
0.68	.5034	.2483	.4966	.2517
0.69	.5098	.2451	.4902	.2549
0.70	.5160	.2420	.4840	.2580
0.71	.5222	.2389	.4778	.2611
0.72	.5284	.2358	.4716	.2642
0.73	.5346	.2327	.4654	.2673
0.74	.5408	.2296	.4592	.2704
0.75	.5468	.2266	.4532	.2734
0.76	.5528	.2236	.4472	.2764
0.77	.5588	.2206	.4412	.2794
0.78	.5646	.2177	.4354	.2823
0.79	.5704	.2148	.4296	.2852
0.80	.5762	.2119	.4238	.2881
0.81	.5820	.2090	.4180	.2910
0.82	.5878	.2061	.4132	.2939
0.83	.5934	.2033	.4066	.2967
0.84	.5990	.2005	.4010	.2995
0.85	.6046	.1977	.3954	.3023
0.86	.6102	.1949	.3898	.3051
0.87	.6156	.1922	.3844	.3078
0.88	.6212	.1894	.3788	.3106
0.89	.6266	.1867	.3734	.3133

z	$\geq -z$ and $\leq z$	$\geq z$ (also use to find $\leq -z$)	$\geq z$ and $\leq -z$	≥ 0 and $\leq z$ (also use to find $\geq -z$ and ≤ 0)
0.90	.6318	.1841	.3682	.3159
0.91	.6372	.1814	.3628	.3186
0.92	.6424	.1788	.3576	.3212
0.93	.6476	.1762	.3524	.3238
0.94	.6528	.1736	.3472	.3264
0.95	.6578	.1711	.3422	.3289
0.96	.6630	.1685	.3370	.3315
0.97	.6680	.1660	.3320	.3340
0.98	.6730	.1635	.3270	.3365
0.99	.6778	.1611	.3222	.3389
1.00	.6826	.1587	.3174	.3413
1.01	.6876	.1562	.3124	.3438
1.02	.6922	.1539	.3078	.3461
1.03	.6970	.1515	.3030	.3485
1.04	.7016	.1492	.2984	.3508
1.05	.7062	.1469	.2938	.3531
1.06	.7108	.1446	.2892	.3554
1.07	.7154	.1423	.2846	.3577
1.08	.7198	.1401	.2802	.3599
1.09	.7242	.1379	.2758	.3621
1.10	.7286	.1357	.2714	.3643
1.11	.7330	.1335	.2670	.3665
1.12	.7372	.1314	.2628	.3686
1.13	.7416	.1292	.2584	.3708
1.14	.7458	.1271	.2542	.3729
1.15	.7498	.1251	.2502	.3749
1.16	.7540	.1230	.2460	.3770
1.17	.7580	.1210	.2420	.3790
1.18	.7620	.1190	.2380	.3810
1.19	.7660	.1170	.2340	.3830
1.20	.7698	.1151	.2302	.3849
1.21	.7738	.1131	.2262	.3869
1.22	.7776	.1112	.2224	.3888
1.23	.7814	.1093	.2186	.3907
1.24	.7850	.1075	.2150	.3925
1.25	.7888	.1056	.2112	.3944
1.26	.7924	.1038	.2076	.3962
1.27	.7960	.1020	.2040	.3980
1.28	.7994	.1003	.2006	.3997
1.29	.8030	.0985	.1970	.4015
1.30	.8064	.0968	.1936	.4032
1.31	.8098	.0951	.1902	.4049
1.32	.8132	.0934	.1868	.4066
1.33	.8164	.0918	.1836	.4082
1.34	.8198	.0901	.1802	.4099

z	≥ −z and ≤ z	≥ z (also use to find ≤ −z)	≥ z and ≤ −z	≥ 0 and ≤ z (also use to find ≥ −z and ≤ 0)
1.35	.8230	.0885	.1770	.4115
1.36	.8262	.0869	.1738	.4131
1.37	.8294	.0853	.1706	.4147
1.38	.8324	.0838	.1676	.4162
1.39	.8354	.0823	.1646	.4177
1.40	.8384	.0808	.1616	.4192
1.41	.8414	.0793	.1586	.4207
1.42	.8444	.0778	.1556	.4222
1.43	.8472	.0764	.1528	.4236
1.44	.8502	.0749	.1498	.4251
1.45	.8530	.0735	.1470	.4265
1.46	.8558	.0721	.1442	.4279
1.47	.8584	.0708	.1416	.4292
1.48	.8612	.0694	.1388	.4306
1.49	.8638	.0681	.1362	.4319
1.50	.8664	.0668	.1336	.4332
1.51	.8690	.0655	.1310	.4345
1.52	.8714	.0643	.1286	.4357
1.53	.8740	.0630	.1260	.4370
1.54	.8764	.0618	.1236	.4382
1.55	.8788	.0606	.1212	.4394
1.56	.8812	.0594	.1188	.4406
1.57	.8836	.0582	.1164	.4418
1.58	.8858	.0571	.1142	.4429
1.59	.8882	.0559	.1118	.4441
1.60	.8904	.0548	.1096	.4452
1.61	.8926	.0537	.1074	.4463
1.62	.8948	.0526	.1052	.4474
1.63	.8968	.0516	.1032	.4484
1.64	.8990	.0505	.1010	.4495
1.65	.9010	.0495	.0990	.4505
1.66	.9030	.0485	.0970	.4515
1.67	.9050	.0475	.0950	.4525
1.68	.9070	.0465	.0930	.4535
1.69	.9090	.0455	.0910	.4545
1.70	.9108	.0446	.0892	.4554
1.71	.9128	.0436	.0872	.4564
1.72	.9146	.0427	.0854	.4573
1.73	.9164	.0418	.0836	.4582
1.74	.9182	.0409	.0818	.4591
1.75	.9198	.0401	.0802	.4599
1.76	.9216	.0392	.0784	.4608
1.77	.9232	.0384	.0764	.4616
1.78	.9250	.0375	.0750	.4625
1.79	.9266	.0367	.0734	.4633

z	$\geq -z$ and $\leq z$	$\geq z$ (also use to find $\leq -z$)	$\geq z$ and $\leq -z$	≥ 0 and $\leq z$ (also use to find $\geq -z$ and ≤ 0)
1.80	.9282	.0359	.0718	.4641
1.81	.9298	.0351	.0702	.4649
1.82	.9312	.0344	.0688	.4656
1.83	.9328	.0336	.0672	.4664
1.84	.9342	.0329	.0658	.4671
1.85	.9356	.0322	.0644	.4678
1.86	.9372	.0314	.0628	.4686
1.87	.9386	.0307	.0614	.4693
1.88	.9398	.0301	.0602	.4699
1.89	.9412	.0294	.0588	.4706
1.90	.9426	.0287	.0574	.4713
1.91	.9438	.0281	.0562	.4719
1.92	.9452	.0274	.0548	.4726
1.93	.9464	.0268	.0536	.4732
1.94	.9476	.0262	.0524	.4738
1.95	.9488	.0256	.0512	.4744
1.96	.9500	.0250	.0500	.4750
1.97	.9512	.0244	.0488	.4756
1.98	.9522	.0239	.0478	.4761
1.99	.9534	.0233	.0466	.4767
2.00	.9544	.0228	.0456	.4772
2.01	.9556	.0222	.0444	.4778
2.02	.9566	.0217	.0434	.4783
2.03	.9576	.0212	.0424	.4788
2.04	.9586	.0207	.0414	.4793
2.05	.9596	.0202	.0404	.4798
2.06	.9606	.0197	.0394	.4803
2.07	.9616	.0192	.0384	.4808
2.08	.9624	.0188	.0376	.4812
2.09	.9634	.0183	.0366	.4817
2.10	.9642	.0179	.0358	.4821
2.11	.9652	.0174	.0348	.4826
2.12	.9660	.0170	.0340	.4830
2.13	.9668	.0166	.0332	.4834
2.14	.9676	.0162	.0324	.4838
2.15	.9684	.0158	.0316	.4842
2.16	.9692	.0154	.0308	.4846
2.17	.9700	.0150	.0300	.4850
2.18	.9708	.0146	.0292	.4854
2.19	.9714	.0143	.0286	.4857
2.20	.9722	.0139	.0278	.4861
2.21	.9728	.0136	.0272	.4864
2.22	.9736	.0132	.0264	.4868
2.23	.9742	.0129	.0258	.4871
2.24	.9750	.0125	.0250	.4875

z	≥ −z and ≤ z	≥ z (also use to find ≤ −z)	≥ z and ≤ −z	≥ 0 and ≤ z (also use to find ≥ −z and ≤ 0)
2.25	.9756	.0122	.0244	.4878
2.26	.9762	.0119	.0238	.4881
2.27	.9768	.0116	.0232	.4884
2.28	.9774	.0113	.0276	.4887
2.29	.9780	.0110	.0220	.4890
2.30	.9786	.0107	.0214	.4893
2.31	.9792	.0104	.0208	.4896
2.32	.9796	.0102	.0204	.4898
2.33	.9802	.0099	.0198	.4901
2.34	.9808	.0096	.0192	.4904
2.35	.9812	.0094	.0188	.4906
2.36	.9818	.0091	.0182	.4909
2.37	.9822	.0089	.0178	.4911
2.38	.9826	.0087	.0174	.4913
2.39	.9832	.0084	.0168	.4916
2.40	.9836	.0082	.0164	.4918
2.41	.9840	.0080	.0160	.4920
2.42	.9844	.0078	.0156	.4922
2.43	.9850	.0075	.0150	.4925
2.44	.9854	.0073	.0146	.4927
2.45	.9858	.0071	.0142	.4929
2.46	.9862	.0069	.0138	.4931
2.47	.9864	.0068	.0136	.4932
2.48	.9868	.0066	.0132	.4934
2.49	.9872	.0064	.0128	.4936
2.50	.9876	.0062	.0124	.4938
2.51	.9880	.0060	.0120	.4940
2.52	.9882	.0059	.0118	.4941
2.53	.9886	.0057	.0114	.4943
2.54	.9890	.0055	.0110	.4945
2.55	.9892	.0054	.0108	.4946
2.56	.9896	.0052	.0104	.4948
2.57	.9898	.0051	.0102	.4949
2.58	.9902	.0048	.0098	.4951
2.59	.9904	.0049	.0096	.4952
2.60	.9906	.0047	.0094	.4953
2.61	.9910	.0045	.0090	.4955
2.62	.9912	.0044	.0088	.4956
2.63	.9914	.0043	.0086	.4957
2.64	.9918	.0041	.0082	.4959
2.65	.9920	.0040	.0080	.4960
2.66	.9922	.0039	.0078	.4961
2.67	.9924	.0038	.0076	.4962
2.68	.9926	.0037	.0074	.4963
2.69	.9928	.0036	.0072	.4964

z	$\geq -z$ and $\leq z$	$\geq z$ (also use to find $\leq -z$)	$\geq z$ and $\leq -z$	≥ 0 and $\leq z$ (also use to find $\geq -z$ and ≤ 0)
2.70	.9930	.0035	.0070	.4965
2.71	.9932	.0034	.0068	.4966
2.72	.9934	.0033	.0066	.4967
2.73	.9936	.0032	.0064	.4968
2.74	.9938	.0031	.0062	.4969
2.75	.9940	.0030	.0060	.4970
2.76	.9942	.0029	.0058	.4971
2.77	.9944	.0028	.0056	.4972
2.78	.9946	.0027	.0054	.4973
2.79	.9948	.0026	.0052	.4974
2.80	.9948	.0026	.0052	.4974
2.81	.9950	.0025	.0050	.4975
2.82	.9952	.0024	.0048	.4976
2.83	.9954	.0023	.0046	.4977
2.84	.9954	.0023	.0046	.4977
2.85	.9956	.0022	.0044	.4978
2.86	.9958	.0021	.0042	.4979
2.87	.9958	.0021	.0042	.4979
2.88	.9960	.0020	.0040	.4980
2.89	.9962	.0019	.0038	.4981
2.90	.9962	.0019	.0038	.4981
2.91	.9964	.0018	.0036	.4982
2.92	.9964	.0018	.0036	.4982
2.93	.9966	.0017	.0034	.4983
2.94	.9968	.0016	.0032	.4984
2.95	.9968	.0016	.0032	.4984
2.96	.9970	.0015	.0030	.4985
2.97	.9970	.0015	.0030	.4985
2.98	.9972	.0014	.0028	.4986
2.99	.9972	.0014	.0028	.4986
3.00	.9974	.0013	.0026	.4987
3.01	.9974	.0013	.0026	.4987
3.02	.9974	.0013	.0026	.4987
3.03	.9976	.0012	.0024	.4988
3.04	.9976	.0012	.0024	.4988
3.05	.9978	.0011	.0022	.4989
3.06	.9978	.0011	.0022	.4989
3.07	.9978	.0011	.0022	.4989
3.08	.9980	.0010	.0020	.4990
3.09	.9980	.0010	.0020	.4990
3.10	.9980	.0010	.0020	.4990
3.11	.9982	.0009	.0018	.4991
3.12	.9982	.0009	.0018	.4991
3.13	.9982	.0009	.0018	.4991
3.14	.9984	.0008	.0016	.4992

z	≥ −z and ≤ z	≥ z (also use to find ≤ −z)	≥ z and ≤ −z	≥ 0 and ≤ z (also use to find ≥ −z and ≤ 0)
3.15	.9984	.0008	.0016	.4992
3.16	.9984	.0008	.0016	.4992
3.17	.9984	.0008	.0016	.4992
3.18	.9986	.0007	.0014	.4993
3.19	.9986	.0007	.0014	.4993
3.20	.9986	.0007	.0014	.4993
3.21	.9986	.0007	.0014	.4993
3.22	.9988	.0006	.0012	.4994
3.23	.9988	.0006	.0012	.4994
3.24	.9988	.0006	.0012	.4994
3.25	.9988	.0006	.0012	.4994
3.30	.9990	.0005	.0010	.4995
3.35	.9992	.0004	.0008	.4996
3.40	.9994	.0003	.0006	.4997
3.45	.9994	.0003	.0006	.4997
3.50	.9996	.0002	.0004	.4998
3.60	.9996	.0002	.0004	.4998
3.70	.9998	.0001	.0002	.4999
3.80	.9998	.0001	.0002	.4999
3.90	.9999	.00005	.00010	.49995
4.00	.99994	.00003	.00006	.49997

Appendix D
Factorials

This table presents the values of $n!$ for numbers from 0 to 20. For example, the value of $7!$ is 5,040.

n	$n!$
0	1
1	1
2	2
3	6
4	24
5	120
6	720
7	5040
8	40320
9	362880
10	3628800
11	39916800
12	479001600
13	6227020800
14	87178291200
15	1307674368000
16	20922789888000
17	355687428096000
18	6402373705728000
19	121645100408832000
20	2432902008176640000

Appendix E
Critical Values for the
t Distribution

The following table presents critical values of *t* for one-tailed and two-tailed tests. The first column is the degrees of freedom associated with the *t* distribution of interest. The entries in the table are the *t* values that define the rejection regions. The top row with the numbers .10, .05, .025, .01, .005, and .0005 represents alpha levels for one-tailed tests. If you were conducting a one-tailed test with alpha = .05 and with 15 degrees of freedom, you would locate the column indicated by .05 and follow down that column until it intersected at 15 in the df column. The entry is 1.753 and this would be your critical *t*.

The row at the top with the numbers .20, .10, .05, .02, .01, and .001 represents alpha levels for two-tailed tests. If you were conducting a two-tailed test with alpha = .05, and with 15 degrees of freedom, you would locate the column indicated by .05 and follow down that column until it intersected at 15 in the df column. The entry is 2.131 and this would define your rejection region.

Source: Table is taken from Table III of Fisher & Yates: *Statistical Tables for Biological, Agricultural and Medical Research*, published by Longman Group Ltd., London (previously published by Oliver & Boyd Ltd., Edinburgh) and by permission of the authors and publishers.

df	Level of significance for one-tailed test					
	.10	.05	.025	.01	.005	.0005
	Level of significance for two-tailed test					
	.20	.10	.05	.02	.01	.001
1	3.078	6.314	12.706	31.821	63.657	636.619
2	1.886	2.920	4.303	6.965	9.925	31.598
3	1.638	2.353	3.182	4.541	5.841	12.941
4	1.533	2.132	2.776	3.747	4.604	8.610
5	1.476	2.015	2.571	3.365	4.032	6.859
6	1.440	1.943	2.447	3.143	3.707	5.959
7	1.415	1.895	2.365	2.998	3.499	5.405
8	1.397	1.860	2.306	2.896	3.355	5.041
9	1.383	1.833	2.262	2.821	3.250	4.781
10	1.372	1.812	2.228	2.764	3.169	4.587
11	1.363	1.796	2.201	2.718	3.106	4.437
12	1.356	1.782	2.179	2.681	3.055	4.318
13	1.350	1.771	2.160	2.650	3.012	4.221
14	1.345	1.761	2.145	2.624	2.977	4.140
15	1.341	1.753	2.131	2.602	2.947	4.073
16	1.337	1.746	2.120	2.583	2.921	4.015
17	1.333	1.740	2.110	2.567	2.898	3.965
18	1.330	1.734	2.101	2.552	2.878	3.922
19	1.328	1.729	2.093	2.539	2.861	3.883
20	1.325	1.725	2.086	2.528	2.845	3.850
21	1.323	1.721	2.080	2.518	2.831	3.819
22	1.321	1.717	2.074	2.508	2.819	3.792
23	1.319	1.714	2.069	2.500	2.807	3.767
24	1.318	1.711	2.064	2.492	2.797	3.745
25	1.316	1.708	2.060	2.485	2.787	3.725
26	1.315	1.706	2.056	2.479	2.779	3.707
27	1.314	1.703	2.052	2.473	2.771	3.690
28	1.313	1.701	2.048	2.467	2.763	3.674
29	1.311	1.699	2.045	2.462	2.756	3.659
30	1.310	1.697	2.042	2.457	2.750	3.646
40	1.303	1.684	2.021	2.423	2.704	3.551
60	1.296	1.671	2.000	2.390	2.660	3.460
120	1.289	1.658	1.980	2.358	2.617	3.373
∞	1.282	1.645	1.960	2.326	2.576	3.291

Appendix F
Power and Sample Size

The following five sets of tables represent sample size tables for purposes of planning investigations in order to achieve desired levels of power. Procedures for using the tables are described in the text for each major statistical technique.

APPENDIX F.1: *t* Test

DIRECTIONAL TEST, ALPHA = 0.05

Power	0.01	0.03	0.05	0.07	0.10	0.15	0.20	0.25	0.30	0.35	0.40	0.45	0.50	0.55	0.60	0.65	0.70	0.75	0.80
															Population Eta Squared				
0.25	48	16	10	7	5	3	3	2	2	2	—	—	—	—	—	—	—	—	—
0.50	136	45	26	19	13	8	6	5	4	3	3	2	2	2	2	—	—	—	—
0.60	181	59	35	25	17	11	8	6	5	4	3	3	3	2	2	2	—	—	—
0.67	216	70	42	29	20	13	9	7	6	5	4	3	3	2	2	2	2	—	—
0.70	236	77	45	32	22	14	10	8	6	5	4	4	3	3	2	2	2	—	—
0.75	270	88	52	36	25	16	11	9	7	6	5	4	3	3	2	2	2	2	—
0.80	310	101	59	42	29	18	13	10	8	6	5	4	4	3	3	2	2	2	—
0.85	360	117	69	48	33	21	15	11	9	7	6	5	4	4	3	3	2	2	2
0.90	429	139	82	58	39	25	18	14	11	9	7	6	5	4	4	3	3	2	2
0.95	542	176	104	73	49	31	22	17	13	11	9	7	6	5	4	4	3	3	2
0.99	789	256	151	105	72	45	32	24	19	15	13	10	9	7	6	5	4	3	3

DIRECTIONAL TEST, ALPHA = 0.01

Power	0.01	0.03	0.05	0.07	0.10	0.15	0.20	0.25	0.30	0.35	0.40	0.45	0.50	0.55	0.60	0.65	0.70	0.75	0.80
															Population Eta Squared				
0.25	138	46	27	20	14	9	7	6	5	4	4	3	3	3	2	2	2	2	2
0.50	272	89	53	37	26	17	12	10	8	7	6	5	4	4	3	3	3	2	2
0.60	334	109	65	46	31	20	15	11	9	8	6	6	5	4	4	3	3	3	2
0.67	382	127	75	53	36	23	17	13	11	9	7	6	5	5	4	4	3	3	2
0.70	408	133	79	56	38	25	18	14	11	9	8	6	6	5	4	4	3	3	2
0.75	452	147	87	61	42	27	20	15	12	10	8	7	6	5	5	4	3	3	3
0.80	503	164	97	68	47	30	22	17	13	11	9	8	7	6	5	5	4	3	3
0.85	567	184	109	77	52	34	24	18	15	12	10	8	7	6	5	5	4	3	3
0.90	652	212	125	88	60	38	27	21	17	14	11	9	8	7	6	5	4	4	3
0.95	790	257	151	106	72	46	33	25	20	16	13	11	9	8	7	6	5	4	3
0.99	1084	352	207	145	99	63	45	34	27	22	18	15	12	10	9	7	6	5	4

NONDIRECTIONAL TEST, ALPHA = 0.05

| Power | Population Eta Squared | | | | | | | | | | | | | | | | | | |
	0.01	0.03	0.05	0.07	0.10	0.15	0.20	0.25	0.30	0.35	0.40	0.45	0.50	0.55	0.60	0.65	0.70	0.75	0.80
0.25	84	28	17	12	8	6	5	3	3	3	2	2	2	2	2	—	—	—	—
0.50	193	63	38	27	18	12	9	7	5	5	4	3	3	3	2	2	2	2	—
0.60	246	80	48	34	23	15	11	8	7	6	5	4	3	3	3	2	2	2	2
0.67	287	93	55	39	27	17	12	10	8	6	5	4	4	3	3	3	2	2	2
0.70	310	101	60	42	29	18	13	10	8	7	6	5	4	4	3	3	2	2	2
0.75	348	113	67	47	32	21	15	11	9	7	6	5	4	4	3	3	2	2	2
0.80	393	128	76	53	36	23	17	13	10	8	7	6	5	4	4	3	3	2	2
0.85	450	146	86	61	41	26	19	14	11	9	8	6	5	5	4	3	3	2	2
0.90	526	171	101	71	48	31	22	17	13	11	9	7	6	5	5	4	3	3	3
0.95	651	211	125	87	60	38	27	21	16	13	11	9	8	6	5	5	4	3	3
0.99	920	298	176	123	84	53	38	29	22	18	15	12	10	9	7	6	5	4	3

NONDIRECTIONAL TEST, ALPHA = 0.01

| Power | Population Eta Squared | | | | | | | | | | | | | | | | | | |
	0.01	0.03	0.05	0.07	0.10	0.15	0.20	0.25	0.30	0.35	0.40	0.45	0.50	0.55	0.60	0.65	0.70	0.75	0.80
0.25	183	60	36	26	18	12	9	7	6	5	4	4	3	3	3	2	2	2	2
0.50	333	109	65	46	31	20	15	11	9	8	6	6	5	4	4	3	3	3	2
0.60	402	131	78	55	38	24	18	14	11	9	8	6	6	5	4	4	3	3	3
0.67	454	148	87	62	42	27	20	15	12	10	8	7	6	5	5	4	3	3	3
0.70	482	157	93	65	45	29	21	16	13	10	9	7	6	5	4	4	3	3	3
0.75	528	172	102	72	49	31	23	17	14	11	9	8	7	6	5	4	4	3	3
0.80	586	190	113	79	54	35	25	19	15	12	10	9	7	6	5	4	4	3	3
0.85	654	213	126	88	60	38	28	21	17	14	11	9	8	7	6	5	4	4	3
0.90	746	242	143	100	69	44	31	24	19	15	13	11	9	8	6	5	5	4	3
0.95	892	290	171	120	82	52	37	28	22	18	15	12	10	9	7	6	5	4	4
0.99	1203	390	230	161	110	70	50	38	30	24	20	16	14	11	10	8	7	6	5

DEGREES OF FREEDOM = 1   ALPHA = 0.05

Power	\multicolumn{19}{c}{Population Eta Squared}
	0.01	0.03	0.05	0.07	0.10	0.15	0.20	0.25	0.30	0.35	0.40	0.45	0.50	0.55	0.60	0.65	0.70	0.75	0.80
0.10	22	8	5	4	3	2	2	2	—	—	—	—	—	—	—	—	—	—	—
0.50	193	63	38	27	18	12	9	7	5	5	4	3	3	3	2	2	2	2	—
0.70	310	101	60	42	29	18	13	10	8	7	6	5	4	4	3	3	2	2	2
0.80	393	128	76	53	36	23	17	13	10	8	7	6	5	4	4	3	3	2	2
0.90	526	171	101	71	48	31	22	17	13	11	9	7	6	5	4	4	3	3	2
0.95	651	211	125	87	60	38	27	21	16	13	11	9	8	6	5	5	4	3	3
0.99	920	298	176	123	84	53	38	29	22	18	15	12	10	9	7	6	5	4	3

DEGREES OF FREEDOM = 2   ALPHA = 0.05

Power	\multicolumn{19}{c}{Population Eta Squared}
	0.01	0.03	0.05	0.07	0.10	0.15	0.20	0.25	0.30	0.35	0.40	0.45	0.50	0.55	0.60	0.65	0.70	0.75	0.80
0.10	22	8	5	4	3	2	2	2	—	—	—	—	—	—	—	—	—	—	—
0.50	165	55	32	23	16	10	8	6	5	4	3	3	3	2	2	2	2	2	—
0.70	255	84	50	35	24	16	11	9	7	6	5	4	4	3	3	2	2	2	2
0.80	319	105	62	44	30	19	14	11	9	7	6	5	4	4	3	3	2	2	2
0.90	417	137	81	57	39	25	18	14	11	9	7	6	5	4	4	3	3	2	2
0.95	511	168	99	69	47	30	22	16	13	11	9	7	6	5	4	4	3	3	2
0.99	708	232	137	96	65	41	29	22	18	14	12	10	8	7	6	5	4	3	3

| | | | | | | | | | Population Eta Squared | | | | | | | | | | |
Power	0.01	0.03	0.05	0.07	0.10	0.15	0.20	0.25	0.30	0.35	0.40	0.45	0.50	0.55	0.60	0.65	0.70	0.75	0.80
0.10	21	7	5	4	3	2	2	2	—	—	—	—	—	—	—	—	—	—	—
0.50	144	48	28	20	14	9	7	5	4	4	3	3	2	2	2	2	2	—	—
0.70	219	72	43	30	21	13	10	8	6	5	4	4	3	3	2	2	2	2	2
0.80	272	90	53	37	26	17	12	9	7	6	5	4	4	3	3	2	2	2	2
0.90	351	115	68	48	33	21	15	12	9	8	6	5	5	4	3	3	3	2	2
0.95	426	140	83	58	40	25	18	14	11	9	7	6	5	5	4	3	3	2	2
0.99	583	191	113	79	54	34	24	19	15	12	10	8	7	6	5	4	4	3	2

| | | | | | | | | | Population Eta Squared | | | | | | | | | | |
Power	0.01	0.03	0.05	0.07	0.10	0.15	0.20	0.25	0.30	0.35	0.40	0.45	0.50	0.55	0.60	0.65	0.70	0.75	0.80
0.10	19	7	5	3	3	2	—	2	—	—	—	—	—	—	—	—	—	—	—
0.50	128	43	25	18	13	8	6	5	4	3	3	3	2	2	2	2	2	—	—
0.70	193	64	38	27	18	12	9	7	6	5	4	3	3	3	2	2	2	2	—
0.80	238	78	46	33	23	15	10	8	7	5	5	4	3	3	3	2	2	2	2
0.90	306	101	59	42	29	18	13	10	8	7	6	5	4	4	3	3	2	2	2
0.95	369	121	72	50	34	22	16	12	10	8	7	6	5	4	3	3	3	2	2
0.99	501	164	97	68	46	30	21	16	13	10	9	7	6	5	4	4	3	3	2

DEGREES OF FREEDOM = 5   ALPHA = 0.05

									Population Eta Squared										
Power	0.01	0.03	0.05	0.07	0.10	0.15	0.20	0.25	0.30	0.35	0.40	0.45	0.50	0.55	0.60	0.65	0.70	0.75	0.80
0.10	18	7	4	3	3	2	—	—	—	—	—	—	—	—	—	—	—	—	—
0.50	117	39	23	17	12	8	6	5	4	3	3	2	2	2	2	2	2	—	—
0.70	174	57	34	24	17	11	8	6	5	4	4	3	3	2	2	2	2	2	—
0.80	213	70	42	29	20	13	9	7	6	5	4	4	3	3	2	2	2	2	2
0.90	273	90	53	37	26	17	12	9	7	6	5	4	4	3	3	2	2	2	2
0.95	328	108	64	45	31	20	14	11	9	7	6	5	4	4	3	3	2	2	2
0.99	442	145	86	60	41	26	19	14	11	9	8	6	5	5	4	3	3	2	2

DEGREES OF FREEDOM = 6   ALPHA = 0.05

									Population Eta Squared										
Power	0.01	0.03	0.05	0.07	0.10	0.15	0.20	0.25	0.30	0.35	0.40	0.45	0.50	0.55	0.60	0.65	0.70	0.75	0.80
0.10	17	6	4	3	2	2	—	—	—	—	—	—	—	—	—	—	—	—	—
0.50	107	36	21	15	11	7	5	4	4	3	3	2	2	2	2	2	—	—	—
0.70	159	53	31	22	15	10	7	6	5	4	3	3	3	2	2	2	2	2	—
0.80	194	64	38	27	19	12	9	7	6	5	4	3	3	3	2	2	2	2	—
0.90	247	81	48	34	23	15	11	8	7	6	5	4	3	3	3	2	2	2	2
0.95	296	97	58	41	28	18	13	10	8	7	5	5	4	3	3	3	2	2	2
0.99	398	131	77	54	37	24	17	13	10	8	7	6	5	4	4	3	3	2	2

DEGREES OF FREEDOM = 8   ALPHA = 0.05

Population Eta Squared

Power	0.01	0.03	0.05	0.07	0.10	0.15	0.20	0.25	0.30	0.35	0.40	0.45	0.50	0.55	0.60	0.65	0.70	0.75	0.80
0.10	16	6	4	3	2	2	—	—	—	—	—	—	—	—	—	—	—	—	—
0.50	94	31	19	13	9	6	5	4	3	3	2	2	2	2	2	2	—	—	—
0.70	137	45	27	19	13	9	6	5	4	4	3	3	2	2	2	2	2	2	—
0.80	167	55	33	23	16	10	8	6	5	4	4	3	3	2	2	2	2	2	2
0.90	211	70	41	29	20	13	9	7	6	5	4	4	3	3	2	2	2	2	2
0.95	251	83	49	35	24	15	11	9	7	6	5	4	4	3	3	2	2	2	2
0.99	335	110	65	46	31	20	14	11	9	7	6	5	4	4	3	3	2	2	2

DEGREES OF FREEDOM = 10   ALPHA = 0.05

Population Eta Squared

Power	0.01	0.03	0.05	0.07	0.10	0.15	0.20	0.25	0.30	0.35	0.40	0.45	0.50	0.55	0.60	0.65	0.70	0.75	0.80
0.10	15	5	4	3	2	2	—	—	—	—	—	—	—	—	—	—	—	—	—
0.50	84	28	17	12	9	6	4	4	3	3	2	2	2	2	2	—	—	—	—
0.70	122	40	24	17	12	8	6	5	4	3	3	2	2	2	2	2	2	—	—
0.80	147	49	29	21	14	9	7	5	4	4	3	3	2	2	2	2	2	—	—
0.90	186	61	36	26	18	12	8	7	5	4	4	3	3	3	2	2	2	2	2
0.95	221	73	43	30	21	14	10	8	6	5	4	4	3	3	2	2	2	2	2
0.99	292	96	57	40	27	18	13	10	8	6	5	5	4	3	3	3	2	2	2

DEGREES OF FREEDOM = 12   ALPHA = 0.05

Population Eta Squared

Power	0.01	0.03	0.05	0.07	0.10	0.15	0.20	0.25	0.30	0.35	0.40	0.45	0.50	0.55	0.60	0.65	0.70	0.75	0.80
0.10	14	5	3	3	2	2	—	—	—	—	—	—	—	—	—	—	—	—	—
0.50	77	26	16	11	8	5	4	3	3	2	2	2	2	2	2	—	—	—	—
0.70	111	37	22	16	11	7	5	4	4	3	3	2	2	2	2	2	—	—	—
0.80	133	44	26	19	13	9	6	5	4	3	3	3	2	2	2	2	2	—	—
0.90	168	55	33	23	16	11	8	6	5	4	4	3	3	2	2	2	2	2	—
0.95	198	65	39	27	19	12	9	7	6	5	4	3	3	3	2	2	2	2	—
0.99	261	86	51	36	25	16	11	9	7	6	5	4	4	3	3	2	2	2	2

DEGREES OF FREEDOM = 15   ALPHA = 0.05

Population Eta Squared

Power	0.01	0.03	0.05	0.07	0.10	0.15	0.20	0.25	0.30	0.35	0.40	0.45	0.50	0.55	0.60	0.65	0.70	0.75	0.80
0.10	13	5	3	3	2	2	—	—	—	—	—	—	—	—	—	—	—	—	—
0.50	68	23	14	10	7	5	4	3	3	2	2	2	2	2	—	—	—	—	—
0.70	98	33	20	14	10	7	5	4	3	3	2	2	2	2	2	2	—	—	—
0.80	118	39	23	17	12	8	6	5	4	3	3	2	2	2	2	2	2	—	—
0.90	147	49	29	21	14	9	7	5	4	4	3	3	2	2	2	2	2	—	—
0.95	174	57	34	24	17	11	8	6	5	4	4	3	3	2	2	2	2	2	—
0.99	227	75	44	31	22	14	10	8	6	5	5	4	3	3	3	2	2	2	2

DEGREES OF FREEDOM = 1   ALPHA = 0.01

Population Eta Squared

Power	0.01	0.03	0.05	0.07	0.10	0.15	0.20	0.25	0.30	0.35	0.40	0.45	0.50	0.55	0.60	0.65	0.70	0.75	0.80
0.10	84	28	17	12	9	6	5	4	3	3	2	2	2	2	2	—	—	—	—
0.50	333	109	65	46	31	20	15	11	9	8	6	6	5	4	4	3	3	3	2
0.70	482	157	93	65	45	29	21	16	13	10	9	7	6	5	5	4	4	3	3
0.80	586	190	113	79	54	35	25	19	15	12	10	9	7	6	5	5	4	3	3
0.90	746	242	143	100	69	44	31	24	19	15	13	11	9	8	6	6	5	4	3
0.95	892	289	171	120	82	52	37	28	22	18	15	12	10	9	7	6	5	4	4
0.99	1203	390	230	161	110	70	50	38	30	24	20	16	14	11	10	8	7	6	5

DEGREES OF FREEDOM = 2   ALPHA = 0.01

Population Eta Squared

Power	0.01	0.03	0.05	0.07	0.10	0.15	0.20	0.25	0.30	0.35	0.40	0.45	0.50	0.55	0.60	0.65	0.70	0.75	0.80
0.10	77	26	16	11	8	5	5	3	3	2	2	2	2	2	2	—	—	—	—
0.50	272	89	53	37	26	16	13	9	7	6	5	4	4	3	3	2	2	2	2
0.70	383	126	74	52	36	23	17	13	10	8	7	6	5	4	4	3	3	2	2
0.80	459	151	89	62	43	27	20	15	12	10	8	7	6	5	4	3	3	3	2
0.90	576	189	111	78	53	34	25	18	15	12	10	8	7	6	5	4	3	3	2
0.95	683	224	132	93	63	40	29	22	17	14	11	9	8	7	6	5	4	3	3
0.99	906	297	175	122	83	53	38	28	22	18	15	12	10	8	7	6	5	4	3

DEGREES OF FREEDOM = 3   ALPHA = 0.01

	Population Eta Squared																		
Power	0.01	0.03	0.05	0.07	0.10	0.15	0.20	0.25	0.30	0.35	0.40	0.45	0.50	0.55	0.60	0.65	0.70	0.75	0.80
0.10	70	23	14	10	7	5	4	3	3	2	2	2	2	2	—	—	—	—	—
0.50	232	76	45	32	22	14	11	8	6	5	4	4	3	3	3	2	2	—	2
0.70	323	106	63	44	30	19	14	11	9	7	6	5	4	4	3	3	2	2	2
0.80	384	126	75	52	36	23	17	13	10	8	7	6	5	4	4	3	3	2	2
0.90	478	157	93	65	44	28	21	15	12	10	8	7	6	5	4	4	3	3	2
0.95	563	184	109	76	52	33	24	18	14	12	10	8	7	6	5	4	3	3	2
0.99	740	242	143	100	68	43	31	23	18	15	12	10	8	7	6	5	4	3	3

DEGREES OF FREEDOM = 4   ALPHA = 0.01

	Population Eta Squared																		
Power	0.01	0.03	0.05	0.07	0.10	0.15	0.20	0.25	0.30	0.35	0.40	0.45	0.50	0.55	0.60	0.65	0.70	0.75	0.80
0.10	64	21	13	9	7	5	4	3	2	2	2	2	2	2	—	—	—	—	—
0.50	204	67	40	28	19	13	10	7	6	5	4	4	3	3	2	2	2	2	2
0.70	280	92	55	38	26	17	13	9	8	6	5	4	4	3	3	3	2	2	2
0.80	333	109	65	46	31	20	15	11	9	7	6	5	4	4	3	3	2	2	2
0.90	412	135	80	56	38	25	18	13	11	9	7	6	5	4	4	3	3	2	2
0.95	483	158	94	66	45	29	21	16	12	10	8	7	6	5	4	4	3	3	2
0.99	631	207	122	86	58	37	27	20	16	13	11	9	7	6	5	4	4	3	3

DEGREES OF FREEDOM = 5   ALPHA = 0.01

Population Eta Squared

Power	0.01	0.03	0.05	0.07	0.10	0.15	0.20	0.25	0.30	0.35	0.40	0.45	0.50	0.55	0.60	0.65	0.70	0.75	0.80
0.10	59	20	12	9	6	4	4	3	2	2	2	2	2	—	—	—	—	—	—
0.50	183	61	36	25	18	11	9	7	5	4	4	3	3	3	2	2	2	2	—
0.70	251	83	49	35	24	15	11	9	7	6	5	4	4	3	3	2	2	2	2
0.80	296	97	58	41	28	18	13	10	8	7	5	5	4	3	3	3	2	2	2
0.90	365	120	71	50	34	22	16	12	10	8	7	5	5	4	3	3	3	2	2
0.95	426	140	83	58	40	25	18	14	11	9	7	6	5	5	4	3	3	2	2
0.99	554	182	107	75	51	33	24	18	14	11	9	8	7	6	5	4	3	3	2

DEGREES OF FREEDOM = 6   ALPHA = 0.01

Population Eta Squared

Power	0.01	0.03	0.05	0.07	0.10	0.15	0.20	0.25	0.30	0.35	0.40	0.45	0.50	0.55	0.60	0.65	0.70	0.75	0.80
0.10	55	19	11	8	6	4	3	3	2	2	2	2	2	—	—	—	—	—	—
0.50	168	55	33	23	16	11	8	6	5	4	4	3	3	2	2	2	2	2	—
0.70	228	75	45	31	22	14	10	8	6	5	4	4	3	3	3	2	2	2	2
0.80	268	88	52	37	25	16	12	9	7	6	5	4	4	3	3	2	2	2	2
0.90	329	108	64	45	31	20	14	11	9	7	6	5	4	4	3	3	2	2	2
0.95	384	126	74	52	36	23	17	13	10	8	7	6	5	4	4	3	3	2	2
0.99	497	163	96	68	46	29	21	16	13	10	9	7	6	5	4	4	3	3	2

DEGREES OF FREEDOM = 8   ALPHA = 0.01

									Population Eta Squared									
Power	0.01	0.03	0.05	0.07	0.10	0.15	0.20	0.25	0.30	0.35	0.40	0.45	0.50	0.55	0.60	0.65	0.70	0.75	0.80
0.10	49	17	10	7	5	4	3	2	2	2	2	2	—	—	—	—	—	—	—
0.50	145	48	29	20	14	9	7	5	4	4	3	3	2	2	2	2	2	—	—
0.70	195	64	38	27	19	12	9	7	6	5	4	3	3	3	2	2	2	2	—
0.80	228	75	45	31	22	14	10	8	6	5	4	4	3	3	3	2	2	2	2
0.90	279	92	54	38	26	17	12	9	8	6	5	4	4	3	3	3	2	2	2
0.95	323	106	63	44	30	19	14	11	9	7	6	5	4	4	3	3	3	2	2
0.99	416	136	81	57	39	25	18	14	11	9	7	6	5	4	4	3	3	2	2

DEGREES OF FREEDOM = 10   ALPHA = 0.01

									Population Eta Squared									
Power	0.01	0.03	0.05	0.07	0.10	0.15	0.20	0.25	0.30	0.35	0.40	0.45	0.50	0.55	0.60	0.65	0.70	0.75	0.80
0.10	45	15	9	7	5	3	3	2	2	2	2	2	—	—	—	—	—	—	—
0.50	128	43	25	18	13	8	6	5	4	3	3	3	2	2	2	2	2	—	—
0.70	172	57	34	24	17	11	8	6	5	4	4	3	3	2	2	2	2	2	—
0.80	201	66	39	28	19	12	9	7	6	5	4	3	3	3	2	2	2	2	2
0.90	244	80	48	34	23	15	11	8	7	6	5	4	3	3	2	2	2	2	2
0.95	283	93	55	39	27	17	12	10	8	6	5	4	4	3	3	3	2	2	2
0.99	361	119	70	49	34	22	16	12	9	8	6	5	5	4	3	3	3	2	2

DEGREES OF FREEDOM = 12   ALPHA = 0.01

Population Eta Squared

Power	0.01	0.03	0.05	0.07	0.10	0.15	0.20	0.25	0.30	0.35	0.40	0.45	0.50	0.55	0.60	0.65	0.70	0.75	0.80
0.10	41	14	9	6	5	3	3	2	2	2	2	—	—	—	—	—	—	—	—
0.50	117	39	23	17	12	8	6	5	4	3	3	2	2	2	2	2	2	—	—
0.70	155	51	31	22	15	10	7	6	5	4	3	3	3	2	2	2	2	2	—
0.80	181	60	35	25	17	11	8	6	5	4	4	3	3	2	2	2	2	2	2
0.90	219	72	43	30	21	13	10	8	6	5	4	4	3	3	2	2	2	2	2
0.95	253	83	49	35	24	15	11	9	7	6	5	4	4	3	3	2	2	2	2
0.99	322	106	63	44	30	19	14	11	9	7	6	5	4	4	3	3	2	2	2

DEGREES OF FREEDOM = 15   ALPHA = 0.01

Population Eta Squared

Power	0.01	0.03	0.05	0.07	0.10	0.15	0.20	0.25	0.30	0.35	0.40	0.45	0.50	0.55	0.60	0.65	0.70	0.75	0.80
0.10	37	13	8	6	4	3	2	2	2	2	2	—	—	—	—	—	—	—	—
0.50	103	34	21	15	10	7	5	4	3	3	3	2	2	2	2	2	—	—	—
0.70	137	45	27	19	13	9	6	5	4	4	3	3	2	2	2	2	2	—	—
0.80	157	52	31	22	15	10	7	6	5	4	3	3	3	2	2	2	2	2	—
0.90	191	63	38	27	18	12	9	7	5	5	4	3	3	3	2	2	2	2	2
0.95	220	73	43	30	21	14	10	8	6	5	4	4	3	3	2	2	2	2	2
0.99	279	92	54	38	26	17	12	9	8	6	5	4	4	3	3	3	2	2	2

DEGREES OF FREEDOM = 2   ALPHA LEVEL = 0.05

Population Eta Squared

Power	0.01	0.03	0.05	0.07	0.10	0.15	0.20	0.25	0.30	0.35	0.40	0.45	0.50	0.55	0.60	0.65	0.70	0.75	0.80
0.10	32	11	7	5	4	3	2	2	2	2	—	—	—	—	—	—	—	—	—
0.50	247	81	48	34	23	15	11	8	7	6	5	4	3	3	3	2	2	2	2
0.70	382	125	74	52	36	23	16	13	10	8	7	6	5	4	4	3	3	2	2
0.80	478	157	93	65	44	28	20	15	12	10	8	7	6	5	4	4	3	3	2
0.90	627	206	121	85	58	37	26	20	16	13	10	9	7	6	5	4	4	3	3
0.95	765	251	148	104	70	45	32	24	19	15	13	10	9	7	6	5	4	4	3
0.99	1060	347	204	143	97	62	44	33	26	21	17	14	12	10	8	7	6	5	4

DEGREES OF FREEDOM = 3   ALPHA LEVEL = 0.05

Population Eta Squared

Power	0.01	0.03	0.05	0.07	0.10	0.15	0.20	0.25	0.30	0.35	0.40	0.45	0.50	0.55	0.60	0.65	0.70	0.75	0.80
0.10	27	9	6	4	3	2	2	2	2	2	—	—	—	—	—	—	—	—	—
0.50	191	63	37	27	18	12	9	7	5	5	4	3	3	3	2	2	2	2	—
0.70	291	96	57	40	27	18	13	10	8	6	5	5	4	3	3	3	2	2	2
0.80	361	118	70	49	34	22	16	12	9	8	6	5	5	4	3	3	3	2	2
0.90	469	154	91	64	44	28	20	15	12	10	8	7	6	5	4	4	3	3	2
0.95	568	186	110	77	53	33	24	18	14	12	10	8	7	6	5	4	3	3	2
0.99	777	254	150	105	72	45	32	25	19	16	13	11	9	7	6	5	4	4	3

DEGREES OF FREEDOM = 4   ALPHA LEVEL = 0.05

												Population Eta Squared							
Power	0.01	0.03	0.05	0.07	0.10	0.15	0.20	0.25	0.30	0.35	0.40	0.45	0.50	0.55	0.60	0.65	0.70	0.75	0.80
0.10	24	8	5	4	3	2	2	2	2	—	—	—	—	—	—	—	—	—	—
0.50	160	53	31	22	15	10	7	6	5	4	3	3	3	2	2	2	2	2	—
0.70	241	79	47	33	23	15	11	8	7	5	5	4	3	3	3	2	2	2	2
0.80	297	98	58	41	28	18	13	10	8	7	5	5	4	3	3	3	2	2	2
0.90	382	126	74	52	36	23	16	13	10	8	7	6	5	4	4	3	3	2	2
0.95	461	151	89	63	43	27	20	15	12	10	8	7	6	5	4	3	3	3	2
0.99	626	205	121	85	58	37	26	20	16	13	10	9	7	6	5	4	4	3	3

DEGREES OF FREEDOM = 5   ALPHA LEVEL = 0.05

												Population Eta Squared							
Power	0.01	0.03	0.05	0.07	0.10	0.15	0.20	0.25	0.30	0.35	0.40	0.45	0.50	0.55	0.60	0.65	0.70	0.75	0.80
0.10	21	8	5	4	3	2	2	2	—	—	—	—	—	—	—	—	—	—	—
0.50	139	46	28	20	14	9	7	5	4	4	3	3	2	2	2	2	2	—	—
0.70	208	69	41	29	20	13	9	7	6	5	4	4	3	3	2	2	2	2	2
0.80	225	84	50	35	24	16	11	9	7	6	5	4	4	3	3	2	2	2	2
0.90	327	108	64	45	31	20	14	11	9	7	6	5	4	4	3	3	2	2	2
0.95	393	129	76	54	37	23	17	13	10	8	7	6	5	4	4	3	3	2	2
0.99	530	174	103	72	49	31	22	17	13	11	9	8	6	5	5	4	3	3	2

DEGREES OF FREEDOM = 6   ALPHA LEVEL = 0.05

Population Eta Squared

Power	0.01	0.03	0.05	0.07	0.10	0.15	0.20	0.25	0.30	0.35	0.40	0.45	0.50	0.55	0.60	0.65	0.70	0.75	0.80
0.10	20	7	5	4	3	2	2	2	—	—	—	—	—	—	—	—	—	—	—
0.50	125	41	25	18	12	8	6	5	4	3	3	3	2	2	2	2	2	—	—
0.70	185	61	36	26	18	12	8	7	5	4	4	3	3	3	2	2	2	2	—
0.80	226	74	44	31	21	14	10	8	6	5	4	4	3	3	3	2	2	2	2
0.90	288	95	56	40	27	17	13	10	8	6	5	5	4	3	3	3	2	2	2
0.95	345	113	67	47	32	21	15	11	9	7	6	5	4	4	3	3	2	2	2
0.99	464	152	90	63	43	27	20	15	12	10	8	7	6	5	4	4	3	3	2

DEGREES OF FREEDOM = 2   ALPHA LEVEL = 0.01

Population Eta Squared

Power	0.01	0.03	0.05	0.07	0.10	0.15	0.20	0.25	0.30	0.35	0.40	0.45	0.50	0.55	0.60	0.65	0.70	0.75	0.80
0.10	115	38	23	16	11	8	6	4	4	3	3	2	2	2	2	2	—	—	—
0.50	406	133	79	55	38	24	17	13	11	9	7	6	5	4	4	3	3	2	2
0.70	574	188	111	78	53	34	24	18	14	12	10	8	7	6	5	4	3	3	2
0.80	688	225	133	93	63	40	29	22	17	14	11	9	8	7	6	5	4	3	3
0.90	864	283	167	117	79	50	36	27	21	17	14	12	10	8	7	6	5	4	3
0.95	1023	335	197	138	94	60	42	32	25	20	16	14	11	9	8	7	5	4	4
0.99	1358	444	261	183	124	79	56	42	33	26	22	18	15	12	10	8	7	6	4

DEGREES OF FREEDOM = 3   ALPHA LEVEL = 0.01

Power	Population Eta Squared																		
	0.01	0.03	0.05	0.07	0.10	0.15	0.20	0.25	0.30	0.35	0.40	0.45	0.50	0.55	0.60	0.65	0.70	0.75	0.80
0.10	92	31	18	13	9	6	5	4	3	3	2	2	2	2	2	—	—	—	—
0.50	308	101	60	42	29	19	13	10	8	7	6	5	4	4	3	3	2	2	2
0.70	429	141	83	58	40	25	18	14	11	9	7	6	5	5	4	3	3	2	2
0.80	511	168	99	69	47	30	22	16	13	11	9	7	6	5	4	4	3	3	2
0.90	636	208	123	86	59	37	27	20	16	13	11	9	7	6	5	4	4	3	3
0.95	749	245	145	101	69	44	31	24	19	15	12	10	9	7	6	5	4	4	3
0.99	985	323	190	133	90	57	41	31	24	19	16	13	11	9	8	6	5	4	3

DEGREES OF FREEDOM = 4   ALPHA LEVEL = 0.01

Power	Population Eta Squared																		
	0.01	0.03	0.05	0.07	0.10	0.15	0.20	0.25	0.30	0.35	0.40	0.45	0.50	0.55	0.60	0.65	0.70	0.75	0.80
0.10	79	26	16	11	8	5	4	3	3	2	2	2	2	2	2	—	—	—	—
0.50	254	84	50	35	24	15	11	9	7	6	5	4	4	3	3	2	2	2	2
0.70	350	115	68	48	33	21	15	12	9	8	6	5	5	4	3	3	3	2	2
0.80	416	136	81	57	39	25	18	14	11	9	7	6	5	4	4	3	3	2	2
0.90	514	169	100	70	48	30	22	17	13	11	9	7	6	5	4	4	3	3	2
0.95	603	198	117	82	56	35	25	19	15	12	10	8	7	6	5	4	4	3	3
0.99	788	258	152	107	73	46	33	25	20	16	13	11	9	8	6	5	4	4	3

DEGREES OF FREEDOM = 5 ALPHA LEVEL = 0.01

Population Eta Squared

Power	0.01	0.03	0.05	0.07	0.10	0.15	0.20	0.25	0.30	0.35	0.40	0.45	0.50	0.55	0.60	0.65	0.70	0.75	0.80
0.10	70	24	14	10	7	5	4	3	3	2	2	2	2	2	—	—	—	—	—
0.50	219	72	43	30	21	14	10	8	6	5	4	4	3	3	2	2	2	2	2
0.70	300	99	58	41	28	18	13	10	8	7	6	5	4	3	3	3	2	2	2
0.80	355	117	69	48	33	21	15	12	9	8	6	5	5	4	3	3	3	2	2
0.90	437	143	85	60	41	26	19	14	11	9	8	6	5	5	4	3	3	2	2
0.95	511	168	99	69	47	30	22	16	13	11	9	7	6	5	4	4	3	3	2
0.99	664	218	128	90	61	39	28	21	17	13	11	9	8	6	5	5	4	3	3

DEGREES OF FREEDOM = 6 ALPHA LEVEL = 0.01

Population Eta Squared

Power	0.01	0.03	0.05	0.07	0.10	0.15	0.20	0.25	0.30	0.35	0.40	0.45	0.50	0.55	0.60	0.65	0.70	0.75	0.80
0.10	64	21	13	9	7	5	4	3	2	2	2	2	2	2	—	—	—	—	—
0.50	195	64	38	27	19	12	9	7	6	5	4	3	3	3	2	2	2	2	—
0.70	265	87	52	36	25	16	12	9	7	6	5	4	4	3	3	2	2	2	2
0.80	312	103	61	43	29	19	14	10	8	7	6	5	4	4	3	3	2	2	2
0.90	383	126	74	52	36	23	16	13	10	8	7	6	5	4	4	3	3	2	2
0.95	447	147	87	61	42	27	19	15	12	9	8	7	6	5	4	3	3	3	2
0.99	579	190	112	79	54	34	24	19	15	12	10	8	7	6	5	4	4	3	2

APPENDIX F.4: Pearson Correlation Coefficient

NONDIRECTIONAL TEST, ALPHA = 0.05

Power	Population Correlation Squared																		
	0.01	0.03	0.05	0.07	0.10	0.15	0.20	0.25	0.30	0.35	0.40	0.45	0.50	0.55	0.60	0.65	0.70	0.75	0.80
0.25	166	56	34	25	17	12	9	8	6	6	5	5	4	4	4	3	3	3	3
0.50	384	127	76	54	38	25	19	15	12	10	9	8	7	6	6	5	5	4	4
0.60	489	162	97	69	48	31	23	18	15	13	11	9	8	7	7	6	5	5	4
0.67	570	188	112	80	55	36	27	21	17	14	12	11	9	8	7	7	6	5	5
0.70	616	203	121	86	59	39	29	23	18	15	13	11	10	9	8	7	6	6	5
0.75	692	228	136	96	67	43	32	25	20	17	14	12	11	10	9	8	7	6	5
0.80	783	258	153	109	75	49	36	28	23	19	16	14	12	11	9	8	7	7	6
0.85	895	294	175	124	85	56	41	32	26	21	18	16	14	12	10	9	8	7	6
0.90	1046	344	204	144	100	65	47	37	30	25	21	18	15	13	12	10	9	8	7
0.95	1308	429	255	180	124	80	58	46	37	30	26	22	19	16	14	12	11	10	8
0.99	1828	599	355	251	172	111	81	63	50	42	35	30	26	22	19	17	14	13	11

NONDIRECTIONAL TEST, ALPHA = 0.01

Power	Population Correlation Squared																		
	0.01	0.03	0.05	0.07	0.10	0.15	0.20	0.25	0.30	0.35	0.40	0.45	0.50	0.55	0.60	0.65	0.70	0.75	0.80
0.25	362	120	72	51	36	24	18	15	12	10	9	7	7	6	5	5	4	4	4
0.50	662	218	130	92	64	42	31	24	19	16	14	12	11	9	8	7	7	6	5
0.60	797	262	156	111	76	50	36	29	23	19	16	14	12	11	9	8	7	7	6
0.67	901	296	176	125	86	56	41	32	26	21	18	16	14	12	10	9	8	7	6
0.70	957	315	187	132	91	59	43	34	27	23	19	17	14	13	11	10	9	8	7
0.75	1052	346	205	145	100	65	47	37	30	25	21	18	16	14	12	10	9	8	7
0.80	1163	382	227	160	110	72	52	41	33	27	23	20	17	15	13	11	10	9	8
0.85	1299	426	253	179	123	80	58	45	36	30	25	22	19	16	14	12	11	9	8
0.90	1480	485	288	203	140	91	66	51	41	34	29	24	21	18	16	14	12	11	9
0.95	1790	587	348	246	169	109	79	62	49	41	34	29	25	22	19	16	14	12	11
0.99	2390	783	464	327	225	145	105	82	65	54	45	38	33	28	24	21	18	16	13

DIRECTIONAL TEST, ALPHA = 0.05

								Population Correlation Squared											
Power	0.01	0.03	0.05	0.07	0.10	0.15	0.20	0.25	0.30	0.35	0.40	0.45	0.50	0.55	0.60	0.65	0.70	0.75	0.80
0.25	99	34	21	15	11	8	6	6	5	4	4	4	3	3	3	3	3	3	2
0.50	277	92	56	40	28	19	14	11	9	8	7	6	6	5	5	4	4	4	3
0.60	368	122	73	52	36	24	18	14	12	10	9	8	7	6	5	5	5	4	4
0.67	430	142	85	61	42	28	21	16	13	11	10	9	8	7	6	5	5	4	4
0.70	470	156	93	66	46	30	22	18	14	12	10	9	8	7	6	6	5	5	4
0.75	537	177	106	75	52	34	25	20	16	14	12	10	9	8	7	6	5	5	5
0.80	618	204	121	86	60	39	29	22	18	15	13	11	10	9	8	7	6	5	5
0.85	727	239	143	101	70	46	33	26	21	18	15	13	11	10	9	8	7	6	5
0.90	864	284	169	120	83	54	39	31	25	21	18	15	13	11	10	9	8	7	6
0.95	1105	363	216	152	105	68	50	39	31	26	22	19	16	14	12	11	10	8	7
0.99	1585	520	308	218	150	97	70	55	44	36	31	26	22	19	17	15	13	11	10

DIRECTIONAL TEST, ALPHA = 0.01

								Population Correlation Squared											
Power	0.01	0.03	0.05	0.07	0.10	0.15	0.20	0.25	0.30	0.35	0.40	0.45	0.50	0.55	0.60	0.65	0.70	0.75	0.80
0.25	273	91	55	39	27	18	14	12	9	8	7	6	6	5	5	4	4	4	3
0.50	540	178	106	76	52	34	25	20	16	14	12	10	9	8	7	6	6	5	5
0.60	663	219	130	92	64	42	31	24	20	16	14	12	11	9	8	7	7	6	5
0.67	757	249	148	105	73	47	35	28	22	18	16	13	12	10	9	8	7	6	6
0.70	809	266	158	112	77	50	37	29	23	19	17	14	12	11	10	8	8	7	6
0.75	897	295	175	124	86	56	41	32	26	21	18	16	14	12	10	9	8	7	6
0.80	998	328	195	138	95	62	45	36	28	24	20	17	15	13	11	10	9	8	7
0.85	1126	370	220	155	107	69	51	40	32	26	22	19	16	14	13	11	10	8	7
0.90	1296	425	253	178	123	80	58	45	36	30	25	22	19	16	14	12	11	9	8
0.95	1585	520	308	218	150	97	70	55	44	36	31	26	22	19	17	15	13	11	10
0.99	2154	706	418	295	203	131	95	74	59	49	41	35	30	26	22	19	17	14	12

APPENDIX F.5: Chi Square Test of Independence

TYPE OF TABLE: 2 × 2   ALPHA = .05

Population Fourfold Point Correlation

Power	.10	.20	.30	.40	.50	.60	.70	.80	.90
.25	165	41	18	10	7	5	3	3	2
.50	384	96	43	24	15	11	8	6	5
.60	490	122	54	31	20	14	10	8	6
.70	617	154	69	39	25	17	13	10	8
.75	694	175	77	43	28	19	14	11	9
.80	785	196	87	49	31	22	16	12	10
.85	898	224	100	56	36	25	18	14	11
.90	1051	263	117	66	42	29	21	16	13
.95	1300	325	144	81	52	36	27	20	16
.99	1837	459	204	115	73	51	37	29	23

TYPE OF TABLE: 2 × 3   ALPHA = .05

Population Value of Cramer's Index

Power	.10	.20	.30	.40	.50	.60	.70	.80	.90
.25	226	56	25	14	9	6	5	4	3
.50	496	124	55	31	20	14	10	8	6
.60	621	155	69	39	25	17	13	10	8
.70	770	193	86	48	31	21	16	12	10
.75	859	215	95	54	34	24	18	13	11
.80	964	241	107	60	39	27	20	15	12
.85	1092	273	121	68	44	30	22	17	13
.90	1265	316	141	79	51	35	26	20	16
.95	1544	386	172	97	62	43	32	24	19
.99	2140	535	238	134	86	59	44	33	26

Appendix F: Power and Sample Size   **A57**

TYPE OF TABLE: 2 × 4   ALPHA = .05

Power	Population Value of Cramer's Index								
	.10	.20	.30	.40	.50	.60	.70	.80	.90
.25	258	65	29	16	10	7	5	4	3
.50	576	144	64	36	23	16	12	9	7
.60	715	179	79	45	29	20	15	11	9
.70	879	220	98	55	35	24	18	14	11
.75	976	244	108	61	39	27	20	15	12
.80	1090	273	121	68	44	30	22	17	13
.85	1230	308	137	77	49	34	25	19	15
.90	1417	354	157	89	57	39	29	22	17
.95	1717	429	191	107	69	48	35	27	21
.99	2352	588	261	147	94	65	48	37	29

TYPE OF TABLE: 3 × 3   ALPHA = 0.05

Power	Population Value of Cramer's Index								
	0.10	0.20	0.30	0.40	0.50	0.60	0.70	0.80	0.90
0.25	154	39	17	10	6	4	3	2	2
0.50	321	80	36	20	13	9	7	5	4
0.60	396	99	44	25	16	11	8	6	5
0.70	484	121	54	30	19	13	10	8	6
0.75	536	134	60	34	21	15	11	8	7
0.80	597	149	66	37	24	17	12	9	7
0.85	671	168	75	42	27	19	14	10	8
0.90	770	193	86	48	31	21	16	12	10
0.95	929	232	103	58	37	26	19	15	11
0.99	1262	316	140	79	50	35	26	20	16

TYPE OF TABLE: 3 × 4   ALPHA = 0.05

Power	Population Value of Cramer's Index								
	0.10	0.20	0.30	0.40	0.50	0.60	0.70	0.80	0.90
0.25	185	46	21	12	7	5	4	3	2
0.50	375	94	42	23	15	10	8	6	5
0.60	460	115	51	29	18	13	9	7	6
0.70	557	139	62	35	22	15	11	9	7
0.75	615	154	68	38	25	17	13	10	8
0.80	681	170	76	43	27	19	14	11	8
0.85	763	191	85	48	31	21	16	12	9
0.90	871	218	97	54	35	24	18	14	11
0.95	1043	261	116	65	42	29	21	16	13
0.99	1403	351	156	88	56	39	29	22	17

TYPE OF TABLE: 4 × 4   ALPHA = 0.05

Power	Population Value of Cramer's Index								
	0.10	0.20	0.30	0.40	0.50	0.60	0.70	0.80	0.90
0.25	148	37	16	9	6	4	3	2	2
0.50	294	73	33	18	12	8	6	5	4
0.60	357	89	40	22	14	10	7	6	4
0.70	430	107	48	27	17	12	9	7	5
0.75	472	118	52	30	19	13	10	7	6
0.80	522	130	58	33	21	14	11	8	6
0.85	582	145	65	36	23	16	12	9	7
0.90	661	165	73	41	26	18	13	10	8
0.95	786	197	87	49	31	22	16	12	10
0.99	1046	262	116	65	42	29	21	16	13

TYPE OF TABLE: 2 × 2   ALPHA = .01

Population Fourfold Point Correlation

Power	.10	.20	.30	.40	.50	.60	.70	.80	.90
.25	362	90	40	23	14	10	7	6	4
.50	664	166	74	41	27	18	14	10	8
.60	800	200	89	50	32	22	16	13	10
.70	961	240	107	60	38	27	20	15	12
.75	1056	264	117	66	42	29	22	17	13
.80	1168	292	130	73	47	32	24	18	14
.85	1305	326	145	82	52	36	27	20	16
.90	1488	372	165	93	60	41	30	23	18
.95	1781	445	198	111	71	49	36	28	22
.99	2403	601	267	150	96	67	49	38	30

TYPE OF TABLE: 2 × 3   ALPHA = .01

Population Value of Cramer's Index

Power	.10	.20	.30	.40	.50	.60	.70	.80	.90
.25	467	117	52	29	19	13	10	7	6
.50	819	205	91	51	33	23	17	13	10
.60	975	244	108	61	39	27	20	15	12
.70	1157	289	129	72	46	32	24	18	14
.75	1264	316	140	79	51	35	26	20	16
.80	1388	347	154	87	56	39	28	22	17
.85	1540	385	171	96	62	43	31	24	19
.90	1743	436	194	109	70	48	36	27	22
.95	2065	516	229	129	83	57	42	32	25
.99	2742	685	305	171	110	76	56	43	34

TYPE OF TABLE: 2 × 4   ALPHA = .01

Power	Population Value of Cramer's Index								
	.10	.20	.30	.40	.50	.60	.70	.80	.90
.25	544	136	60	34	22	15	11	8	7
.50	931	233	103	58	37	26	19	15	11
.60	1101	275	122	69	44	31	22	17	14
.70	1297	324	144	81	52	36	26	20	16
.75	1412	353	157	88	56	39	29	22	17
.80	1546	386	172	97	62	43	32	24	19
.85	1709	427	190	107	68	47	35	27	21
.90	1925	481	214	120	77	53	39	30	24
.95	2267	567	252	142	91	63	46	35	28
.99	2983	746	331	186	119	83	61	47	37

TYPE OF TABLE: 3 × 3   ALPHA = 0.01

Power	Population Value of Cramer's Index								
	0.10	0.20	0.30	0.40	0.50	0.60	0.70	0.80	0.90
0.25	304	76	34	19	12	8	6	5	4
0.50	512	128	57	32	20	14	10	8	6
0.60	602	151	67	38	24	17	12	9	7
0.70	706	177	78	44	28	20	14	11	9
0.75	767	192	85	48	31	21	16	12	9
0.80	824	206	92	52	33	23	17	13	10
0.85	924	231	103	58	37	26	19	14	11
0.90	1037	259	115	65	41	29	21	16	13
0.95	1217	304	135	76	49	34	25	19	15
0.99	1590	398	177	99	64	44	32	25	20

TYPE OF TABLE: 3 × 4   ALPHA = 0.01

Population Value of Cramer's Index

Power	0.10	0.20	0.30	0.40	0.50	0.60	0.70	0.80	0.90
0.25	357	89	40	22	14	10	7	6	4
0.50	588	147	65	37	24	16	12	9	7
0.60	687	172	76	43	27	19	14	11	8
0.70	801	200	89	50	32	22	16	13	10
0.75	867	217	96	54	35	24	18	14	11
0.80	944	236	105	59	38	26	19	15	12
0.85	1037	259	115	65	41	29	21	16	13
0.90	1159	290	129	72	46	32	24	18	14
0.95	1352	338	150	85	54	38	28	21	17
0.99	1751	438	195	109	70	49	36	27	22

TYPE OF TABLE: 4 × 4   ALPHA = 0.01

Population Value of Cramer's Index

Power	0.10	0.20	0.30	0.40	0.50	0.60	0.70	0.80	0.90
0.25	280	70	31	18	11	8	6	4	3
0.50	453	113	50	28	18	13	9	7	6
0.60	526	132	58	33	21	15	11	8	6
0.70	610	153	68	38	24	17	12	10	8
0.75	658	165	73	41	26	18	13	10	8
0.80	714	179	79	45	29	20	15	11	9
0.85	782	196	87	49	31	22	16	12	10
0.90	871	218	97	54	35	24	18	14	11
0.95	1010	253	112	63	40	28	21	16	12
0.99	1296	324	144	81	52	36	26	20	16

Appendix G
Critical Values for the
F Distribution

The following table presents critical values for an *F* distribution for alpha levels of .05 and .01. The .05 critical values are in roman type and the .01 critical values are in bold-face type. The "degrees of freedom in the numerator," listed as columns, corresponds to the "degrees of freedom between" in traditional analysis of variance. The "degrees of freedom in the denominator," listed in rows, corresponds to the "degrees of freedom within" in traditional analysis of variance. For example, the critical value of *F* for $\alpha = .05$, df between = 2, and df within = 11, is 3.98.

Source: Reprinted by permission from *Statistical Methods* by George W. Snedecor and William G. Cochran, Seventh Edition. © 1980 by The Iowa State University Press, Ames, Iowa 50010.

df denom	1	2	3	4	5	6	7	8	9	10	11	12	14	16	20	24	30	40	50	75	100	200	500	∞
1	161 4052	200 4999	216 5403	225 5625	230 5764	234 5859	237 5928	239 5981	241 6022	242 6056	243 6082	244 6106	245 6142	246 6169	248 6208	249 6234	250 6258	251 6286	252 6302	253 6323	253 6334	254 6352	254 6361	254 6366
2	18.51 98.49	19.00 99.01	19.16 99.17	19.25 99.25	19.30 99.30	19.33 99.33	19.36 99.34	19.37 99.36	19.38 99.38	19.39 99.40	19.40 99.41	19.41 99.42	19.42 99.43	19.43 99.44	19.44 99.45	19.45 99.46	19.46 99.47	19.47 99.48	19.47 99.48	19.48 99.49	19.49 99.49	19.49 99.49	19.50 99.50	19.50 99.50
3	10.13 34.12	9.55 30.81	9.28 29.46	9.12 28.71	9.01 28.24	8.94 27.91	8.88 27.67	8.84 27.49	8.81 27.34	8.78 27.23	8.76 27.13	8.74 27.05	8.71 26.92	8.69 26.83	8.66 26.69	8.64 26.60	8.62 26.50	8.60 26.41	8.58 26.30	8.57 26.27	8.56 26.23	8.54 26.18	8.54 26.14	8.53 26.12
4	7.71 21.20	6.94 18.00	6.59 16.69	6.39 15.98	6.26 15.52	6.16 15.21	6.09 14.98	6.04 14.80	6.00 14.66	5.96 14.54	5.93 14.45	5.91 14.37	5.87 14.24	5.84 14.15	5.80 14.02	5.77 13.93	5.74 13.83	5.71 13.74	5.70 13.69	5.68 13.61	5.66 13.57	5.65 13.52	5.64 13.48	5.63 13.46
5	6.61 16.26	5.79 13.27	5.41 12.06	5.19 11.39	5.05 10.97	4.95 10.67	4.88 10.45	4.82 10.27	4.78 10.15	4.74 10.05	4.70 9.96	4.68 9.89	4.64 9.77	4.60 9.68	4.56 9.55	4.53 9.47	4.50 9.38	4.46 9.29	4.44 9.24	4.42 9.17	4.40 9.13	4.38 9.07	4.37 9.04	4.36 9.02
6	5.99 13.74	5.14 10.92	4.76 9.78	4.53 9.15	4.39 8.75	4.28 8.47	4.21 8.26	4.15 8.10	4.10 7.98	4.06 7.87	4.03 7.79	4.00 7.72	3.96 7.60	3.92 7.52	3.87 7.39	3.84 7.31	3.81 7.23	3.77 7.14	3.75 7.09	3.72 7.02	3.71 6.99	3.69 6.94	3.68 6.90	3.67 6.88
7	5.59 12.25	4.74 9.55	4.35 8.45	4.12 7.85	3.97 7.46	3.87 7.19	3.79 7.00	3.73 6.84	3.68 6.71	3.63 6.62	3.60 6.54	3.57 6.47	3.52 6.35	3.49 6.27	3.44 6.15	3.41 6.07	3.38 5.98	3.34 5.90	3.32 5.85	3.29 5.78	3.28 5.75	3.25 5.70	3.24 5.67	3.23 5.65
8	5.32 11.26	4.46 8.65	4.07 7.59	3.84 7.01	3.69 6.63	3.58 6.37	3.50 6.19	3.44 6.03	3.39 5.91	3.34 5.82	3.31 5.74	3.28 5.67	3.23 5.56	3.20 5.48	3.15 5.36	3.12 5.28	3.08 5.20	3.05 5.11	3.03 5.06	3.00 5.00	2.98 4.96	2.96 4.91	2.94 4.88	2.93 4.86
9	5.12 10.56	4.26 8.02	3.86 6.99	3.63 6.42	3.48 6.06	3.37 5.80	3.29 5.62	3.23 5.47	3.18 5.35	3.13 5.26	3.10 5.18	3.07 5.11	3.02 5.00	2.98 4.92	2.93 4.80	2.90 4.73	2.86 4.64	2.82 4.56	2.80 4.51	2.77 4.45	2.76 4.41	2.73 4.36	2.72 4.33	2.71 4.31
10	4.96 10.04	4.10 7.56	3.71 6.55	3.48 5.99	3.33 5.64	3.22 5.39	3.14 5.21	3.07 5.06	3.02 4.95	2.97 4.85	2.94 4.78	2.91 4.71	2.86 4.60	2.82 4.52	2.77 4.41	2.74 4.33	2.70 4.25	2.67 4.17	2.64 4.12	2.61 4.05	2.59 4.01	2.56 3.96	2.55 3.93	2.54 3.91
11	4.84 9.65	3.98 7.20	3.59 6.22	3.36 5.67	3.20 5.32	3.09 5.07	3.01 4.88	2.95 4.74	2.90 4.63	2.86 4.54	2.82 4.46	2.79 4.40	2.74 4.29	2.70 4.21	2.65 4.10	2.61 4.02	2.57 3.94	2.53 3.86	2.50 3.80	2.47 3.74	2.45 3.70	2.42 3.66	2.41 3.62	2.40 3.60
12	4.75 9.33	3.88 6.93	3.49 5.95	3.26 5.41	3.11 5.06	3.00 4.82	2.92 4.65	2.85 4.50	2.80 4.39	2.76 4.30	2.72 4.22	2.69 4.16	2.64 4.05	2.60 3.98	2.54 3.86	2.50 3.78	2.46 3.70	2.42 3.61	2.40 3.56	2.36 3.49	2.35 3.46	2.32 3.41	2.31 3.38	2.30 3.36
13	4.67 9.07	3.80 6.70	3.41 5.74	3.18 5.20	3.02 4.86	2.92 4.62	2.84 4.44	2.77 4.30	2.72 4.19	2.67 4.10	2.63 4.02	2.60 3.96	2.55 3.85	2.51 3.78	2.46 3.67	2.42 3.59	2.38 3.51	2.34 3.42	2.32 3.37	2.28 3.30	2.26 3.27	2.24 3.21	2.22 3.18	2.21 3.16
14	4.60 8.86	3.74 6.51	3.34 5.56	3.11 5.03	2.96 4.69	2.85 4.46	2.77 4.28	2.70 4.14	2.65 4.03	2.60 3.94	2.56 3.86	2.53 3.80	2.48 3.70	2.44 3.62	2.39 3.51	2.35 3.43	2.31 3.34	2.27 3.26	2.24 3.21	2.21 3.14	2.19 3.11	2.16 3.06	2.14 3.02	2.13 3.00
15	4.54 8.68	3.68 6.36	3.29 5.42	3.06 4.89	2.90 4.56	2.79 4.32	2.70 4.14	2.64 4.00	2.59 3.89	2.55 3.80	2.51 3.73	2.48 3.67	2.43 3.56	2.39 3.48	2.33 3.36	2.29 3.29	2.25 3.20	2.21 3.12	2.18 3.07	2.15 3.00	2.12 2.97	2.10 2.92	2.08 2.89	2.07 2.87

Degrees of freedom for denominator

(continued)

df	1	2	3	4	5	6	7	8	9	10	11	12	13	14	15	16	17	18	19	20	21	22	23	24
16	2.01/2.75	2.02/2.77	2.04/2.80	2.07/2.86	2.09/2.89	2.13/2.96	2.16/3.01	2.20/3.10	2.24/3.18	2.28/3.25	2.33/3.37	2.37/3.45	2.42/3.55	2.45/3.61	2.49/3.69	2.54/3.78	2.59/3.89	2.66/4.03	2.74/4.20	2.85/4.44	3.01/4.77	3.24/5.29	3.63/6.23	4.49/8.53
17	1.96/2.65	1.97/2.67	1.99/2.70	2.02/2.76	2.04/2.79	2.08/2.86	2.11/2.92	2.15/3.00	2.19/3.08	2.23/3.16	2.29/3.27	2.33/3.35	2.38/3.45	2.41/3.52	2.45/3.59	2.50/3.68	2.55/3.79	2.62/3.93	2.70/4.10	2.81/4.34	2.96/4.67	3.20/5.18	3.59/6.11	4.45/8.40
18	1.92/2.57	1.93/2.59	1.95/2.62	1.98/2.68	2.00/2.71	2.04/2.78	2.07/2.83	2.11/2.91	2.15/3.00	2.19/3.07	2.25/3.19	2.29/3.27	2.34/3.37	2.37/3.44	2.41/3.51	2.46/3.60	2.51/3.71	2.58/3.85	2.66/4.01	2.77/4.25	2.93/4.58	3.16/5.09	3.55/6.01	4.41/8.28
19	1.88/2.49	1.90/2.51	1.91/2.54	1.94/2.60	1.96/2.63	2.00/2.70	2.02/2.76	2.07/2.84	2.11/2.92	2.15/3.00	2.21/3.12	2.26/3.19	2.31/3.30	2.34/3.36	2.38/3.43	2.43/3.52	2.48/3.63	2.55/3.77	2.63/3.94	2.74/4.17	2.90/4.50	3.13/5.01	3.52/5.93	4.38/8.18
20	1.84/2.42	1.85/2.44	1.87/2.47	1.90/2.53	1.92/2.56	1.96/2.63	1.99/2.69	2.04/2.77	2.08/2.86	2.12/2.94	2.18/3.05	2.23/3.13	2.28/3.23	2.31/3.30	2.35/3.37	2.40/3.45	2.45/3.56	2.52/3.71	2.60/3.87	2.71/4.10	2.87/4.43	3.10/4.94	3.49/5.85	4.35/8.10
21	1.81/2.36	1.82/2.38	1.84/2.42	1.87/2.47	1.90/2.51	1.93/2.58	1.96/2.63	2.00/2.72	2.05/2.80	2.09/2.88	2.15/2.99	2.20/3.07	2.25/3.17	2.28/3.24	2.32/3.31	2.37/3.40	2.42/3.51	2.49/3.65	2.57/3.81	2.68/4.04	2.84/4.37	3.07/4.87	3.47/5.78	4.32/8.02
22	1.78/2.31	1.80/2.33	1.81/2.37	1.84/2.42	1.87/2.46	1.91/2.53	1.93/2.58	1.98/2.67	2.03/2.75	2.07/2.83	2.13/2.94	2.18/3.02	2.23/3.12	2.26/3.18	2.30/3.26	2.35/3.35	2.40/3.45	2.47/3.59	2.55/3.76	2.66/3.99	2.82/4.31	3.05/4.82	3.44/5.72	4.30/7.94
23	1.76/2.26	1.77/2.28	1.79/2.32	1.82/2.37	1.84/2.41	1.88/2.48	1.91/2.53	1.96/2.62	2.00/2.70	2.04/2.78	2.10/2.89	2.14/2.97	2.20/3.07	2.24/3.14	2.28/3.21	2.32/3.30	2.38/3.41	2.45/3.54	2.53/3.71	2.64/3.94	2.80/4.26	3.03/4.76	3.42/5.66	4.28/7.88
24	1.73/2.21	1.74/2.23	1.76/2.27	1.80/2.33	1.82/2.36	1.86/2.44	1.89/2.49	1.94/2.58	1.98/2.66	2.02/2.74	2.09/2.85	2.13/2.93	2.18/3.03	2.22/3.09	2.26/3.17	2.30/3.25	2.36/3.36	2.43/3.50	2.51/3.67	2.62/3.90	2.78/4.22	3.01/4.72	3.40/5.61	4.26/7.82
25	1.71/2.17	1.72/2.19	1.74/2.23	1.77/2.29	1.80/2.32	1.84/2.40	1.87/2.45	1.92/2.54	1.96/2.62	2.00/2.70	2.06/2.81	2.11/2.89	2.16/2.99	2.20/3.05	2.24/3.13	2.28/3.21	2.34/3.32	2.41/3.46	2.49/3.63	2.60/3.86	2.76/4.18	2.99/4.68	3.38/5.57	4.24/7.77
26	1.69/2.13	1.70/2.15	1.72/2.19	1.76/2.25	1.78/2.28	1.82/2.36	1.85/2.41	1.90/2.50	1.95/2.58	1.99/2.66	2.05/2.77	2.10/2.86	2.15/2.96	2.18/3.02	2.22/3.09	2.27/3.17	2.32/3.29	2.39/3.42	2.47/3.59	2.59/3.82	2.74/4.14	2.96/4.64	3.37/5.53	4.22/7.72
27	1.67/2.10	1.68/2.12	1.71/2.16	1.74/2.21	1.76/2.25	1.80/2.33	1.84/2.38	1.88/2.47	1.93/2.55	1.97/2.63	2.03/2.74	2.08/2.83	2.13/2.93	2.16/2.98	2.20/3.06	2.25/3.14	2.30/3.26	2.37/3.39	2.46/3.56	2.57/3.79	2.73/4.11	2.95/4.60	3.35/5.49	4.21/7.68
28	1.65/2.06	1.67/2.09	1.69/2.13	1.72/2.18	1.75/2.22	1.78/2.30	1.81/2.35	1.87/2.44	1.91/2.52	1.94/2.60	2.02/2.71	2.06/2.80	2.12/2.90	2.15/2.95	2.19/3.03	2.24/3.11	2.29/3.23	2.36/3.36	2.44/3.53	2.56/3.76	2.71/4.07	2.93/4.57	3.34/5.45	4.20/7.64
29	1.64/2.03	1.65/2.06	1.68/2.10	1.71/2.15	1.73/2.19	1.77/2.27	1.80/2.32	1.85/2.41	1.90/2.49	1.93/2.57	2.00/2.68	2.05/2.77	2.10/2.87	2.14/2.92	2.18/3.00	2.22/3.08	2.28/3.20	2.35/3.32	2.43/3.50	2.54/3.73	2.70/4.04	2.93/4.54	3.33/5.52	4.18/7.60
30	1.62/2.01	1.64/2.03	1.66/2.07	1.69/2.13	1.72/2.16	1.76/2.24	1.79/2.29	1.84/2.38	1.89/2.47	1.93/2.55	1.99/2.66	2.04/2.74	2.09/2.84	2.12/2.90	2.16/2.98	2.21/3.06	2.27/3.17	2.34/3.30	2.42/3.47	2.53/3.70	2.69/4.02	2.92/4.51	3.32/5.39	4.17/7.56

Degrees of freedom for denominator

Degrees of freedom for numerator

df (denom)	1	2	3	4	5	6	7	8	9	10	11	12	14	16	20	24	30	40	50	75	100	200	500	∞
32	4.15 / 7.50	3.30 / 5.34	2.90 / 4.46	2.67 / 3.97	2.51 / 3.66	2.40 / 3.42	2.32 / 3.25	2.25 / 3.12	2.19 / 3.01	2.14 / 2.94	2.10 / 2.86	2.07 / 2.80	2.02 / 2.70	1.97 / 2.62	1.91 / 2.51	1.86 / 2.42	1.82 / 2.34	1.76 / 2.25	1.74 / 2.20	1.69 / 2.12	1.67 / 2.08	1.64 / 2.02	1.61 / 1.98	1.59 / 1.96
34	4.13 / 7.44	3.28 / 5.29	2.88 / 4.42	2.65 / 3.93	2.49 / 3.61	2.38 / 3.38	2.30 / 3.21	2.23 / 3.08	2.17 / 2.97	2.12 / 2.89	2.08 / 2.82	2.05 / 2.76	2.00 / 2.66	1.95 / 2.58	1.89 / 2.47	1.84 / 2.38	1.80 / 2.30	1.74 / 2.21	1.71 / 2.15	1.67 / 2.08	1.64 / 2.04	1.61 / 1.98	1.59 / 1.94	1.57 / 1.91
36	4.11 / 7.39	3.26 / 5.25	2.86 / 4.38	2.63 / 3.89	2.48 / 3.58	2.36 / 3.35	2.28 / 3.18	2.21 / 3.04	2.15 / 2.94	2.10 / 2.86	2.06 / 2.78	2.03 / 2.72	1.98 / 2.62	1.93 / 2.54	1.87 / 2.43	1.82 / 2.35	1.78 / 2.26	1.72 / 2.17	1.69 / 2.12	1.65 / 2.04	1.62 / 2.00	1.59 / 1.94	1.56 / 1.90	1.55 / 1.87
38	4.10 / 7.35	3.25 / 5.21	2.85 / 4.34	2.62 / 3.86	2.46 / 3.54	2.35 / 3.32	2.26 / 3.15	2.19 / 3.02	2.14 / 2.91	2.09 / 2.82	2.05 / 2.75	2.02 / 2.69	1.96 / 2.59	1.92 / 2.51	1.85 / 2.40	1.80 / 2.32	1.76 / 2.22	1.71 / 2.14	1.67 / 2.08	1.63 / 2.00	1.60 / 1.97	1.57 / 1.90	1.54 / 1.86	1.53 / 1.84
40	4.08 / 7.31	3.23 / 5.18	2.84 / 4.31	2.61 / 3.83	2.45 / 3.51	2.34 / 3.29	2.25 / 3.12	2.18 / 2.99	2.12 / 2.88	2.07 / 2.80	2.04 / 2.73	2.00 / 2.66	1.95 / 2.56	1.90 / 2.49	1.84 / 2.37	1.79 / 2.29	1.74 / 2.20	1.69 / 2.11	1.66 / 2.05	1.61 / 1.97	1.59 / 1.94	1.55 / 1.88	1.53 / 1.84	1.51 / 1.81
42	4.07 / 7.27	3.22 / 5.15	2.83 / 4.29	2.59 / 3.80	2.44 / 3.49	2.32 / 3.26	2.24 / 3.10	2.17 / 2.96	2.11 / 2.86	2.06 / 2.77	2.02 / 2.70	1.99 / 2.64	1.94 / 2.54	1.89 / 2.46	1.82 / 2.35	1.78 / 2.26	1.73 / 2.17	1.68 / 2.08	1.64 / 2.02	1.60 / 1.94	1.57 / 1.91	1.54 / 1.85	1.51 / 1.80	1.49 / 1.78
44	4.06 / 7.24	3.21 / 5.12	2.82 / 4.26	2.58 / 3.78	2.43 / 3.46	2.31 / 3.24	2.23 / 3.07	2.16 / 2.94	2.10 / 2.84	2.05 / 2.75	2.01 / 2.68	1.98 / 2.62	1.92 / 2.52	1.88 / 2.44	1.81 / 2.32	1.76 / 2.24	1.72 / 2.15	1.66 / 2.06	1.63 / 2.00	1.58 / 1.92	1.56 / 1.88	1.52 / 1.82	1.50 / 1.78	1.48 / 1.75
46	4.05 / 7.21	3.20 / 5.10	2.81 / 4.24	2.57 / 3.76	2.42 / 3.44	2.30 / 3.22	2.22 / 3.05	2.14 / 2.92	2.09 / 2.82	2.04 / 2.73	2.00 / 2.66	1.97 / 2.60	1.91 / 2.50	1.87 / 2.42	1.80 / 2.30	1.75 / 2.22	1.71 / 2.13	1.65 / 2.04	1.62 / 1.98	1.57 / 1.90	1.54 / 1.86	1.51 / 1.80	1.48 / 1.76	1.46 / 1.72
48	4.04 / 7.19	3.19 / 5.08	2.80 / 4.22	2.56 / 3.74	2.41 / 3.42	2.30 / 3.20	2.21 / 3.04	2.14 / 2.90	2.08 / 2.80	2.03 / 2.71	1.99 / 2.64	1.96 / 2.58	1.90 / 2.48	1.86 / 2.40	1.79 / 2.28	1.74 / 2.20	1.70 / 2.11	1.64 / 2.02	1.61 / 1.96	1.56 / 1.88	1.53 / 1.84	1.50 / 1.78	1.47 / 1.73	1.45 / 1.70
50	4.03 / 7.17	3.18 / 5.06	2.79 / 4.20	2.56 / 3.72	2.40 / 3.41	2.29 / 3.18	2.20 / 3.02	2.13 / 2.88	2.07 / 2.78	2.02 / 2.70	1.98 / 2.62	1.95 / 2.56	1.90 / 2.46	1.85 / 2.39	1.78 / 2.26	1.74 / 2.18	1.69 / 2.10	1.63 / 2.00	1.60 / 1.94	1.55 / 1.86	1.52 / 1.82	1.48 / 1.76	1.46 / 1.71	1.44 / 1.68
55	4.02 / 7.12	3.17 / 5.01	2.78 / 4.16	2.54 / 3.68	2.38 / 3.37	2.27 / 3.15	2.18 / 2.98	2.11 / 2.85	2.05 / 2.75	2.00 / 2.66	1.97 / 2.59	1.93 / 2.53	1.88 / 2.43	1.83 / 2.35	1.76 / 2.23	1.72 / 2.15	1.67 / 2.06	1.61 / 1.96	1.58 / 1.90	1.52 / 1.82	1.50 / 1.78	1.46 / 1.71	1.43 / 1.66	1.41 / 1.64
60	4.00 / 7.08	3.15 / 4.98	2.76 / 4.13	2.52 / 3.65	2.37 / 3.34	2.25 / 3.12	2.17 / 2.95	2.10 / 2.82	2.04 / 2.72	1.99 / 2.63	1.95 / 2.56	1.92 / 2.50	1.86 / 2.40	1.81 / 2.32	1.75 / 2.20	1.70 / 2.12	1.65 / 2.03	1.59 / 1.93	1.56 / 1.87	1.50 / 1.79	1.48 / 1.74	1.44 / 1.68	1.41 / 1.63	1.39 / 1.60
65	3.99 / 7.04	3.14 / 4.95	2.75 / 4.10	2.51 / 3.62	2.36 / 3.31	2.24 / 3.09	2.15 / 2.93	2.08 / 2.79	2.02 / 2.70	1.98 / 2.61	1.94 / 2.54	1.90 / 2.47	1.85 / 2.37	1.80 / 2.30	1.73 / 2.18	1.68 / 2.09	1.63 / 2.00	1.57 / 1.90	1.54 / 1.84	1.49 / 1.76	1.46 / 1.71	1.42 / 1.64	1.39 / 1.60	1.37 / 1.56
70	3.98 / 7.01	3.13 / 4.92	2.74 / 4.08	2.50 / 3.60	2.35 / 3.29	2.23 / 3.07	2.14 / 2.91	2.07 / 2.77	2.01 / 2.67	1.97 / 2.59	1.93 / 2.51	1.89 / 2.45	1.84 / 2.35	1.79 / 2.28	1.72 / 2.15	1.67 / 2.07	1.62 / 1.98	1.56 / 1.88	1.53 / 1.82	1.47 / 1.74	1.45 / 1.69	1.40 / 1.62	1.37 / 1.56	1.35 / 1.53
80	3.96 / 6.96	3.11 / 4.88	2.72 / 4.04	2.48 / 3.56	2.33 / 3.25	2.21 / 3.04	2.12 / 2.87	2.05 / 2.74	1.99 / 2.64	1.95 / 2.55	1.91 / 2.48	1.88 / 2.41	1.82 / 2.32	1.77 / 2.24	1.70 / 2.11	1.65 / 2.03	1.60 / 1.94	1.54 / 1.84	1.51 / 1.78	1.45 / 1.70	1.42 / 1.65	1.38 / 1.57	1.35 / 1.52	1.32 / 1.49

Degrees of freedom for denominator

A66 *Appendix G: Critical Values for the F Distribution*

Critical values for the F distribution, each cell shows two stacked values (upper / lower).

Degrees of freedom for denominator																								
100	1.28 / 1.43	1.30 / 1.46	1.34 / 1.51	1.39 / 1.59	1.42 / 1.64	1.48 / 1.73	1.51 / 1.79	1.57 / 1.89	1.63 / 1.98	1.68 / 2.06	1.75 / 2.19	1.79 / 2.26	1.84 / 2.36	1.88 / 2.43	1.92 / 2.51	1.97 / 2.59	2.03 / 2.69	2.10 / 2.82	2.19 / 2.99	2.30 / 3.20	2.46 / 3.51	2.70 / 3.98	3.09 / 4.82	3.94 / 6.90
125	1.25 / 1.37	1.27 / 1.40	1.31 / 1.46	1.36 / 1.54	1.39 / 1.59	1.45 / 1.68	1.49 / 1.75	1.55 / 1.85	1.60 / 1.94	1.65 / 2.03	1.72 / 2.15	1.77 / 2.23	1.83 / 2.33	1.86 / 2.40	1.90 / 2.47	1.95 / 2.56	2.01 / 2.65	2.08 / 2.79	2.17 / 2.95	2.29 / 3.17	2.44 / 3.47	2.68 / 3.94	3.07 / 4.78	3.92 / 6.84
150	1.22 / 1.33	1.25 / 1.37	1.29 / 1.43	1.34 / 1.51	1.37 / 1.56	1.44 / 1.66	1.47 / 1.72	1.54 / 1.83	1.59 / 1.91	1.64 / 2.00	1.71 / 2.12	1.76 / 2.20	1.82 / 2.30	1.85 / 2.37	1.89 / 2.44	1.94 / 2.53	2.00 / 2.62	2.07 / 2.76	2.16 / 2.92	2.27 / 3.13	2.43 / 3.44	2.67 / 3.91	3.06 / 4.75	3.91 / 6.81
200	1.19 / 1.28	1.22 / 1.33	1.26 / 1.39	1.32 / 1.48	1.35 / 1.53	1.42 / 1.62	1.45 / 1.69	1.52 / 1.79	1.57 / 1.88	1.62 / 1.97	1.69 / 2.09	1.74 / 2.17	1.80 / 2.28	1.83 / 2.34	1.87 / 2.41	1.92 / 2.50	1.98 / 2.60	2.05 / 2.73	2.14 / 2.90	2.26 / 3.11	2.41 / 3.41	2.65 / 3.88	3.04 / 4.71	3.89 / 6.76
400	1.13 / 1.19	1.16 / 1.24	1.22 / 1.32	1.28 / 1.42	1.32 / 1.47	1.38 / 1.57	1.42 / 1.64	1.49 / 1.74	1.54 / 1.84	1.60 / 1.92	1.67 / 2.04	1.72 / 2.12	1.78 / 2.23	1.81 / 2.29	1.85 / 2.37	1.90 / 2.46	1.96 / 2.55	2.03 / 2.69	2.12 / 2.85	2.23 / 3.06	2.39 / 3.36	2.62 / 3.83	3.02 / 4.66	3.86 / 6.70
1000	1.08 / 1.11	1.13 / 1.19	1.19 / 1.28	1.26 / 1.38	1.30 / 1.44	1.36 / 1.54	1.41 / 1.61	1.47 / 1.71	1.53 / 1.81	1.58 / 1.89	1.65 / 2.01	1.70 / 2.09	1.76 / 2.20	1.80 / 2.26	1.84 / 2.34	1.89 / 2.43	1.95 / 2.53	2.02 / 2.66	2.10 / 2.82	2.22 / 3.04	2.38 / 3.34	2.61 / 3.80	3.00 / 4.62	3.85 / 6.66
∞	1.00 / 1.00	1.11 / 1.15	1.17 / 1.25	1.24 / 1.36	1.28 / 1.41	1.35 / 1.52	1.40 / 1.59	1.46 / 1.69	1.52 / 1.79	1.57 / 1.87	1.64 / 1.99	1.69 / 2.07	1.75 / 2.18	1.79 / 2.24	1.83 / 2.32	1.88 / 2.41	1.94 / 2.51	2.01 / 2.64	2.09 / 2.80	2.21 / 3.02	2.37 / 3.32	2.60 / 3.78	2.99 / 4.60	3.84 / 6.64

Appendix H
Critical Values for the F_{MAX} Statistic

The following is a table of critical values for the F_{MAX} statistic. The first column is $n - 1$ where n is the number of subjects within any given group. If the analysis of variance had unequal numbers of subjects in each group, use the largest n (assuming the group sample sizes are not too discrepant). The second column is the alpha level for the test, either .05 or .01. Columns 3 through 13 represent the number of groups, k.

Consider the case of an analysis of variance with three groups, and $n = 7$ in each group. The critical value of F_{MAX} for $n - 1 = 7 - 1 = 6$ and $k = 3$, $\alpha = .05$, would be 8.38.

Source: Table 31 in *Biometrika Tables for Statisticians*, E. Pearson and O. Hartley, eds., New York: Cambridge, 1958. Reprinted by permission of Imperial College of Science and Technology.

$n-1$	α	2	3	4	5	6	7	8	9	10	11	12
						k = number of groups						
4	.05	9.60	15.5	20.6	25.2	29.5	33.6	37.5	41.4	44.6	48.0	51.4
	.01	23.2	37.	49.	59.	69.	79.	89.	97.	106.	113.	120.
5	.05	7.15	10.8	13.7	16.3	18.7	20.8	22.9	24.7	26.5	28.2	29.9
	.01	14.9	22.	28.	33.	38.	42.	46.	50.	54.	57.	60.
6	.05	5.82	8.38	10.4	12.1	13.7	15.0	16.3	17.5	18.6	19.7	20.7
	.01	11.1	15.5	19.1	22.	25.	27.	30.	32.	34.	36.	37.
7	.05	4.99	6.94	8.44	9.70	10.8	11.8	12.7	13.5	14.3	15.1	15.8
	.01	8.89	12.1	14.5	16.5	18.4	20.	22.	23.	24.	26.	27.
8	.05	4.43	6.00	7.18	8.12	9.03	9.78	10.5	11.1	11.7	12.2	12.7
	.01	7.50	9.9	11.7	13.2	14.5	15.8	16.9	17.9	18.9	19.8	21.
9	.05	4.03	5.34	6.31	7.11	7.80	8.41	8.95	9.45	9.91	10.3	10.7
	.01	6.54	8.5	9.9	11.1	12.1	13.1	13.9	14.7	15.3	16.0	16.6
10	.05	3.72	4.85	5.67	6.34	6.92	7.42	7.87	8.28	8.66	9.01	9.34
	.01	5.85	7.4	8.6	9.6	10.4	11.1	11.8	12.4	12.9	13.4	13.9
12	.05	3.28	4.16	4.79	5.30	5.72	6.09	6.42	6.72	7.00	7.25	7.48
	.01	4.91	6.1	6.9	7.6	8.2	8.7	9.1	9.5	9.9	10.2	10.6
15	.05	2.86	3.54	4.01	4.37	4.68	4.95	5.19	5.40	5.59	5.77	5.93
	.01	4.07	4.9	5.5	6.0	6.4	6.7	7.1	7.3	7.5	7.8	8.0
20	.05	2.46	2.95	3.29	3.54	3.76	3.94	4.10	4.24	4.37	4.49	4.59
	.01	3.32	3.8	4.3	4.6	4.9	5.1	5.3	5.5	5.6	5.8	5.9
30	.05	2.07	2.40	2.61	2.78	2.91	3.02	3.12	3.21	3.29	3.36	3.39
	.01	2.63	3.0	3.3	3.4	3.6	3.7	3.8	3.9	4.0	4.1	4.2
60	.05	1.67	1.85	1.96	2.04	2.11	2.17	2.22	2.26	2.30	2.33	2.36
	.01	1.96	2.2	2.3	2.4	2.4	2.5	2.5	2.6	2.6	2.7	2.7

Appendix I
Studentized Range Values

The following table presents studentized range values (q) for alpha levels of .05 and .01. The first column contains the value of the degrees of freedom within or, as it is also called, "degrees of freedom error." The second column contains the alpha level to be used. The remaining columns refer to k or the number of groups in the experiment (that is, the number of values on the independent variable). Suppose we want to determine q for an experiment comparing the mean scores of three groups with five subjects per group. In this study, $k = 3$, df within = 12. For an $\alpha = .05$, the value of q is 3.77 and for $\alpha = .01$, it is 5.05.

Source: R. Kirk, *Experimental Design: Procedures for the Behavioral Sciences.* Monterey, California: Brooks/Cole, 1967.

Percentage Points of the Studentized Range

Error df	α	k = number of groups									
		2	3	4	5	6	7	8	9	10	11
5	.05	3.64	4.60	5.22	5.67	6.03	6.33	6.58	6.80	6.99	7.17
	.01	5.70	6.98	7.80	8.42	8.91	9.32	9.67	9.97	10.24	10.48
6	.05	3.46	4.34	4.90	5.30	5.63	5.90	6.12	6.32	6.49	6.65
	.01	5.24	6.33	7.03	7.56	7.97	8.32	8.61	8.87	9.10	9.30
7	.05	3.34	4.16	4.68	5.06	5.36	5.61	5.82	6.00	6.16	6.30
	.01	4.95	5.92	6.54	7.01	7.37	7.68	7.94	8.17	8.37	8.55
8	.05	3.26	4.04	4.53	4.89	5.17	5.40	5.60	5.77	5.92	6.05
	.01	4.75	5.64	6.20	6.62	6.96	7.24	7.47	7.68	7.86	8.03
9	.05	3.20	3.95	4.41	4.76	5.02	5.24	5.43	5.59	5.74	5.87
	.01	4.60	5.43	5.96	6.35	6.66	6.91	7.13	7.33	7.49	7.65
10	.05	3.15	3.88	4.33	4.65	4.91	5.12	5.30	5.46	5.60	5.72
	.01	4.48	5.27	5.77	6.14	6.43	6.67	6.87	7.05	7.21	7.36
11	.05	3.11	3.82	4.26	4.57	4.82	5.03	5.20	5.35	5.49	5.61
	.01	4.39	5.15	5.62	5.97	6.25	6.48	6.67	6.84	6.99	7.13
12	.05	3.08	3.77	4.20	4.51	4.75	4.95	5.12	5.27	5.39	5.51
	.01	4.32	5.05	5.50	5.84	6.10	6.32	6.51	6.67	6.81	6.94
13	.05	3.06	3.73	4.15	4.45	4.69	4.88	5.05	5.19	5.32	5.43
	.01	4.26	4.96	5.40	5.73	5.98	6.19	6.37	6.53	6.67	6.79
14	.05	3.03	3.70	4.11	4.41	4.64	4.83	4.99	5.13	5.25	5.36
	.01	4.21	4.89	5.32	5.63	5.88	6.08	6.26	6.41	6.54	6.66
15	.05	3.01	3.67	4.08	4.37	4.59	4.78	4.94	5.08	5.20	5.31
	.01	4.17	4.84	5.25	5.56	5.80	5.99	6.16	6.31	6.44	6.55
16	.05	3.00	3.65	4.05	4.33	4.56	4.74	4.90	5.03	5.15	5.26
	.01	4.13	4.79	5.19	5.49	5.72	5.92	6.08	6.22	6.35	6.46
17	.05	2.98	3.63	4.02	4.30	4.52	4.70	4.86	4.99	5.11	5.21
	.01	4.10	4.74	5.14	5.43	5.66	5.85	6.01	6.15	6.27	6.38
18	.05	2.97	3.61	4.00	4.28	4.49	4.67	4.82	4.96	5.07	5.17
	.01	4.07	4.70	5.09	5.38	5.60	5.79	5.94	6.08	6.20	6.31
19	.05	2.96	3.59	3.98	4.25	4.47	4.65	4.79	4.92	5.04	5.14
	.01	4.05	4.67	5.05	5.33	5.55	5.73	5.89	6.02	6.14	6.25
20	.05	2.95	3.58	3.96	4.23	4.45	4.62	4.77	4.90	5.01	5.11
	.01	4.02	4.64	5.02	5.29	5.51	5.69	5.84	5.97	6.09	6.19
24	.05	2.92	3.53	3.90	4.17	4.37	4.54	4.68	4.81	4.92	5.01
	.01	3.96	4.55	4.91	5.17	5.37	5.54	5.69	5.81	5.92	6.02
30	.05	2.89	3.49	3.85	4.10	4.30	4.46	4.60	4.72	4.82	4.92
	.01	3.89	4.45	4.80	5.05	5.24	5.40	5.54	5.65	5.76	5.85
40	.05	2.86	3.44	3.79	4.04	4.23	4.39	4.52	4.63	4.73	4.82
	.01	3.82	4.37	4.70	4.93	5.11	5.26	5.39	5.50	5.60	5.69
60	.05	2.83	3.40	3.74	3.98	4.16	4.31	4.44	4.55	4.65	4.73
	.01	3.76	4.28	4.59	4.82	4.99	5.13	5.25	5.36	5.45	5.53
120	.05	2.80	3.36	3.68	3.92	4.10	4.24	4.36	4.47	4.56	4.64
	.01	3.70	4.20	4.50	4.71	4.87	5.01	5.12	5.21	5.30	5.37
∞	.05	2.77	3.31	3.63	3.86	4.03	4.17	4.29	4.39	4.47	4.55
	.01	3.64	4.12	4.40	4.60	4.76	4.88	4.99	5.08	5.16	5.23

	k = number of groups										Error
12	13	14	15	16	17	18	19	20	α	df
7.32	7.47	7.60	7.72	7.83	7.93	8.03	8.12	8.21	.05	5
10.70	10.89	11.08	11.24	11.40	11.55	11.68	11.81	11.93	.01	
6.79	6.92	7.03	7.14	7.24	7.34	7.43	7.51	7.59	.05	6
9.48	9.65	9.81	9.95	10.08	10.21	10.32	10.43	10.54	.01	
6.43	6.55	6.66	6.76	6.85	6.94	7.02	7.10	7.17	.05	7
8.71	8.86	9.00	9.12	9.24	9.35	9.46	9.55	9.65	.01	
6.18	6.29	6.39	6.48	6.57	6.65	6.73	6.80	6.87	.05	8
8.18	8.31	8.44	8.55	8.66	8.76	8.85	8.94	9.03	.01	
5.98	6.09	6.19	6.28	6.36	6.44	6.51	6.58	6.64	.05	9
7.78	7.91	8.03	8.13	8.23	8.33	8.41	8.49	8.57	.01	
5.83	5.93	6.03	6.11	6.19	6.27	6.34	6.40	6.47	.05	10
7.49	7.60	7.71	7.81	7.91	7.99	8.08	8.15	8.23	.01	
5.71	5.81	5.90	5.98	6.06	6.13	6.20	6.27	6.33	.05	11
7.25	7.36	7.46	7.56	7.65	7.73	7.81	7.88	7.95	.01	
5.61	5.71	5.80	5.88	5.95	6.02	6.09	6.15	6.21	.05	12
7.06	7.17	7.26	7.36	7.44	7.52	7.59	7.66	7.73	.01	
5.53	5.63	5.71	5.79	5.86	5.93	5.99	6.05	6.11	.05	13
6.90	7.01	7.10	7.19	7.27	7.35	7.42	7.48	7.55	.01	
5.46	5.55	5.64	5.71	5.79	5.85	5.91	5.97	6.03	.05	14
6.77	6.87	6.96	7.05	7.13	7.20	7.27	7.33	7.39	.01	
5.40	5.49	5.57	5.65	5.72	5.78	5.85	5.90	5.96	.05	15
6.66	6.76	6.84	6.93	7.00	7.07	7.14	7.20	7.26	.01	
5.35	5.44	5.52	5.59	5.66	5.73	5.79	5.84	5.90	.05	16
6.56	6.66	6.74	6.82	6.90	6.97	7.03	7.09	7.15	.01	
5.31	5.39	5.47	5.54	5.61	5.67	5.73	5.79	5.84	.05	17
6.48	6.57	6.66	6.73	6.81	6.87	6.94	7.00	7.05	.01	
5.27	5.35	5.43	5.50	5.57	5.63	5.69	5.74	5.79	.05	18
6.41	6.50	6.58	6.65	6.73	6.79	6.85	6.91	6.97	.01	
5.23	5.31	5.39	5.46	5.53	5.59	5.65	5.70	5.75	.05	19
6.34	6.43	6.51	6.58	6.65	6.72	6.78	6.84	6.89	.01	
5.20	5.28	5.36	5.43	5.49	5.55	5.61	5.66	5.71	.05	20
6.28	6.37	6.45	6.52	6.59	6.65	6.71	6.77	6.82	.01	
5.10	5.18	5.25	5.32	5.38	5.44	5.49	5.55	5.59	.05	24
6.11	6.19	6.26	6.33	6.39	6.45	6.51	6.56	6.61	.01	
5.00	5.08	5.15	5.21	5.27	5.33	5.38	5.43	5.47	.05	30
5.93	6.01	6.08	6.14	6.20	6.26	6.31	6.36	6.41	.01	
4.90	4.98	5.04	5.11	5.16	5.22	5.27	5.31	5.36	.05	40
5.76	5.83	5.90	5.96	6.02	6.07	6.12	6.16	6.21	.01	
4.81	4.88	4.94	5.00	5.06	5.11	5.15	5.20	5.24	.05	60
5.60	5.67	5.73	5.78	5.84	5.89	5.93	5.97	6.01	.01	
4.71	4.78	4.84	4.90	4.95	5.00	5.04	5.09	5.13	.05	120
5.44	5.50	5.56	5.61	5.66	5.71	5.75	5.79	5.83	.01	
4.62	4.68	4.74	4.80	4.85	4.89	4.93	4.97	5.01	.05	∞
5.29	5.35	5.40	5.45	5.49	5.54	5.57	5.61	5.65	.01	

A72 *Appendix I: Studentized Range Values*

Appendix J
Critical Values for
Pearson *r*

The following table presents values of *r* that would be viewed as being statistically significant based on different alpha levels for $\rho = 0$. The entries in the table are Pearson correlation coefficients. The first column represents the degrees of freedom associated with the correlation ($N - 2$). The remaining columns represent alpha levels of .10, .05, .02, and .01, respectively, for a two-tailed test or .05, .025, .01, and .005 for a one-tailed test. If we observed a correlation of .466 with 18 degrees of freedom, we can see by inspecting the row for 18 df that this would lead to the rejection of the null hypothesis that $\rho = 0$ for alpha levels of .10 or .05 but not for alpha levels of .02 or .01 (assuming a two-tailed test). The statistical significance of negative correlations is found by ignoring the sign of the coefficient and using the table as described above.

Source: G. A. Ferguson, *Statistical Analysis in Psychology and Education*, McGraw-Hill, 1976. Used with permission.

Critical values of the correlation coefficient

	Level of significance for one-tailed test			
	.05	.025	.01	.005
	Level of significance for two-tailed test			
df	.10	.05	.02	.01
1	.988	.997	.9995	.9999
2	.900	.950	.980	.990
3	.805	.878	.934	.959
4	.729	.811	.882	.917
5	.669	.754	.833	.874
6	.622	.707	.789	.834
7	.582	.666	.750	.798
8	.549	.632	.716	.765
9	.521	.602	.685	.735
10	.497	.576	.658	.708
11	.476	.553	.634	.684
12	.458	.532	.612	.661
13	.441	.514	.592	.641
14	.426	.497	.574	.623
15	.412	.482	.558	.606
16	.400	.468	.542	.590
17	.389	.456	.528	.575
18	.378	.444	.516	.561
19	.369	.433	.503	.549
20	.360	.423	.492	.537
21	.352	.413	.482	.526
22	.344	.404	.472	.515
23	.337	.396	.462	.505
24	.330	.388	.453	.496
25	.323	.381	.445	.487
26	.317	.374	.437	.479
27	.311	.367	.430	.471
28	.306	.361	.423	.463
29	.301	.355	.416	.456
30	.296	.349	.409	.449
35	.275	.325	.381	.418
40	.257	.304	.358	.393
45	.243	.288	.338	.372
50	.231	.273	.322	.354
60	.211	.250	.295	.325
70	.195	.232	.274	.303
80	.183	.217	.256	.283
90	.173	.205	.242	.267
100	.164	.195	.230	.254

Appendix K
Fisher's Transformation
of Pearson *r*

The following table presents Fisher's transformation of Pearson correlation coefficients for testing null hypotheses other than $\rho = 0$. The column labeled *r* represents the value of the relevant correlation coefficient, and the adjacent column labeled *r'* is the transformed value. For example, a correlation of .200 would be transformed to a value of .203.

Source: G. A. Ferguson, *Statistical Analysis in Psychology and Education*. New York: McGraw-Hill, 1976.

r	r'	r	r'	r	r'	r	r'	r	r'
.000	.000	.200	.203	.400	.424	.600	.693	.800	1.099
.005	.005	.205	.208	.405	.430	.605	.701	.805	1.113
.010	.010	.210	.213	.410	.436	.610	.709	.810	1.127
.015	.015	.215	.218	.415	.442	.615	.717	.815	1.142
.020	.020	.220	.224	.420	.448	.620	.725	.820	1.157
.025	.025	.225	.229	.425	.454	.625	.733	.825	1.172
.030	.030	.230	.234	.430	.460	.630	.741	.830	1.188
.035	.035	.235	.239	.435	.466	.635	.750	.835	1.204
.040	.040	.240	.245	.440	.472	.640	.758	.840	1.221
.045	.045	.245	.250	.445	.478	.645	.767	.845	1.238
.050	.050	.250	.255	.450	.485	.650	.775	.850	1.256
.055	.055	.255	.261	.455	.491	.655	.784	.855	1.274
.060	.060	.260	.266	.460	.497	.660	.793	.860	1.293
.065	.065	.265	.271	.465	.504	.665	.802	.865	1.313
.070	.070	.270	.277	.470	.510	.670	.811	.870	1.333
.075	.075	.275	.282	.475	.517	.675	.820	.875	1.354
.080	.080	.280	.288	.480	.523	.680	.829	.880	1.376
.085	.085	.285	.293	.485	.530	.685	.838	.885	1.398
.090	.090	.290	.299	.490	.536	.690	.848	.890	1.422
.095	.095	.295	.304	.495	.543	.695	.858	.895	1.447
.100	.100	.300	.310	.500	.549	.700	.867	.900	1.472
.105	.105	.305	.315	.505	.556	.705	.877	.905	1.499
.110	.110	.310	.321	.510	.563	.710	.887	.910	1.528
.115	.116	.315	.326	.515	.570	.715	.897	.915	1.557
.120	.121	.320	.332	.520	.576	.720	.908	.920	1.589
.125	.126	.325	.337	.525	.583	.725	.918	.925	1.623
.130	.131	.330	.343	.530	.590	.730	.929	.930	1.658
.135	.136	.335	.348	.535	.597	.735	.940	.935	1.697
.140	.141	.340	.354	.540	.604	.740	.950	.940	1.738
.145	.146	.345	.360	.545	.611	.745	.962	.945	1.783
.150	.151	.350	.365	.550	.618	.750	.973	.950	1.832
.155	.156	.355	.371	.555	.626	.755	.984	.955	1.886
.160	.161	.360	.377	.560	.633	.760	.996	.960	1.946
.165	.167	.365	.383	.565	.640	.765	1.008	.965	2.014
.170	.172	.370	.388	.570	.648	.770	1.020	.970	2.092
.175	.177	.375	.394	.575	.655	.775	1.033	.975	2.185
.180	.182	.380	.400	.580	.662	.780	1.045	.980	2.298
.185	.187	.385	.406	.585	.670	.785	1.058	.985	2.443
.190	.192	.390	.412	.590	.678	.790	1.071	.990	2.647
.195	.198	.395	.418	.595	.685	.795	1.085	.995	2.994

Appendix L
Critical Values for the
Chi Square Distribution

The following table presents critical values of χ^2 for different alpha levels. The first column presents the degrees of freedom associated with the χ^2. The remaining columns represent different alpha levels. For a χ^2 with 5 degrees of freedom, and $\alpha = .05$, the critical value of χ^2 would be 11.07. For a χ^2 with 5 degrees of freedom and $\alpha = .01$, the critical value of χ^2 would be 15.09. For a χ^2 with 10 degrees of freedom, and $\alpha = .05$, the critical value of χ^2 would be 18.31.

Source: R. P. Runyon and A. Haber, *Fundamentals of Behavioral Statistics*, Addison-Wesley, 1976. Used with permission.

Degrees of freedom (df)	α = .99	.98	.95	.90	.80	.70	.50	.30	.20	.10	.05	.02	.01
1	.000157	.000628	.00393	.0158	.0642	.148	.455	1.074	1.642	2.706	3.841	5.412	6.635
2	.0201	.0404	.103	.211	.446	.713	1.386	2.408	3.219	4.605	5.991	7.824	9.210
3	.115	.185	.352	.584	1.005	1.424	2.366	3.665	4.642	6.251	7.815	9.837	11.341
4	.297	.429	.711	1.064	1.649	2.195	3.357	4.878	5.989	7.779	9.488	11.668	13.277
5	.554	.752	1.145	1.610	2.343	3.000	4.351	6.064	7.289	9.236	11.070	13.388	15.086
6	.872	1.134	1.635	2.204	3.070	3.828	5.348	7.231	8.558	10.645	12.592	15.033	16.812
7	1.239	1.564	2.167	2.833	3.822	4.671	6.346	8.383	9.803	12.017	14.067	16.622	18.475
8	1.646	2.032	2.733	3.490	4.594	5.527	7.344	9.524	11.030	13.362	15.507	18.168	20.090
9	2.088	2.532	3.325	4.168	5.380	6.393	8.343	10.656	12.242	14.684	16.919	19.679	21.666
10	2.558	3.059	3.940	4.865	6.179	7.267	9.342	11.781	13.442	15.987	18.307	21.161	23.209
11	3.053	3.609	4.575	5.578	6.989	8.148	10.341	12.899	14.631	17.275	19.675	22.618	24.725
12	3.571	4.178	5.226	6.304	7.807	9.034	11.340	14.011	15.812	18.549	21.026	24.054	26.217
13	4.107	4.765	5.892	7.042	8.634	9.926	12.340	15.119	16.985	19.812	22.362	25.472	27.688
14	4.660	5.368	6.571	7.790	9.467	10.821	13.339	16.222	18.151	21.064	23.685	26.873	29.141
15	5.229	5.985	7.261	8.547	10.307	11.721	14.339	17.322	19.311	22.307	24.996	28.259	30.578
16	5.812	6.614	7.962	9.312	11.152	12.624	15.338	18.418	20.465	23.542	26.296	29.633	32.000
17	6.408	7.255	8.672	10.085	12.002	13.531	16.338	19.511	21.615	24.769	27.587	30.995	33.409
18	7.015	7.906	9.390	10.865	12.857	14.440	17.338	20.601	22.760	25.989	28.869	32.346	34.805
19	7.633	8.567	10.117	11.651	13.716	15.352	18.338	21.689	23.900	27.204	30.144	33.687	36.191
20	8.260	9.237	10.851	12.443	14.578	16.266	19.337	22.775	25.038	28.412	31.410	35.020	37.566
21	8.897	9.915	11.591	13.240	15.445	17.182	20.337	23.858	26.171	29.615	32.671	36.343	38.932
22	9.542	10.600	12.338	14.041	16.314	18.101	21.337	24.939	27.301	30.813	33.924	37.659	40.289
23	10.196	11.293	13.091	14.848	17.187	19.021	22.337	26.018	28.429	32.007	35.172	38.968	41.638
24	10.856	11.992	13.848	15.659	18.062	19.943	23.337	27.096	29.553	33.196	36.415	40.270	42.980
25	11.524	12.697	14.611	16.473	18.940	20.867	24.337	28.172	30.675	34.382	37.652	41.566	44.314
26	12.198	13.409	15.379	17.292	19.820	21.792	25.336	29.246	31.795	35.563	38.885	42.856	45.642
27	12.877	14.125	16.151	18.114	20.703	22.719	26.336	30.319	32.912	36.741	40.113	44.140	46.963
28	13.565	14.847	16.928	18.939	21.588	23.647	27.336	31.391	34.027	37.916	41.337	45.419	48.278
29	14.256	15.574	17.708	19.768	22.475	24.577	28.336	32.461	35.139	39.087	42.557	46.693	49.588
30	14.953	16.306	18.493	20.599	23.364	25.508	29.336	33.530	36.250	40.256	43.773	47.962	50.892

Appendix M
Critical Values for the
Mann-Whitney U Test

The following tables present critical values of U for purposes of performing a Mann-Whitney U test. The first row presents the sample size for group 1 and the first column presents the sample size for group 2. The critical value of U is where the sample sizes intersect. For example, if an experiment had ten subjects in group 1 and five subjects in group 2, the critical value of U would be 8, for $\alpha = .05$, two-tailed test.

Source: R. E. Kirk, *Introductory Statistics.* Copyright © 1978 by Wadsworth Publishing Company, Inc. Reprinted by permission of the publisher, Brooks/Cole Publishing Company, Monterey, California.

Critical Values of the Mann–Whitney U.[a] For a one-tailed test at $\alpha = 0.01$ (roman type) and $\alpha = 0.005$ (boldface type) and for a two-tailed test at $\alpha = 0.02$ (roman type) and $\alpha = 0.01$ (boldface type).

n_2 \ n_1	1	2	3	4	5	6	7	8	9	10	11	12	13	14	15	16	17	18	19	20
1	—[b]	—	—	—	—	—	—	—	—	—	—	—	—	—	—	—	—	—	—	—
2	—	—	—	—	—	—	—	—	—	—	—	—	0	0	0	0	0	0	1	1
	—	—	—	—	—	—	—	—	—	—	—	—	—	—	—	—	—	—	0	0
3	—	—	—	—	—	—	0	0	1	1	1	2	2	2	3	3	4	4	4	5
	—	—	—	—	—	—	—	—	0	0	0	1	1	1	2	2	2	2	3	3
4	—	—	—	—	0	1	1	2	3	3	4	5	5	6	7	7	8	9	9	10
	—	—	—	—	—	0	0	1	1	2	2	3	3	4	5	5	6	6	7	8
5	—	—	—	0	1	2	3	4	5	6	7	8	9	10	11	12	13	14	15	16
	—	—	—	—	0	1	1	2	3	4	5	6	7	7	8	9	10	11	12	13
6	—	—	—	1	2	3	4	6	7	8	9	11	12	13	15	16	18	19	20	22
	—	—	—	0	1	2	3	4	5	6	7	9	10	11	12	13	15	16	17	18
7	—	—	0	1	3	4	6	7	9	11	12	14	16	17	19	21	23	24	26	28
	—	—	—	0	1	3	4	6	7	9	10	12	13	15	16	18	19	21	22	24
8	—	—	0	2	4	6	7	9	11	13	15	17	20	22	24	26	28	30	32	34
	—	—	—	1	2	4	6	7	9	11	13	15	17	18	20	22	24	26	28	30
9	—	—	1	3	5	7	9	11	14	16	18	21	23	26	28	31	33	36	38	40
	—	—	0	1	3	5	7	9	11	13	16	18	20	22	24	27	29	31	33	36
10	—	—	1	3	6	8	11	13	16	19	22	24	27	30	33	36	38	41	44	47
	—	—	0	2	4	6	9	11	13	16	18	21	24	26	29	31	34	37	39	42
11	—	—	1	4	7	9	12	15	18	22	25	28	31	34	37	41	44	47	50	53
	—	—	0	2	5	7	10	13	16	18	21	24	27	30	33	36	39	42	45	48
12	—	—	2	5	8	11	14	17	21	24	28	31	35	38	42	46	49	53	56	60
	—	—	1	3	6	9	12	15	18	21	24	27	31	34	37	41	44	47	51	54
13	—	0	2	5	9	12	16	20	23	27	31	35	39	43	47	51	55	59	63	67
	—	—	1	3	7	10	13	17	20	24	27	31	34	38	42	45	49	53	56	60
14	—	0	2	6	10	13	17	22	26	30	34	38	43	47	51	56	60	65	69	73
	—	—	1	4	7	11	15	18	22	26	30	34	38	42	46	50	54	58	63	67
15	—	0	3	7	11	15	19	24	28	33	37	42	47	51	56	61	66	70	75	80
	—	—	2	5	8	12	16	20	24	29	33	37	42	46	51	55	60	64	69	73
16	—	0	3	7	12	16	21	26	31	36	41	46	51	56	61	66	71	76	82	87
	—	—	2	5	9	13	18	22	27	31	36	41	45	50	55	60	65	70	74	79
17	—	0	4	8	13	18	23	28	33	38	44	49	55	60	66	71	77	82	88	93
	—	—	2	6	10	15	19	24	29	34	39	44	49	54	60	65	70	75	81	86
18	—	0	4	9	14	19	24	30	36	41	47	53	59	65	70	76	82	88	94	100
	—	—	2	6	11	16	21	26	31	37	42	47	53	58	64	70	75	81	87	92
19	—	1	4	9	15	20	26	32	38	44	50	56	63	69	75	82	88	94	101	107
	—	0	3	7	12	17	22	28	33	39	45	51	56	63	69	74	81	87	93	99
20	—	1	5	10	16	22	28	34	40	47	53	60	67	73	80	87	93	100	107	114
	—	0	3	8	13	18	24	30	36	42	48	54	60	67	73	79	86	92	99	105

[a] To be significant for any given n_1 and n_2, obtained U must be *equal to* or *less than* the value shown in the table.

[b] Dashes in the body of the table indicate that no decision is possible at the stated level of significance.

Critical values for a one-tailed test at $\alpha = 0.05$ (roman type) and $\alpha = 0.025$ (boldface type) and for a two-tailed test at $\alpha = 0.10$ (roman type) and $\alpha = 0.05$ (boldface type).

n_2 \ n_1	1	2	3	4	5	6	7	8	9	10	11	12	13	14	15	16	17	18	19	20
1	—	—	—	—	—	—	—	—	—	—	—	—	—	—	—	—	—	—	0	0
	—	—	—	—	—	—	—	—	—	—	—	—	—	—	—	—	—	—	—	—
2	—	—	—	—	0	0	0	1	1	1	1	2	2	2	3	3	3	4	4	4
	—	—	—	—	—	—	—	**0**	**0**	**0**	**0**	**1**	**1**	**1**	**1**	**1**	**2**	**2**	**2**	**2**
3	—	—	0	0	1	2	2	3	3	4	5	5	6	7	7	8	9	9	10	11
	—	—	—	—	**0**	**1**	**1**	**2**	**2**	**3**	**3**	**4**	**4**	**5**	**5**	**6**	**6**	**7**	**7**	**8**
4	—	—	0	1	2	3	4	5	6	7	8	9	10	11	12	14	15	16	17	18
	—	—	—	**0**	**1**	**2**	**3**	**4**	**4**	**5**	**6**	**7**	**8**	**9**	**10**	**11**	**11**	**12**	**13**	**13**
5	—	0	1	2	4	5	6	8	9	11	12	13	15	16	18	19	20	22	23	25
	—	—	**0**	**1**	**2**	**3**	**5**	**6**	**7**	**8**	**9**	**11**	**12**	**13**	**14**	**15**	**17**	**18**	**19**	**20**
6	—	0	2	3	5	7	8	10	12	14	16	17	19	21	23	25	26	28	30	32
	—	—	**1**	**2**	**3**	**5**	**6**	**8**	**10**	**11**	**13**	**14**	**16**	**17**	**19**	**21**	**22**	**24**	**25**	**27**
7	—	0	2	4	6	8	11	13	15	17	19	21	24	26	28	30	33	35	37	39
	—	—	**1**	**3**	**5**	**6**	**8**	**10**	**12**	**14**	**16**	**18**	**20**	**22**	**24**	**26**	**28**	**30**	**32**	**34**
8	—	1	3	5	8	10	13	15	18	20	23	26	28	31	33	36	39	41	44	47
	—	**0**	**2**	**4**	**6**	**8**	**10**	**13**	**15**	**17**	**19**	**22**	**24**	**26**	**29**	**31**	**34**	**36**	**38**	**41**
9	—	1	3	6	9	12	15	18	21	24	27	30	33	36	39	42	45	48	51	54
	—	**0**	**2**	**4**	**7**	**10**	**12**	**15**	**17**	**20**	**23**	**26**	**28**	**31**	**34**	**37**	**39**	**42**	**45**	**48**
10	—	1	4	7	11	14	17	20	24	27	31	34	37	41	44	48	51	55	58	62
	—	**0**	**3**	**5**	**8**	**11**	**14**	**17**	**20**	**23**	**26**	**29**	**33**	**36**	**39**	**42**	**45**	**48**	**52**	**55**
11	—	1	5	8	12	16	19	23	27	31	34	38	42	46	50	54	57	61	65	69
	—	**0**	**3**	**6**	**9**	**13**	**16**	**19**	**23**	**26**	**30**	**33**	**37**	**40**	**44**	**47**	**51**	**55**	**58**	**62**
12	—	2	5	9	13	17	21	26	30	34	38	42	47	51	55	60	64	68	72	77
	—	**1**	**4**	**7**	**11**	**14**	**18**	**22**	**26**	**29**	**33**	**37**	**41**	**45**	**49**	**53**	**57**	**61**	**65**	**69**
13	—	2	6	10	15	19	24	28	33	37	42	47	51	56	61	65	70	75	80	84
	—	**1**	**4**	**8**	**12**	**16**	**20**	**24**	**28**	**33**	**37**	**41**	**45**	**50**	**54**	**59**	**63**	**67**	**72**	**76**
14	—	2	7	11	16	21	26	31	36	41	46	51	56	61	66	71	77	82	87	92
	—	**1**	**5**	**9**	**13**	**17**	**22**	**26**	**31**	**36**	**40**	**45**	**50**	**55**	**59**	**64**	**67**	**74**	**78**	**83**
15	—	3	7	12	18	23	28	33	39	44	50	55	61	66	72	77	83	88	94	100
	—	**1**	**5**	**10**	**14**	**19**	**24**	**29**	**34**	**39**	**44**	**49**	**54**	**59**	**64**	**70**	**75**	**80**	**85**	**90**
16	—	3	8	14	19	25	30	36	42	48	54	60	65	71	77	83	89	95	101	107
	—	**1**	**6**	**11**	**15**	**21**	**26**	**31**	**37**	**42**	**47**	**53**	**59**	**64**	**70**	**75**	**81**	**86**	**92**	**98**
17	—	3	9	15	20	26	33	39	45	51	57	64	70	77	83	89	96	102	109	115
	—	**2**	**6**	**11**	**17**	**22**	**28**	**34**	**39**	**45**	**51**	**57**	**63**	**67**	**75**	**81**	**87**	**93**	**99**	**105**
18	—	4	9	16	22	28	35	41	48	55	61	68	75	82	88	95	102	109	116	123
	—	**2**	**7**	**12**	**18**	**24**	**30**	**36**	**42**	**48**	**55**	**61**	**67**	**74**	**80**	**86**	**93**	**99**	**106**	**112**
19	0	4	10	17	23	30	37	44	51	58	65	72	80	87	94	101	109	116	123	130
	—	**2**	**7**	**13**	**19**	**25**	**32**	**38**	**45**	**52**	**58**	**65**	**72**	**78**	**85**	**92**	**99**	**106**	**113**	**119**
20	0	4	11	18	25	32	39	47	54	62	69	77	84	92	100	107	115	123	130	138
	—	**2**	**8**	**13**	**20**	**27**	**34**	**41**	**48**	**55**	**62**	**69**	**76**	**83**	**90**	**98**	**105**	**112**	**119**	**127**

Appendix N
Critical Values for the
Wilcoxon Signed-Rank Test

The following table presents critical values of the T statistic for the Wilcoxon signed-rank test. The first column represents the sample size. The remaining columns represent different alpha levels for one- and two-tailed tests. As an example, the critical value of T for a sample size 10 ($N = 10$), $\alpha = .05$, two-tailed test, is 8.0.

Source: Wilcoxon, Katti, and Wilcox, "Critical Values and Probability Levels of the Wilcoxon Rank Sum Test and the Wilcoxon Signed Rank Test." Reproduced with the permission of American Cyanamid Company.

The symbol T denotes the smaller sum of ranks associated with differences that are all of the same sign. For any given N (number of ranked differences), the obtained T is significant at a given level if it is equal to or <u>less than</u> the value shown in the table.

	Level of significance for one-tailed test					Level of significance for one-tailed test			
	.05	.025	.01	.005		.05	.025	.01	.005
	Level of significance for two-tailed test					Level of significance for two-tailed test			
N	.10	.05	.02	.01	N	.10	.05	.02	.01
5	0	--	--	--	28	130	116	101	91
6	2	0	--	--	29	140	126	110	100
7	3	2	0	--	30	151	137	120	109
8	5	3	1	0	31	163	147	130	118
9	8	5	3	1	32	175	159	140	128
10	10	8	5	3	33	187	170	151	138
11	13	10	7	5	34	200	182	162	148
12	17	13	9	7	35	213	195	173	159
13	21	17	12	9	36	227	208	185	171
14	25	21	15	12	37	241	221	198	182
15	30	25	19	15	38	256	235	211	194
16	35	29	23	19	39	271	249	224	207
17	41	34	27	23	40	286	264	238	220
18	47	40	32	27	41	302	279	252	233
19	53	46	37	32	42	319	294	266	247
20	60	52	43	37	43	336	310	281	261
21	67	58	49	42	44	353	327	296	276
22	75	65	55	48	45	371	343	312	291
23	83	73	62	54	46	389	361	328	307
24	91	81	69	61	47	407	378	345	322
25	100	89	76	68	48	426	396	362	339
26	110	98	84	75	49	446	415	379	355
27	119	107	92	83	50	466	434	397	373

(Slight discrepancies will be found between the critical values appearing in the table above and in Table 2 of the 1964 revision of F. Wilcoxon, and R.A. Wilcox, <u>Some Rapid Approximate Statistical Procedures</u>, New York, Lederle Laboratories, 1964. The disparity reflects the latter's policy of selecting the critical value nearest a given significance level, occasionally overstepping that level. For example, for N = 8,

the probability of a T of 3 = 0.0390 (two-tail)

and

the probability of a T of 4 = 0.0546 (two-tail).

Wilcoxon and Wilcox select a T of 4 as the critical value at the 0.05 level of significance (two-tail), whereas this table reflects a more conservative policy by setting a T of 3 as the critical value at this level.)

Appendix O
Critical Values for
Spearman's r

The following table presents values of the Spearman rank order correlation coefficient that would be rejected at the .10, .05, .02, and .01 alpha levels for a nondirectional test, and alpha levels of .05, .025, .01, and .005 for directional tests, ρ_s = 0. The first column is the sample size, N. The remaining columns present the correlation coefficients for different alpha levels. If the absolute value of a correlation exceeds or is equal to the appropriate value in the table, then the null hypothesis that ρ_s = 0 would be rejected. For example, if N = 10, any Spearman correlation greater than or equal to .648 would lead to the rejection of the null hypothesis for α = .05, two-tailed test. For negative values of r_s, the sign of the coefficient is ignored and the table is used as described above.

	Level of significance for one-tailed test			
	.05	.025	.01	.005
	Level of significance for two-tailed test			
N	.10	.05	.02	.01
5	.900	1.000	1.000	--
6	.829	.886	.943	1.000
7	.714	.786	.893	.929
8	.643	.738	.833	.881
9	.600	.683	.783	.833
10	.564	.648	.746	.794
12	.506	.591	.712	.777
14	.456	.544	.645	.715
16	.425	.506	.601	.665
18	.399	.475	.564	.625
20	.377	.450	.534	.591
22	.359	.428	.508	.562
24	.343	.409	.485	.537
26	.329	.392	.465	.515
28	.317	.377	.448	.496
30	.306	.364	.432	.478

Source: G. Ferguson, *Statistical Analysis in Psychology and Education.* New York: McGraw-Hill, 1976.

Glossary of Symbols

a	Point where the regression line intersects the Y axis (13)	n	Number of observations in a subgroup (9)
b	Slope of a regression line (13)	$n!$	The factorial of a number, n (5)
nCr	Combination of n objects taken r at a time (5)	O	In chi square analysis, an observed frequency (14)
cf	Cumulative frequency (2)	nPr	Permutation of n objects taken r at a time (5)
CD	Critical difference in the Tukey HSD test (11)	p	Probability
CP	Cross products (13)	pdf	Probability density function (2)
D	The difference between two correlated scores (10)	$p(A)$	Probability of event A (5)
$\overline{D}$	The mean of the difference between two correlated scores (10)	$p(A\|B)$	Conditional probability of event A given event B (5)
df	Degrees of freedom (6)	$p(A,B)$	Probability of events A and B (5)
E	In chi square, the expected frequency (14)	$p(A \text{ or } B)$	Probability of event A or B (5)
$\overline{E}$	Expected sum of ranks in the Wilcoxon rank sum test (15)	q	Value of the studentized statistic (11)
Eta^2	The eta squared statistic (9)	r	Sample Pearson correlation coefficient (13)
E_R^2	Epsilon squared for sample data (15)	r_g	Glass rank biserial correlation coefficient (15)
f	Frequency of a score (2)	r_c	Matched pairs rank biserial correlation coefficient (15)
F	A variance ratio (11)	r_s	Spearman's rank order correlation for a sample (15)
H	Kruskal-Wallis test statistic (15)	r'	Fisher's log transform of r (13)
H_0	A null hypothesis (7)	rf	Relative frequency (2)
H_1	An alternative hypothesis (7)	s	Sample standard deviation (3)
HSD	Tukey's honest significant difference test (11)	$\hat{s}$	Estimate of the population standard deviation (6)
L	Lower limit of a numerical category (4)		
Md	Median (3)	s^2	Sample variance (3)
MS	Mean Square (11)	$\hat{s}^2$	Estimate of the population variance (6)
N	Number of observations in an experiment (1)	$\hat{s}_Y$	Estimate of the standard error of the mean (6)

$\hat{s}_{\bar{Y}_1 - \bar{Y}_2}$	Estimate of the standard error of the difference between two independent means.
$\hat{s}_{\bar{D}}$	Estimate of the standard error of the difference between two correlated means (10)
$\hat{s}_p^2$	Pooled estimate of a population variance (9)
SS	Sum of squares (3)
S_{XY}	Standard error of estimate (13)
T	A test statistic in Wilcoxon's matched pairs test
t	Student's t score (7)
U	Test statistic in the Mann-Whitney U test (15)

X	A variable (1)
X_P	A percentile (4)
PR_X	A percentile rank (4)
$\overline{X}$	A sample mean for variable X (3)
V	Cramer's statistic or the fourfold point correlation (14)
Y	A variable (1)
$\overline{Y}$	A sample mean for variable Y (3)
z	A standard score in a normal distribution (4)

GREEK LETTERS (Number in parentheses indicates chapter where the symbol is introduced and discussed)

α	The probability of a type I error (7)
β	The probability of a type II error (7)
μ	The mean of a population (6)
μ_D	The population mean of a set of difference scores (10)
ρ	Pearson product moment correlation for a population (13)
ρ_s	Spearman's rank order correlation for a population (15)
Σ	Summation: An instruction to add (1)

σ	Standard deviation of a population (3)
σ^2	Variance of a population (3)
$\sigma_{\overline{X}}$	Standard error of the mean (6)
$\sigma_{\bar{Y}_1 - \bar{Y}_2}$	Standard error of the difference between two independent means (9)
σ_R	Standard error of the R statistic in Wilcoxon's rank sum test (15)
σ_T	Standard error of Wilcoxon's T statistic (15)
χ^2	The chi square statistic (14)

References

Ainsworth, M. S., Blehar, M. C., Waters, E., and Wall, S. *Patterns of Attachment: A Psychological Study of the Strange Situation.* Hillsdale, N.J.: Erlbaum, 1978.

Anderson, N. H. Functional measurement and psychophysical judgment. *Psychological Review,* 1970, *77,* 153–170.

Aronson, E., Turner, J. A., and Carlsmith, J. Communicator credibility and communication discrepancy as a determinant of opinion change. *Journal of Abnormal and Social Psychology,* 1963, *67,* 31–36.

Barcus, F. E. *Saturday Children's Television: A Report of TV Programming and Advertising on Boston Commercial Television.* Newton, Mass.: Action for Children's Television, 1971.

Barron, F. The psychology of creativity. In Newcomb, T. (Ed), *New Directions in Psychology: II.* New York: Holt, Rinehart, and Winston, 1965.

Bennett, E. L., Krech, D., and Rosenzweig, M. R. Reliability and regional specificity of cerebral effects of environmental complexity and training. *Journal of Comparative and Physiological Psychology,* 1964, *57,* 440–441.

Bohrnstedt, G. W. and Carter, T. M. Robustness in regression analysis. *Sociological Methodology,* 1971, 118–146.

Boneau, C. A. The effects of violations of assumptions underlying the *t* test. *Psychological Bulletin,* 1960, *57,* 49–64.

Borden, R. Environmental attitudes and beliefs in technology. Unpublished manuscript. Department of Psychological Sciences, Purdue University, 1978.

Box, G. E. and Anderson, S. L. Permutation theory in the derivation of robust criteria and the study of departures from assumption. *Journal of the Royal Statistical Society, Series B,* 1955, *17,* 1–26.

Brehm, J. Post-decision changes in desirability of alternatives. *Journal of Abnormal and Social Psychology,* 1956, *52,* 384–389.

Carrol, R. M. and Nordholm, L. A. Sampling characteristics of Kelley's η^2 and Hays' ω^2. *Educational and Psychological Measurement,* 1975, *35,* 541–554.

Casler, L. The effects of hypnosis on GESP. *Journal of Parapsychology,* 1964, *28,* 126–134.

Cohen, J. An alternative to Marascuilo's large sample multiple comparisons for proportions. *Psychological Bulletin,* 1967, *67,* 199–201.

Cohen, J. *Statistical Power Analysis for the Behavioral Sciences.* New York: Academic Press, 1969 (Second edition, 1977).

Collier, R. D., Baker, F. B., Mandeville, G. K., and Hayes, T. F. Estimates of test size for several test procedures based on conventional variance ratios in the repeated measure design. *Psychometrika,* 1967, *32,* 339–353.

Conover, W. J. Rejoinder. *Journal of the American Statistical Association,* 1974, *69,* 382. (a)

Conover, W. J. Some reasons for not using the Yates continuity correction on 2×2 contingency tables. *Journal of the American Statistical Association,* 1974, *69,* 374-376. (b)

Cox, C. *Genetic Studies of Genius.* Stanford, Calif. Stanford University Press, 1926.

Deaux, K. and Emswiller, T. Explanations of successful performance on sex-linked tasks: What is skill for the male is luck for the female. *Journal of Personality and Social Psychology,* 1974, *29,* 80–85.

Dunn, O. J. Multiple comparisons using rank sums. *Technometrics,* 1964, *6.*

Eron, L. D. Relationship of TV viewing habits and aggressive behavior in children. *Journal of Abnormal and Social Psychology,* 1963, *67,* 193–196.

Feldman-Summers, S. and Ashworth, C. D. Factors related to intentions to report a rape. Unpublished manuscript. Department of Psychology, University of Washington, Seattle, 1980.

Ferguson, G. A. *Statistical Analysis in Psychology and Education.* New York: McGraw-Hill, 1976.

Festinger, L. and Carlsmith, J. M. Cognitive consequences of forced compliance. *Journal of Abnormal and Social Psychology*, 1959, *58*, 203–210.

Fiedler, F. *A Theory of Leadership Effectiveness*. New York: McGraw-Hill, 1967.

Fisher, R. *Statistical Methods for Research Workers*. New York: Hafner, 1950.

Fleishman, A. I. Confidence intervals for correlation ratios. *Educational and Psychological Measurement*, 1980, *40*, 659–670.

Friedman, H. *Introduction to Statistics*. New York: Random House, 1972.

Frieze, I. H., Parsons, J. E., Johnson, P. B., Ruble, D. N., and Zellman, G. L. *Women and Sex Roles: A Social Psychological Perspective*. New York: Norton, 1978.

Gallup, G. *The Sophisticated Poll Watcher's Guide*. Princeton, N. J.: Princeton Opinion Press, 1976.

Glass, G. V. and Hakstian, A. R. Measures of association in comparative experiments: Their development and interpretation. *American Educational Research Journal*, 1969, *6*, 403–414.

Glass, G., Peckham, P. D., and Sanders, J. R. Consequences of failure to meet assumptions underlying the fixed effects analyses of variance and covariance. *Review of Educational Research*, 1974, *42*, 237–288.

Glass, G. and Stanley, J. C. *Statistical Methods in Education and Psychology*. Englewood Cliffs, N.J.: Prentice-Hall, 1970.

Goldberg, P. Are women prejudiced against women? *Transaction*, 1968, *5*, 28–30.

Goodman, L. A. Simultaneous confidence intervals for contrasts among multinomial populations. *Annals of Mathematical Statistics*, 1964, *35*, 716–725.

Greenwald, A. Consequences of prejudice against the null hypothesis. *Psychological Bulletin*, 1975, *82*, 1–20.

Grizzle, J. E. Continuity correction in the χ^2 test for 2 × 2 tables. *American Statistician*, 1967, *21*, 28–32.

Guilford, J. *Fundamental Statistics in Psychology and Education*. New York: McGraw-Hill, 1965.

Haggard, E. A. *Intraclass Correlation and Analysis of Variance*. New York: Dryden Press, 1958.

Harvath, J. Problem solving performance and music. *American Journal of Psychiatry*, 1943, *22*, 211–212.

Havilcek, L. and Peterson, N. Effect of violation of assumptions upon significance levels of the Pearson *r*. *Psychological Bulletin*, 1977, *84*, 373–377.

Hays, W. L. *Statistics for Psychologists*. New York: Holt, Rinehart, and Winston, 1963.

Hays, W. L. and Winkler, R. L. *Statistics: Probability, Inference, and Decision*. New York: Holt, Rinehart, and Winston, 1971.

Hsu, T. C. and Feldt, L. S. The effect of limitations on the number of criterion score values on the significance level of the *F*-test. *American Educational Research Journal*, 1969, *6*, 515–527.

Huck, S. W. and Sandler, H. M. *Rival Hypotheses: Alternative Interpretations of Data-Based Conclusions*. New York: Harper and Row, 1979.

Huff, D. *How to Lie with Statistics*. New York: Norton, 1954.

Hurlock, E. An evaluation of certain incentives used in schoolwork. *Journal of Educational Psychology*, 1925, *16*, 145–159.

Jaccard, J. Factors affecting the acceptance of male oral contraceptives. Unpublished manuscript. Department of Psychology, Purdue University, 1980.

Jensen, A. R. *Educability and Group Differences*. New York: Basic Books, 1973.

Johnson, M. K. and Liebert, R. M. *Statistics*. Englewood Cliffs, N.J.: Prentice-Hall, 1977.

Johnson, T. Luck versus ability: A replication. Unpublished manuscript, 1976.

Kelman, H. C. and Hovland, C. I. Reinstatement of the communicator in delayed measurement of opinion change. *Journal of Abnormal and Social Psychology*, 1953, *48*, 327–335.

Kennedy, J. J. The eta coefficient in complex ANOVA designs. *Educational and Psychological Measurement*, 1970, *30*, 885–889.

Kerlinger, F. N. *Foundations of Behavioral Research*. New York: Holt, Rinehart, and Winston, 1973.

Kesselman, H. J. A Monte Carlo investigation of three estimates of treatment magnitude: Epsilon squared, eta squared, and omega squared. *Canadian Psychological Review*, 1975, *16*, 44–48.

Kessleman, H. J., Games, P. A., and Rogan, J. C. Protecting the overall rate of Type I errors for pairwise comparisons with an omnibus test statistic. *Psychological Bulletin*, 1979, *86*, 884–888.

Kirk, R. E. *Experimental Design: Procedures for the Behavioral Sciences*. Monterey, Calif.: Brooks-Cole, 1968.

Kirk, R. E. *Statistical Issues: A Reader for the Behavioral Sciences.* Monterey, Calif.: Brooks-Cole, 1972.

Kirk, R. E. *Introductory Statistics.* Monterey, Calif.: Brooks-Cole, 1978.

Kruskal, W. H. and Wallis, W. A. Use of ranks in one criterion variance analysis. *Journal of the American Statistical Association,* 1952, *47,* 583–621.

Lissitz, R. W. and Chardos, S. A study of the effect of the violation of the assumption of independent sampling upon the type I error rate of the two group *t* test. *Educational and Psychological Measurement,* 1975, *35,* 353–359.

Lord, F. M. On the statistical treatment of football numbers. *American Psychologist,* 1953, *8,* 750–751.

Lunney, G. H. Using analysis of variance with a dichotomous dependent variable: An empirical study. *Journal of Educational Measurement,* 1970, *7,* 263–269.

Mantel, N. Comment and suggestion. *Journal of the American Statistical Association,* 1974, *69,* 378–380.

Marascuilo, L. A. and McSweeney, M. *Nonparametric and Distribution-Free Methods for the Social Sciences.* Monterey, Calif.: Brooks-Cole, 1977.

McArthur, L. Z. and Eisen, S. V. Television and sex role stereotyping. *Journal of Applied Social Psychology,* 1976, *6,* 329–351.

McCall, R. B. *Fundamental Statistics for Psychology.* New York: Harcourt, Brace and World, 1980.

McCall, R. and Appelbaum, M. Bias in the analysis of repeated measures designs: Some alternative approaches. *Child Development,* 1973, *44,* 401–415.

McConnell, J. V. New evidence for the transfer of training effect in planaria. Symposium on the biological basis of memory traces, International Congress of Psychology, Moscow, 1966.

McNemar, Q. *Psychological Statistics.* New York: Wiley, 1962.

Miettinen, D. S. Comment. *Journal of the American Statistical Association,* 1974, *69,* 380–382.

Miller, D. and Swanson, G. *Inner Conflict and Defense.* New York: Holt, Rinehart, and Winston, 1960.

Miller, R. G. *Simultaneous Statistical Inference.* New York: McGraw-Hill, 1966.

Minium, E. *Statistical Reasoning in Psychology and Education.* New York: Wiley, 1970.

Morrow, F. and Davidson, D. Race and family size decisions. Unpublished manuscript. Department of Psychology, Purdue University, 1976.

Myers, J. L. *Fundamentals of Experimental Design.* Boston: Allyn and Bacon, 1972.

Nezlek, J. Social behavior and diaries. Unpublished manuscript. Department of Psychology, College of William and Mary, Williamsburg, Va., 1978.

Norris, R. C. and Hjelm, H. F. Non-normality and product moment correlation. *Journal of Experimental Education,* 1961, *29,* 261–270.

Pearson, E. S. and Please, N. W. Relations between the shape of the population distribution of four simple test statistics. *Biometrika,* 1975, *62,* 223–241.

Plackett, R. L. The continuity correction in 2×2 tables. *Biometrika,* 1964, *51,* 237–337.

Rubovits, P. C. and Maehr, M. L. Pygmalion black and white. *Journal of Personality and Social Psychology,* 1973, *25,* 210–218.

Ryan, T. A. Multiple comparisons in psychological research. *Psychological Bulletin,* 1959, *56,* 26–47.

Scheffé, H. *The Analysis of Variance.* New York: Wiley, 1959.

Sears, D. O. Political behavior. In G. Lindzey and E. Aronson (Eds.), *The Handbook of Social Psychology.* Reading, Mass.: Addison-Wesley, 1969.

Smith, W. L., Phillipus, M. J., and Guard, H. L. Psychometric study of children with learning problems and 14-6 positive spike EEG patterns, treated with ethosuximide (zarontin) and placebo. *Archives of Disease in Childhood,* 1968, *43,* 616–619.

Sroufe, L. A. and Waters, E. Attachment as an organizational construct. *Child Development,* 1977, *48,* 1184–1199.

Steel, R. G. A rank sum test for comparing all pairs of treatments. *Technometrics,* 1960, *2,* 197–207.

Steiner, I. D. *Group Process and Productivity.* New York: Academic Press, 1972.

Stevens, S. S. Mathematics, measurement, and psychophysics. In S. S. Stevens (Ed.) *Handbook of Experimental Psychology,* New York: Wiley, 1951.

Subrahmaniam, K., Subrahmaniam, K., and Messeri, J. On the robustness of some tests of significance in sampling from a compound normal population. *Journal of the American Statistical Association,* 1975, *70,* 435–438.

Tapp, J. L., and Levin, F. J. *Law, Justice, and the Individual in Society: Psychological and Legal Issues.* New York: Holt, Rinehart, and Winston, 1977.

Thorndike, R. Regression fallacies in the matched groups experiment. *Psychometrika,* 1942, *7,* 85–102.

Touhey, J. C. Effects of additional women professionals on ratings of occupational prestige and desirability. *Journal of Personality and Social Psychology,* 1974, *29,* 86–89.

Tukey, J. W. The problem of multiple comparisons. Unpublished manuscript. Princeton University, 1953.

Warkov, S. and Greeley, A. Parochial school origins and education achievement. *American Sociological Review,* 1966, *31,* 406–414.

Wechsler, D. *The Measurement and Appraisal of Adult Intelligence.* Baltimore: Williams and Wilkens, 1958.

Weil, A. T., Zinberg, N. E., and Nelson, J. Clinical and psychological effects of marihuana in man. *Science,* 1968, *162,* 1234–1242.

Wiggins, J. S. *Personality and Prediction: Principles of Personality Assessment.* Reading, Mass.: Addison-Wesley, 1973.

Wike, E. L. *Data Analysis: A Statistical Primer for Psychology Students.* Chicago: Aldine, 1971.

Wilcoxon, F. *Some Rapid Approximate Statistical Procedures.* New York: American Cyanamid Company, 1949.

Willingham, W. W. Predicting success in graduate education. *Science,* 1974, *183,* 275–278.

Winer, B. J. *Statistical Principles in Experimental Design.* New York: McGraw-Hill, 1971.

Witte, R. S. *Statistics.* New York: Holt, Rinehart, and Winston, 1980.

Zeisel, H. and Kalvin, H. Parking tickets and missing women: Statistics and the law. In J. Janur et al. (Eds.) *Statistics: A Guide to the Unknown.* San Francisco: Holden-Day, 1972.

Zelazo, P. R., Zelazo, N. A., and Kolb, S. Walking in the newborn. *Science,* 1972, *176,* 314–315.

Answers to Selected Exercises

CHAPTER ONE

1. The independent variable is the preference for aggressive television shows. The dependent variable is the peer rating of aggression. Both are quantitative in nature.

3. The independent variable is group size and the dependent variable is the time until problem solution. Both are quantitative in nature.

5. (a) quantitative   (c) quantitative

    (e) qualitative

8. (a) 37   (c) 15   (e) 128

    (g) 3.75   (i) 227   (k) -117

    (m) 185   (o) 1110

10. (a) 36   (b) 36   (c) 32

    (d) 32   (e) $k\Sigma X = \Sigma Xk$

12. (a) 4.89   (c) 1.42   (e) 6.25

    (g) 6.32   (i) 1.00   (k) 12.25

    (m) 2.00

14. The probability is .05. My decision would depend on the nature of the operation and the consequences of failure. The information about probability of success is not sufficient, by and of itself, to make a decision.

16. All members of the population have an equal chance of being included in the sample.

CHAPTER TWO

1–2.

Score	f	rf	cf	crf
8	2	.10	20	1.00
7	4	.20	18	.90
6	10	.50	14	.70
5	0	.00	4	.20
4	2	.10	4	.20
3	2	.10	2	.10

3. 8 days = 10%; more than 6 days = 30%; less than 5 days = 20%

4. Score of 8 = .10; score of 6 or 8 = .60; less than 7 = .70

8–9.

Score	f	rf	cf	crf
120–129	5	.10	50	1.00
110–119	10	.20	45	.90
100–109	20	.40	35	.70
90–99	10	.20	15	.30
80–89	5	.10	5	.10

16.

Score	f	rf	cf	crf
5	351	.19	1850	1.00
4	574	.31	1499	.81
3	185	.10	925	.50
2	407	.22	740	.40
1	333	.18	333	.18

CHAPTER THREE

1. Four-year-olds: $\overline{X} = 7.0$, Md $= 7.0$, Mode $= 7.0$, SS $= 40.00$, $s = 1.58$
   Twelve-year-olds: $\overline{X} = 20.0$, Md $= 20.0$, Mode $= 20.0$, SS $= 46.00$, $s = 1.70$

3. Mean of first set $= 12.0$; Mean of second set $= 15.0$; Mean of third set $= 2.0$. If a constant, k, is added to each score, X, then the mean of the new set of scores will equal $\overline{X} + k$. If k is subtracted from each X, then the mean of the transformed scores will be $\overline{X} - k$.

7. Set A, because it contains one very extreme score

9. Company A: $\overline{X} = 4.0$, $s = 1.29$
   Company B: $\overline{X} = 4.0$, $s = 0.00$
   Although both companies have the same mean level of satisfaction, employees of Company A exhibit more variability in their satisfaction than those of Company B.

11. $\overline{X} = 181.2$, $s = 10.3$

13. (a) Yes, positive skew

    (b) Yes, negative skew

    (c) No

19. All measures of variability are zero, since all of the scores are the same.

CHAPTER FOUR

1. The standard score is 4.00, implying the response to the critical question was 4 standard deviations above the mean. This is quite unique and suggests that the individual may be lying.

3. (a) 2.99   (c) 4.25   (e) 7.67

4. (a) 94.40   (c) 20.20

5. (a) 30.0   (c) 65.0   (e) 50.0

6. (a) .1587   (e) .5000

9. (a) 120   (c) 100   (e) 115

11. (a) .33   (c) $-.67$ (e) -1.67

14. Distribution C

15. (a) 104.2   (c) 94.8

CHAPTER FIVE

1. $p(\text{favors}) = .514$; $p(\text{opposes}) = .486$

3. No, because $p(B|A) \neq p(B)$

5. Four: $p(\text{male, favors})$, $p(\text{female, favors})$, $p(\text{male, opposes})$, $p(\text{female, opposes})$

8. .80 and .58, respectively

10. .08 and .32, respectively

12. .12

13. (a) 10   (c) 6

14. (a) 60   (c) 720

17. 6 and 4, respectively

19. (a) .010   (c) .172

21. $\mu = 124.8$ $\sigma = 5$

CHAPTER SIX

5. Population A, because $\sigma = 5$ is less than $\sigma = 7$, yielding standard errors of .91 versus 1.28 for A and B, respectively.

7. They will all equal the population mean.

10. N and σ

13. $\overline{X} = 5.0$, $\hat{s}_{\overline{X}} = .408$

16. 90% $= 72.75$ to 76.05
    95% $= 72.44$ to 76.36
    99% $= 71.82$ to 76.98

CHAPTER SEVEN

2. (b) 2.093   (c) 1.833

4. When $N > 30$

9. (a) $t = 10.00$, the null hypothesis is rejected

   (b) $t = 1.00$, we fail to reject the null hypothesis

   (c) $t = 5.00$, the null hypothesis is rejected

   (a) is different from (b) because of the larger N, (b) is different from (c) because of the larger $\hat{s}$

12. 95%: 2.78 to 3.44
    99%: 2.68 to 3.54

CHAPTER EIGHT

3. The cause of the spurious relationship is a person's gender. Males tend to have shorter hair than females. Males tend to be taller than females.

5. The independent variable is race (qualitative) and the dependent variable is the family size decision (quantitative). The independent variable is between-subjects in nature.

7. The independent variable is the presence-absence of others in a problem-solving situation (qualitative). The dependent variable is time until solution. The independent variable is within-subjects in nature.

CHAPTER NINE

3. Scores in the populations are normally distributed, with equal variances. The samples are independently and randomly selected from their respective populations.

5. (a) 1.734   (c) 2.101

6. $t = 6.71$, df = 8. The null hypothesis is rejected.

7. $G = 5.50$, $SS_{TOTAL} = 26.50$

8. $T_M = 1.5$, $T_F = -1.5$

9. Male scores: 6.5, 5.5, 5.5, 5.5, 4.5
Female scores: 6.5, 5.5, 5.5, 5.5, 4.5

10. $SS_{ERROR} = 4.00$, $Eta^2 = .85$

19. $n = 36$ per group, or $N = 72$

CHAPTER TEN

3. (a) 1.833   (b) 2.262

4. Condition A: 12, 12, 12, 11, 13
Condition B: 14, 14, 14, 15, 13

9. Independent groups t test

11. Correlated groups t test

13. Correlated groups t test

CHAPTER ELEVEN

3. (a) For $\alpha = .05$, critical value of $F = 3.59$
(c) For $\alpha = .05$, critical value of $F = 3.35$

6. A versus B, A versus C, A versus D, B versus C, B versus D, C versus D

7. (a) $T_X = -2.0$ $T_Y = 0.0$ $T_Z = 2.0$

(b)

Source	SS	df	MS	F
Between	40.00	2	20.00	10.00
Within	24.00	14	2.00	
Total	64.00	14		

For $\alpha = .05$, F critical is 3.88. Since 10.00 exceeds this, the null hypothesis is rejected. Eta squared = .625. HSD test yields:

Null Hypothesis Tested	Absolute Difference Between Sample Means	Value of CD	Is Null Hypothesis Rejected?
$\mu_X = \mu_Y$	2	2.38	No
$\mu_X = \mu_Z$	4	2.38	Yes
$\mu_Y = \mu_Z$	2	2.38	No

8.

Source	SS	df	MS	F
Between	54.00	3	18.00	3.60
Within	100.00	20	5.00	
Total	154.00	23		

(a) $N = 24$ (b) Unexplained variance = .65

17. (c)

19. (a)

CHAPTER TWELVE

3.

Source	SS	df	MS	F
Treatments	20	2	10	5
Error	116	58	2	
Within Subjects	136	60		

5. (a) For alpha = .05, critical $F = 3.25$
(c) For alpha = .05, critical $F = 3.17$

7. $\overline{Y}_A = 6.0$, $\overline{Y}_B = 6.0$, $\overline{Y}_C = 6.0$. The adjusted data means are the same as the original means because we have removed the effects of a disturbance variable, not a confounding variable. This, by definition, affects within-group variability and not between-group variability.

12. (c)

14. (d)

CHAPTER THIRTEEN

4. It indicates whether the relationship is direct or inverse. This is also true of the sign of a correlation coefficient.

5. (a) $SS_X = 50$ $SS_Y = 60$ $CP = 10$

   (b) $r = .18$ $a = 5.00$ $b = .20$

   (c) $Y = 5.00 + .20X$

   (d)

Y	$\hat{Y}$	$Y - \hat{Y}$
7	5.6	1.4
9	6.6	2.4
3	5.6	-2.6
8	5.4	2.6
8	6.2	1.8
9	6.2	2.8
6	6.6	$-.6$
4	6.0	-2.0
2	6.4	-4.4
4	5.4	-1.4

6. 6 units

10. $-.65$

13. Pearson's correlation coefficient. Both variables are quantitative in nature, are measured on at least an interval level, and are probably well described by a linear relationship.

15. One-way repeated measures analysis of variance. The independent variable is the type of aspirin and it has four values. It is within-subjects in nature. The dependent variable is the preference rating and it is quantitative in nature.

CHAPTER FOURTEEN

3. (a) For alpha = .05, critical value is 3.841
   (c) For alpha = .05, critical value is 5.991

5. 3 rows and 4 columns

7. When df = 1

9. When df = 1

11. $\chi^2 = 13.19$, the null hypothesis is rejected for alpha = .05. Cramer's $V = .16$.

13. Correlated groups t test. The independent variable is qualitative in nature (pretest versus posttest), is within-subjects in nature, and has two values. The dependent measure is quantitative in nature (score on reading skill test).

15. One-way between-subjects analysis of variance. The independent variable, religion, is qualitative, between-subjects in nature, and has three values. The dependent variable, number of children desired, is quantitative in nature.

CHAPTER FIFTEEN

2. A Wilcoxon rank sum test was applied to analyze the relationship between ownership of car and G.P.A. The difference in ranks for owners versus nonowners was statistically nonsignificant ($z = .394$, ns).

3. A Wilcoxon signed-rank test was used to analyze recognition times as a function of the marijuana, no-marijuana conditions. The test indicated a statistically nonsignificant difference ($T = 17$, ns).

7. A Spearman rank order correlation was performed on the ranked scores for crime rate and size of police force. The observed correlation, $r = .738$, was statistically significant ($t = 4.37$, df = 16, $p < .05$). As the crime rate increased, so did the size of the police force.

9. A Kruskal-Wallis test was applied to the ranked recall data as a function of the three types of words. The resulting H statistic was statistically significant ($H = 9.26$, df = 2, $p < .05$). The strength of the relationship, as indexed by epsilon squared, was .605.

A post hoc test suggested by Dunn (1964) indicated that all three groups were statistically significantly different from each other.

11. A Spearman rank order correlation was performed on the ranked scores for conservatism and attitude toward gun control. The observed correlation of .984 was statistically significant ($t = 23.43$, df $= 18$, $p = .05$). As conservatism increased, attitudes in favor of gun control also increased.

CHAPTER SIXTEEN

1. The independent variable is time of the year (December versus May) and it is a qualitative variable with two values. The dependent variable is scores on the depression scale and it is a quantitative variable. This is a case II situation, and the statistical test used is the correlated groups t test. This is because the independent variable is within-subjects in nature.

3. The independent variable is the presence versus absence of music, and the dependent variable is the growth in inches of the plants. The former is a qualitative variable and the latter is a quantitative variable. This is a case II situation. An independent groups t test would be used because the independent variable has two values and is between-subjects in nature.

5. The independent variable is social class and the dependent variable is dogmatism. Both are quantitative in nature. This is a case IV situation. Assuming a linear relationship, Pearson's correlation coefficient could be applied to the data. However, it would be important to examine the scatterplot for the possible existence of a nonlinear relationship.

7. The independent variable is the presence or absence of hypnosis, and the dependent variable is the temperature of the hand. The former is a qualitative variable, and the latter is a quantitative variable, yielding a case II situation. An independent groups t test would be applied since the independent variable is between-subjects in nature and has two values.

9. The independent variable is the time of morning (first wake up versus three hours later), and the dependent variable is grip strength. The former is a qualitative variable (although it could be conceptualized as a quantitative variable), and the latter is a quantitative variable. This yields a case II situation. A correlated groups t test would be used since the independent variable is within-subjects in nature and has two values.

11. The independent (predictor) variable is gender, and the dependent variable is use of marijuana (used versus not used). Both variables are qualitative in nature and a chi square test of independence is used to analyze the data.

13. The independent variable is time spent on an exam, and the dependent variable is performance on the exam. Both are quantitative in nature. The investigator classified the independent measure into three groups in order to compare their performance. Consequently, a Kruskal-Wallis test would be most appropriate. The independent variable is between-subjects in nature and has three values.

15. The independent variable is quantity of publications, and the dependent variable is reputation of the department. Both are quantitative in nature. Spearman's rank order correlation could be used to analyze this relationship.

CHAPTER SEVENTEEN

1. Two and three, respectively.

3. The independent variables are gender (male versus female), and social class (low versus medium versus high). The dependent variable is prejudicial attitudes toward women.

5.

Source	SS	df	MS	F
A	20	2	10	5.0
B	45	3	15	7.5
A × B	60	6	10	5.0
Within	216	108	2	
Total	341	119		

7. $N = 60$. Number of groups $= 6$. Number of subjects per group $= 10$.

9. Eta^2 for $A = .125$. Eta^2 for $B = .10$. Eta^2 for $A \times B = .10$.

11.

	B_1	B_2	B_3
A_1	6	5	4
A_2	1	2	3

13.

Source	SS	df	MS	F
A	180.00	1	180.00	21.77
B	45.00	1	45.00	5.44
$A \times B$	5.00	1	5.00	.61
Within	132.32	16	8.27	
Total	362.32	19		

15. Both involve the case of unequal n per group. With unweighted means analysis of variance, the unequal n are *not* theoretically meaningful, whereas in least squares analysis of variance they are.

17. The two main effects are statistically significant (alpha = .05) but the interaction effect is not. Eta squared for the main effects of A and B is .14 and .54, respectively. The nature of the relationship for A is such that subjects in A_2 tend to have higher scores than those in A_1. For B, an HSD test is necessary. This reveals that all of the differences between means are statistically significant, alpha = .05.

Index

Statistical Tests for Analyzing Relationships Between Variables

Parametric Test	Nonparametric Counterpart
Independent Groups t test	*Mann-Whitney U Test/ Wilcoxon Test*
Test of null hypothesis — *t* statistic	Test of null hypothesis — *U* statistic/*z* statistic
Strength of effect — Eta squared	Strength of effect — Glass rank biserial correlation coefficient
Nature of effect — Inspection of group means	Nature of effect — Inspection of rank sums
Correlated Groups t Test	*Wilcoxon Signed Rank Test*
Test of null hypothesis — *t* statistic	Test of null hypothesis — *T* statistic/*z* statistic
Strength of effect — Eta squared	Strength of effect — Matched pairs ranked biserial coefficient
Nature of effect — Inspection of group means	Nature of effect — Inspection of rank sums
One-Way Between-Subjects Analysis of Variance	*Kruskal-Wallis Test*
Test of null hypothesis — *F* ratio	Test of null hypothesis — *H* statistic
Strength of effect — Eta squared	Strength of effect — Epsilon squared
Nature of effect — Tukey HSD test	Nature of effect — Dunn's procedure